OTHER WRITINGS BY TODD SILER

Metaphorms: Forms of Metaphor

The Art of Thought

Architectonics of Thought:
    A Symbolic Model of Neuropsychological Processes

"Neurocosmology:
    Ideas and Images Toward an Art-Science-Technology
    Synthesis," in *Leonardo*

The Biomirror

Cerebreactors

BREAKING

THE MIND BARRIER

The Artscience of Neurocosmology

TODD SILER

SIMON AND SCHUSTER

New York • London • Toronto • Sydney • Tokyo • Singapore

SIMON AND SCHUSTER
Simon & Schuster Building
Rockefeller Center
1230 Avenue of the Americas
New York, New York 10020

Copyright © 1990 by Todd Siler
All rights reserved
including the right of reproduction
in whole or in part in any form.
SIMON AND SCHUSTER and colophon are
registered trademarks of Simon & Schuster Inc.
Designed by Edith Fowler
Manufactured in the United States of America

10 9 8 7 6 5 4 3 2 1

Library of Congress Cataloging in Publication Data

Siler, Todd.
    Breaking the mind barrier: the artscience of
neurocosmology / Todd Siler.
        p.   cm.
    Includes bibliographical references and index.
    1. Philosophical anthropology.   2. Thought
and thinking.   3. Brain—Evolution.
4. Creative ability.   I. Title.
BD450.S497   1990
128′.2—dc20   90-44049
                                                    CIP
ISBN 0-671-69097-3

In memory of my brother-in-law,
Christopher Lyon, one of nature's gifts:
architect, naturalist, and human being

Acknowledgments

The roots of this book—where mind, nature, life, cosmos became one integrated reality—first formed in the summer of 1969. As a fifteen-year-old visiting Jerusalem with my brother, Paul, I approached the Wailing Wall awed by the rituals which charged this special space. Absorbed by some excavation that revealed the depth of the Wall's history, it occurred to me that the structure I was looking at and its symbolism were considerably larger than what my eyes sized for comprehension. Suddenly, the Wall expanded beyond its historical significance, breaking barriers of culture, knowledge, and experience. It represented to me a vestige of nature itself and of our collective mind. The countless messages wedged in the crevices of this metaphorical wall reminded me of humankind's great capacity to embrace itself as a unified force in searching for, or requesting, guidance from nature—from ourselves.

In hindsight, I've learned how *all* centers of intensified human reflection, searching, and inspired creation possess the potential to enlighten. These centers of thought are enriched learning environments that convey the wisdom and creative reach of civilizations.

To those who have helped me further other critical realizations that have led to self-discovery, I acknowledge your important contributions. I am especially appreciative of the generous time I've spent at the Massachusetts Institute of Technology discussing my work with many exceptional thinkers who include Dr. Walle J. H. Nauta and Dr. Gerald Schneider of the Fleischmann Center for Neuroscience; Dr. Stephan Chorover of the Brain and Cognitive Science Department; Otto Piene, director of the Center for Advanced Visual Studies, and the Fellows of the Center; Dr. Kosta Tsipis, Principal Research Scientist in the Department of Science, Technology and Society; Dr. David Gossard, director of the Computer-Aided Design Laboratory in the Department of Mechanical Engineering; Dr. Shai Haran, a graduate of the Department of Pure Mathematics; Dr. Stanford Anderson, chairman of the History, Theory, and Criticism Program in the Department of Architecture; Dr. Kenneth Manning, head of the Writing Program and Professor of History of Science; and at Harvard University, Dr. James Ackerman, the Arthur Kingsley Porter Professor of Fine Arts; also, Dr. Eric Schwartz of the Brain Research Laboratory at New York University Medical Center. In addition, I am indebted to Ron Feldman, director of the Ronald Feldman Fine Arts Gallery in New York City, for his continuous support and sustained belief in my work—and earlier, to Thomas Cornell of Bowdoin College and Leonard Baskin of Smith College. All of these individuals have enriched my ways of seeing the world. Although many may not share my points of view, they have nonetheless been instrumental in shaping my explorations, making the process of learning a positive, transformative experience.

I extend my gratitude to the many organizations that have seeded my work, among them the I.B.M. Thomas J. Watson Foundation, the Fulbright Foundation, the Massachusetts Council on the Arts and Humanities/Artists Foundation Fellowship Program, the Council for the Arts at M.I.T. and the Meitec Fellowship Program in Japan. Their fellowships arrived at just the right moment—when fundamentally new experiences were necessary for growth.

To Andra Akers, founder of International Synergy, I offer bountiful appreciation for recognizing my quest and pointing me toward John Brockman, my literary agent. Without Brockman's vision and perspicacity, this book would never be. To my editor at Simon and Schuster, Robert Asahina, whose sensitive steering has been most productive, my thankfulness. I am also obliged to Bonnie Derman, a graduate of the physics department at M.I.T., whose timely discussions of this book over many years helped maintain an overview of its contents. The critical comments by Eric Begleiter and Sharon Gallagher are also much appreciated.

To my wife, Lees Ruoff, for her love and nurturance, my lasting gratefulness. And to my tribe of grandparents, parents, brothers, and sisters who have always given me their unconditional love: Your confidence and respect have allowed me the freedom to challenge myself in pursuing the ideas in this book. This is perhaps the most powerful agent in the realization of any dream—the applied freedom of mind and wonderment. To be free in this sense is to venture seeing not only *what is* in the world, but *what is possible.*

Contents

NEUROCOSMOLOGY

Discovering the Brain-Universe

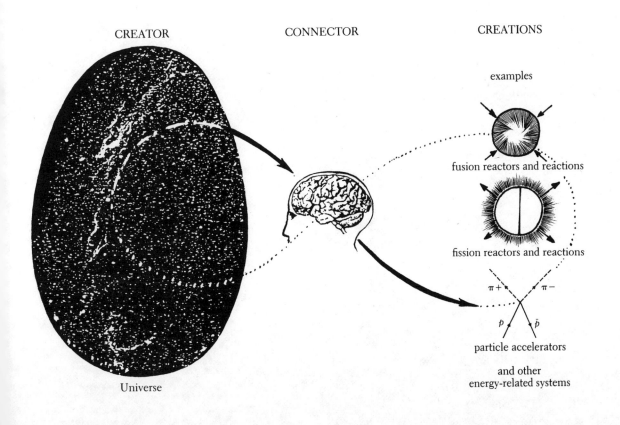

CREATOR

CONNECTOR

CREATIONS

examples

fusion reactors and reactions

fission reactors and reactions

$\pi+$ $\pi-$

p $\bar{p}$

particle accelerators

and other
energy-related systems

Universe

Discoveries are largely a function of the methods used.
—SANTIAGO RAMÓN Y CAJAL

Only connect.—E. M. FORSTER

Introduction

AN OCEAN OF INTERRELATED THEMES

The messages of this book are basic. The universe imparts its creative processes to us. We, in turn, impart our creative processes to the things we create. Our creations reveal the nature of our minds directly and so the universe indirectly. This is the great current of influences that changes our lives in accord with the lifeful changes of the universe.

Breaking the Mind Barrier ventures into this ocean current with one premise: In decoding the brain, we decode the universe—and vice versa. In many ways, the brain *is* what the brain *creates*. Its workings reflect the workings of its creations. This implies the brain imposes its dynamics on *everything* it makes—from concepts of the universe to the techniques used to test or represent those concepts; from chaos and catastrophe theories to technological tools we use to probe our knowledge or record our experiences. In our quest to learn how the brain works, we need to look comparatively at our objects of creation because they are *the brain's physical reflections*. These reflections can be as metaphorical or as literal as you choose to interpret them.

This book is allegorical in many respects. Even though you're presented with numerous literal connections, I entreat you to explore these as ideas. You could say that the explorers of this story want to understand the great whale, nature—to intimately know this creature—which rarely surfaces long enough to be befriended. And so we journey into opaque seas and deep space in search of nature's brain-universe.

The expression *brain-universe* has two distinct meanings. One refers to the universe(s) *of* the brain, or the inner cosmos, which is generally interpreted independently of the outer cosmos. The subtitle of this book could correctly imply a study of the origin and evolution of the human nervous system and mind.

The second meaning refers to the human brain *in relation to* the universe. This consideration is the subject of *neurocosmology*. The book presents a vision of nature as a world of what I call "processmorphs" (in Chapter 3)—things that are alike in process but not necessarily alike in form: from heavenly bodies to human brains. Neurocosmology combines neuroscience, cosmology, and art, among other disciplines, in discovering the connectedness of the brain and the universe. It does not dwell on the well-trodden comparisons between microcosmic and macrocosmic structures—where the form of one thing dominates another in scale and complexity. Instead, it illuminates similarities between structureless processes of stars, minds, and other seemingly diverse forms of matter.

The interpretation of "process" presented here expands beyond the notions of process discussed in this century in the context of philosophy, art, deep structures in linguistics, and dissipative structures in thermodynamics. In comparing the processes of things and thoughts, this discussion de-emphasizes the issues of size, form, magnitude, and time. The emphasis here is on showing similarities of processes in biological, physical, psychological, and symbolic systems, *regardless of the scale on which these processes exist, their proportions, or their applications*. The reader is invited to discover how processmorphs connect us with everything we create and are influcnced by.

This book examines why we must completely rethink our relationship with nature, seeing beyond nature's seductive surfaces and structures. These infinitely diverse structures hypnotically hold our fascination, while confusing our comprehension of nature's wholeness. *Breaking the Mind Barrier* suggests how we might peer more thoroughly into the processes of nature, which seem to remain the same as forms change.

MANY VESSELS, MANY EXPLORERS

Neuroscience is the study of the whole nervous system, its organization, processes, pathology, and development.

Cosmology is the study of the whole physical universe, its processes, structures, and evolution based on both observational data and theoretical constructs.

Art is a means of communicating thoughts and experiences in a mostly *personal* way. It's a process of communication that partially conceals its process of communicating. Contemporary art recognizes ever-widening definitions of art and aesthetics, representing all artforms and affirming all approaches to art-making. In time, the same may be true for the making of science. We are close to realizing that art is what scientists do when they hypothesize and create, while science is what artists do when their expressions challenge what they hypothesize and create.

AN ARTSCIENCE OF MODELING NATURE

Neurocosmology is neither art nor science; it is both. Or rather it is something in between. It seeks to reverse the isolation between our kingdoms of disciplines and our ways of exploring and knowing the world. It also encourages us to recognize and acknowledge all forms of self-imposed limitations in order to transgress them. One of the most serious limitations in creating more complete models involves the cloistering of knowledge in the sciences and technology, arts and humanities.

The exploration I propose requires art's natural partnership with numerous disciplines. Its principal means of exploring the world—its *modus operandi*—is the process of "metaphorming," or forming metaphors, which I describe in Chapter 1. Through this process—through metaphorms—we relate information from one discipline to another, connecting potentially all sources and forms of information. Metaphorms are expressions of nature's unity.

The tone of neurocosmology is suggestive, not explanatory. It doesn't offer detailed understandings; it offers insights that lead to understanding some details about nature and ourselves. In examining the ideas in this book, I invite you to take things out of a limited context and put them back in their original broadest context—that is, nature's integrated state. This process of contexturing information involves forming ideas about the brain-universe, without forsaking other ideas and methods of inquiry just because they clash with one's own. We need to become "re-formed splitters." After dividing and differentiating knowledge, we need to stand back and reintegrate it. Neurocosmology emphasizes that we think of "unity *with* unity of mind," in order to become as unified in viewpoint as the knowledge and forces we seek to unify. However, this must be accomplished without sacrificing the diversity of disciplinary knowledge.

By integrating these fields of knowledge, we can attain more encompassing and meaningful models of nature that take into account the whole of human experience. Irrespective of how we analyze or represent the *neurocosmic* connection, what's important is that it be discussed in both the broadest and the narrowest contexts—adopting the humanist vision of Renaissance thinkers.

WITH BOUNDLESS SHORELINES AND VISTAS

The concepts that unfold in the pages ahead were most likely born many hundreds of thousands of years ago in the imaginations of our ancestors who saw our world as I now see it: wondrously poised between the weighted symmetries and asymmetries of nature—those invariances and variances of life that make up the whole of reality. Our great ancestors may have had no words for symmetry and asymmetry, order and chaos, structure and process, or other complementary concepts, but their sense of these concepts is likely if they shared our present functional neuroanatomy—which they did. Our nervous systems operate according to principles of complementarity, like the cosmos. This book explores the similarities between these seemingly unlike systems of matter—brain and cosmos—challenging us to see this relationship anew as we recognize how these entities work as an integrated whole.

As we begin to understand our evolution in the context of the cosmos, we will consider questions like these, all of which are interrelated:

Can we learn anything about the functioning of the brain through the physical descriptions of the universe?

Can the logic and the ideas that come out of our analyses of the behavior of stars, galaxies, and clusters contribute to the ideas needed to understand the behavior or evolution of nerve cells, nuclei, and nervous systems?

Can the brain be thought of as a "biological star"—an energy-burning star of consciousness—whose systems have a dynamic resemblance to the fusion and fission processes that form and shape our universe?

Could there be a merging of brain functions ("cerebral fusion") at the instant of intuition and a splitting of these functions ("cerebral fission") in moments of reasoning?

In what way do the physical laws of nature exist independently of the brain? How does the brain influence these laws through its interpretations?

How is our sense of nature-reality-life influenced by these physical laws? *Does every detail of nature detail our nature?*

QUESTIONS AS INSTRUMENTS FOR EXPLORING NATURE

These and hundreds of similar questions are the primary instruments of neurocosmology. They are used to interpret brain and cosmic processes. Through works of *artscience*—or art that combines science—and writing, neurocosmologists pose questions. What are *we* that *it* (the universe) is not? Are we both the mirrors and the mirrors' reflections of the universe? Or is "the mirror" one of the mind's mythic creations? Do patterns of matter *reflect* patterns of mind? Or are these patterns one and the same thing—both physical and nonphysical in nature? (See "reflectionism" in Chapter 6.)

As we read in ancient cosmogonic myths of man-world origins, there remains an unbroken introspection not only on our place in the stars but on the stars' place in us. Going beyond explanations in astrophysics of a star's power production, we must allow ourselves to wonder: In what ways is the brain's production of energy similar to a star's production of energy? What are the details of these similarities and how can they be investigated metaphorically and literally? Going beyond the descriptions in high-energy physics of interactive forces: What are the forces of interaction between and within our minds? How can they be visualized mathematically or spoken of descriptively other than through poetry and philosophy? Beyond speculations of consciousness, are elements of the human mind, such as ideas and sensations, artifacts of the universe? What tools other than questions can be used to examine these "artifacts"? By examining the brain, do we examine everything?

COMBINING ALL SENSES

As soon as the phrase "total universe" is introduced in a philosophical discussion of nature or reality or life, the whole sentence and thought suddenly become fuzzy. It's as if the concept embedded in this phrase is too large for our present imaginations. Or it's too expansive for our current vocabulary, which feels most secure with its succinct, crystalline definitions. Perhaps this notion figuratively takes up too much room in our brains. The universe can in fact include or refer to *everything*—from concepts about chaos to physical entropy, from the idea of gravitons to the reality of acceleration (or gravitation). Its constituent elements may make up the entirety of human experience and all of nature. Gertrude Stein's luminous words—"Art isn't everything. It's just about everything."—come to mind when I contemplate what the universe is about. However, in contrast to Stein, I believe that art, like the universe, is everything! It is a uniquely open-ended system of infinite possibilities.

Since the age of the Scientific Revolution, some four hundred years ago, the discipline of science has been perceived as the citadel of objectivity, the single path to tangible truth. During this age, the procurement of data and the proclamation of facts were regarded as strictly rational enterprises. Scientists were intolerant of transient inner truths derived from the wellspring of mental life. Science had a completely empirical program equipped with inductive reasoning. These regulatory mechanisms filtered out lots of falsities and half-truths about the external world. Unfortunately, they also filtered out much of their imagination. Science's methods of inquiry were fairly straightforward: Some aspect of nature was observed and teased apart by sleuths bearing hypotheses and experiments for distilling the facts. The idea that the observer actually influences the observed was not needed given the state of Newtonian physics. This way of interpreting phenomena became increasingly problematic in modern quantum physics as researchers looked for the fundamental building blocks of matter and the language of nature. Particle physicists were halted by a wall of quandaries, many of which were related to the problem of observation; the detector was no longer a neutral bystander but an influential participant. Any external system, such as a detector or the mind, always exercises its indirect influences on the things it observes (and creates).

In hindsight, it's not so difficult to see how certain limitations in the scientific method may have arisen from such a rigid method of inquiry in the formative stages. Some historians of science have dealt with these limitations in the same way Cervantes's Don Quixote met his fate with the mirror-carrying soldiers of reality who revealed the apparent paucity of his endeavors. Science has been exposed repeatedly to the realities of change and uncertainty, "subjective objectivity," and relativism. Awareness of these and other realities continues to stimulate the pulse of science, challenging entrenched ideas so that they never stagnate into dogmas. The idea that there are unchanging laws of anything, no less an unchanging nature, has unlodged itself from our anxious visions of omniscience.

The traditions of art, which can be confidently stated as having existed from the beginning of recorded human history, rarely attempted to answer how the language of either art or nature worked—or how nature worked through art. This self-inquiry is a more recent phenomenon that was perhaps stimulated by artists' awareness of the sciences' passion for self-examination. The arts' earlier pursuits were primarily of beauty, or the truth of beauty, which can only be known through private experience, as John Keats reminds us in his famous poem "Ode on a Grecian Urn." Beauty, for Keats as for the arts, includes both the joyously sensual and "the agonies, the strife/ Of human hearts," which Shakespeare spoke of. Beauty recognizes that familiar sorrow we feel in sensing that we may never truly know ourselves. And yet art and metaphor can speak with the voice of truth as spoken from nature's mouth and as informed by nature's eyes.

SEEING THE WHOLE WITHOUT PARTS AND HOLES

The type of creative exploration advocated in this book preserves the best of both the scientific method and artistic traditions. It attempts to move beyond (yet also along with) both domains in its search for a unified world: the world of the undivided brain-universe. We begin with what the fifteenth-century theologian, mathematician, and statesman Cardinal Nicholas of Cusa called "learned ignorance." We start from that humble point and shortly thereafter conclude: There are definitive truths —but only in limited frames of reference. We are privileged with modest insights which we use to illuminate nature. The discovery and interpretation of facts and truths depend on the exploratory perspectives, beliefs, contexts, assumptions, and unique prejudices of the interpreter. There's just no way around this reality. There's only so far we can climb up Jacob's ladder before we realize it's both suspended from the ether of our imaginations and grounded in its truth.

The explorations herein entertain the possibility that we may ultimately never understand ourselves in a strictly biological or physical sense. That is, this knowledge may continually escape us as we try to grasp the modeler's place or part in the models of nature. We will almost certainly discover that our present descriptions of both brain and cosmos fall light-years short of their true properties and behavioral characteristics.

To date, I am skeptical that even the finest researchers in neuroscience and theoretical physics can now move beyond artful speculation about the interactions of biological processes and subatomic physical processes. This is in great part because of the fragmentary ways in which the human brain and human behavior continue to be observed. It also has to do with the manner in which the universe is studied. Neurophysiology and psychology are seen as occupying opposite corners of one room. Meanwhile theoretical physics, astrophysics, astronomy, and cosmology are given the corners of yet another room. What neurocosmologists propose are ways of working these and many more disciplines into one room. Also proposed are means of encouraging researchers to talk to one another about these connections, including the perspectives of informed laypersons.

To those seeking a neatly defined agenda for the exploration introduced here—complete with axioms, algorithms, and terse, mathematical expressions—I offer this statement by the pioneering neurophysiologist Warren McCulloch: "Don't bite my finger, look where I am pointing."

FROM PHILOSOPHY TO PHYSICS TO METAPHYSICS TO ART

It is critical that we understand the origin and evolution of the brain in the context of the birth and growth of the universe. The brain appears to have evolved out of the cosmos; it is a product of the producer. As such, it bears the marks of its creation. Neurocosmology endeavors to discern these "marks" and to interpret their implications.

To see ourselves in the evolutionary context of the universe is to recognize how our minds *are* nature, in every detail and behavior. We are "processmorphs" of volcanic eruptions, earthquakes, wind storms, geological and atmospheric disturbances, etc. Even though we don't look like any of these things in our outward appearances, the processes of our thoughts, feelings, and actions resemble these and other phenomena. What we cannot identify as facets of nature either in our character or in our mental abstractions, we can identify in the things we create. As Samuel Butler expressed it: "A hen is merely an egg's way of making another egg." Humankind is merely nature's way of making more nature. But perhaps humankind is also a way for nature to know itself, to become aware, to develop, through history, a freedom!

Exploring this relationship, we can discover how our thoughts touch and shape everything. And how our mental evolution and welfare hinge on understanding the nature of our minds as they're embedded in our creations. To the degree that we remain unaware of the dynamics of this relation, we will continue to ignore the hazardous consequences of realizing our more malignant ideas. The fallout from ignorance can be as environmentally catastrophic as nuclear winter or the greenhouse effect. *Ignorance has a longer half-life than plutonium.*

Through our joined perspectives, we have the ability to create a world in which one person's life experiences and beliefs may be respected, shared, and understood by others. I feel this is imperative for human development and survival—a cultural necessity, not an intellectual luxury. We want to build communication between all peoples, advancing the collective wisdom of our civilizations. Either we all learn to communicate and to cooperate with one another or we perish. Period. This message is stated with the same earnestness as the visual notation (shown below) which appeared in Benjamin Franklin's *Pennsylvania Gazette* 237 years ago.

"Join, or Die." The first known American cartoon, published by Benjamin Franklin in his Pennsylvania Gazette, *1754, to support his plan for colonial union presented at the Albany Congress. (The Granger Collection)*

Only by art can we get outside ourselves; instead of seeing only one world, our own, we see it under multiple forms.
—MARCEL PROUST

Nature is an artist that works from within instead of from without.—JOHN DEWEY

1

Art and Metaphorms: Methods of Inquiry

The idea of relating the human body to celestial bodies is older than recorded history. The Paleolithic cave painters, the ancient Egyptian and Mayan pyramid architects, the Celtic builders of megalithic monuments, the early Eastern Indian rock-cut temple sculptors, and North American Indian artisans—*all* must have felt the influence of the cosmos on their lives, as their artforms manifest. Scholars of Eastern mythology point out that the universe was conceived as a "great mother" who bore many worlds—and in whose womb will form other worlds, in a continuous circle of novel creations. The Chinese Tao reveals a similar insight in which all human activity and natural events are believed to exist in accord with the ways of the universe, and the various systems of Indian philosophy have expounded on the body-mind-self-universe relation for millennia.

One masterwork of metaphor which touches on many of the concerns raised in neurocosmology is the Upanishad. This ancient treatise, conceived in the eighth to sixth century B.C. from Hindu philosophy, poetically speculates on the merging of matter and *spirit* (their word for *energy*). It does so with such foresight that one might think its authors had anticipated our contemporary concept of the equivalence of mass and energy. Moreover, it pictures the evolutionary process by which matter grows toward and away from order. In the Maitrayani Upanishad, for instance, the term "quiddity" is used to describe the creative process by which the cosmos—like all living creatures—is stimulated to evolve from states of unbalance or chaos to states of balance and order and back again continuously.

Elsewhere, the pre-Socratic philosophers in the sixth century B.C. intimated many of our current notions about consciousness, matter, and the fundamental principles of the world. The Greek astronomer Anaximander, for instance, conceived of reality as a whole whose parts are all interdependent. And the philosophers Empedocles and Anaxagoras explored the opposite point of view; they theorized that reality is composed of many independent parts with a multitude of "ultimate principles." Also, there were the philosophers Leucippus, Democritus, and Epicurus, who proposed that all matter is made up of finite indivisible elements.

With these precedents, I say that the neural principles and cosmological concepts of today are seedlings grown in part from the artful speculations of the ancients, in the same way as the seeds of modern algebraic geometry were sown by the Arabs, Hindus, Chinese, and Greeks sometime around A.D. 1100. The history of these ideas is a lengthy and imaginative one which has influenced our lives in the deepest sense. Analyzed, it documents how the ancients formulated concepts that later shaped the modern "nuclear mind," setting in motion chain reactions of reason that continue to change our sensibilities and perspectives. To this day we remain impressed by the magnitude of the ancients' imagination in the service of their curiosity. They provided the necessary conceptual tools and "thought technologies" for studying ourselves. Two of the most compelling and provocative tools they used in their explorations are art and metaphor. It is toward the development of these instruments of knowledge that this chapter is directed.

Metaphor—like art—is about *experience*. One has to experience it in imagination, before analyzing its expressions for their meaning. When we are asked to identify metaphors in our speech, or consider their meanings, invariably some uncomfortable silence fills the neurospace like fog before it lifts with our broken response. And yet, we use metaphors as unconsciously and freely as we use water. We cannot speak without speaking metaphorically—as our spoken languages are constructed from symbolic building materials that are metaphorical in nature. Words are only symbols for their meanings. One symbol, sign, word, object, or idea *represents* another.

Our symbolic minds work by implicitly relating different things and processes. Through metaphors, we may relate something we know to something we don't. A rich metaphor can be a gold mine for hypotheses relating the properties of different systems. Such is the case of the German astronomer Johannes Kepler's concept of the "music of the spheres" that helped the science of astronomy listen anew to its descriptions of the universe. Kepler's metaphorical views of the cosmos, as expressed in his earliest book *Mysterium Cosmographicum* (1595) and in his more mature work *Harmonices Mundi* (1619), identified a number of unsuspected relations between familiar things: arithmetic, geometry, music, and astronomy. Kepler approached astronomy through metaphor, as he intended to "uncover the magic of mere numbers and to demonstrate the music of the spheres,"[1] the harmonious nature of the cosmos. His speculations, which were based on the rigorous astronomical observations of Nicolaus Copernicus and Tycho Brahe, provided the seminal thinking for Isaac Newton's theory of gravitation. And this theory opened the door to modern science—a door that Kepler's contemporary the Renaissance scientist Galileo had built from his belief in systematic, experimental evidence. Here the work of metaphor makes sense within the context of scientific research investigating unity in nature.

Metaphors were particularly valuable in these and other discoveries in that they demonstrated how similar models are often shared by dissimilar things. Models of the way one thing works can be used to describe the workings of very different things: Such were the comparisons relating a pump's mechanics to those of the human heart (Da Vinci); comparisons turning a clock's mechanics into the dynamics of the cosmos (Descartes); comparisons linking a factory to the operations of the human mind (Leibniz); and comparisons figuring the computational engine of a computer in relation to the brain (Turing).

There are innumerable examples of commonplace metaphors that have yielded the source material for extraordinary ideas. Two that are most confidently cited in the domain of science are the discoveries of

Isaac Newton and Albert Einstein. The twenty-two-year-old Newton observed the falling of an apple from a tree. The same phenomenon may already have been casually witnessed by millions of other people, some of whom, historians tell us, already conceived of such a force as gravity. Yet Newton was the first to dare conjecture that there must be a similar force acting on the stars and planets. This conjecture has since served to lift humankind into space. It may also prove to be useful in helping us conceptualize the influences of gravitational forces at work within and between human minds. No doubt as we continue to define the ambiguous nature of gravity we will, in turn, learn about the gravity of ambiguity —discovering how the mind gravitates towards certain matters of thought and creations.

As we've read in countless reports, the thirteen-year-old Einstein engaged in the process of metaphor when he envisioned himself riding on a beam of light; in effect, he become part of the beam as he "experienced" the system he was attempting to describe. This kinesthetic experience led Einstein to conceptualize relativity and to formulate his mathematical speculations on the relationships between matter and energy, time and space. It's as if the sensoriums of these two discoverers were completely open to the journey and reach of metaphor.

The history of human creations—from 14,000-year-old Magdalenian cave paintings at Altamira to the cerebral pursuits of theoretical physics and contemporary art—demonstrates that metaphors are universe-makers. With pleasant redundancy, the rhythms of metaphor drum the message that there are as many ways of interpreting the world as there are ways of representing our perceptions and responses.[2] Each representation touches on some aspect of the thing or process represented. The mental images we form in response to metaphoric concepts have meanings that extend beyond our descriptions of the concepts themselves. One could say they're as light or dense as the things to which they refer.

In describing the processes of art and metaphor, I use the word "metaphorm." A metaphorm is an object, image, concept, or process that we compare to something else. For example, when we say that the mind is a machine, the machine becomes a metaphorm which represents the mechanical aspects of the human mind. And vice versa: Mind becomes a metaphorm representing the organic or biological aspects of machines. Both concepts are expanded by this new association.

Neurocosmology insists that one consider *every* object, image, concept, and process as a metaphorm. All things are intrinsically metaphorms, whether or not one uses them metaphorically. Regardless of the context in which a thing or process exists, our minds can connect it to something else, while discriminating between connections. *Metaphorming* encompasses all forms of metaphor including analogy, allegory, allusion, symbolism, and trope, or figure of speech. Moreover, it can involve all of our physical senses, insofar as it implicates every mode of thinking, feeling, and creating. Many of the metaphorms in this book tend to be visual objects or ideas even though they're derived from nonvisual sources.

The words *metaphorm* and *metaphorming* are both verbs and nouns. They are the ways and means of implying likeness between things. In metaphorming something, we can traverse the constraints of logic and verbal thought, transferring or relating from one object to another a new meaning, pattern, or set of associations. Like the language of pure mathematics, which can describe abstract *n*th-dimensional processes and forms, the symbolic language of metaphorms is also multidimensional. It operates simultaneously on many planes of associations, nuances, and meanings. Metaphorms invoke the idea of forming, connecting, shaping *some thing* (or information) in our mind's eyes and hands.

In the next few pages we see how metaphorms are nature's instruments for uniting things and thoughts. They convey ideas that direct our awareness to relationships that elude full descriptive analyses. Metaphorms 1 through 5, for example, intimate that the arts and the sciences, universe and brain, share certain relationships—without being specific or defending this intimation.

METAPHORM 1. The Parallel Worlds of Art and Science. *Although they sometimes share similar interests and perceptions, they're essentially different in their representations. Science is generally thought of (by those outside this field) as a linear progression of events, research, etc. This linearity is often indicated by these types of visual notations. The world of art is shown as moving in the opposite direction towards infinity: As we see here the arts and the sciences parallel one another and come together (on some issues), although they never really meet on one point. If we were to substitute the words* brain *and* universe *for the words* science *and* art, *respectively, the meaning of this metaphorm changes abruptly.*

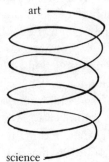

METAPHORM 2. Two Views of the Same Thing or Process. *Each view implies something else regarding levels or hierarchies, perspectives and motion. Again, substitute the words* brain *and* universe *for the words* science *and* art; *consider the implications of these changes.*

Those of you who are not satisfied with this simple visual notation, and who regard this relationship as being more complex, might indicate this by creating two lines or curves paralleling each other, or two planes intersecting, or this relationship can be represented as a matrix form, where systems of discrete points (representing science) are connected by a fairly organized system of lines (symbolizing art). To add a sense of action to these static geometrical forms, you might call upon the active symbol of the spiral to express this idea (Metaphorm 2).

These diagrams may still not satisfy those who see the arts' relationship to the sciences as being more complementary, closely interrelated, and, in fact, naturally intertwining. These individuals might be inclined to visually express this relationship in terms of a biological metaphorm; they might select, for example, the macromolecule of the nucleic acid DNA, which is present in all living cells (Metaphorm 3).

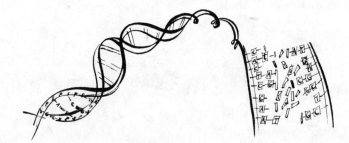

METAPHORM 3. Four Nucleotides Composing the Twisted Ladder of DNA. *Represented here are different combinations of interrelationships between art and science. With some knowledge regarding the anatomy and physiology of this extraordinary structure, you can translate the dynamics of DNA and its role in gene construction and chromosomal activity in terms of this metaphorm. Simply, you can understand this image on whatever plane of complexity you choose. If you're particularly imaginative, such things as the synthesis and multiplication of DNA will mean something in the context of relationships between the arts and the sciences.*

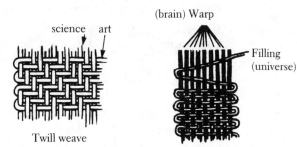

METAPHORM 4. Art and Science as Weave Constructions. *Each weave symbolically represents either science or art, brain or universe.*

For a nonbiological version of this scheme, we might adopt the example of weaves (Metaphorm 4).

If we were to generalize about fabric or DNA, and strip it of symbolism, we would reduce our experience of these objects to ordinary perception. They would also have little use in these pedagogical exercises.

Finally, the metaphorms above do not speak of the search for truth or meaning in the scienses and arts. The dynamics of this search might be described by using an astronomical analogy (Metaphorm 5).

We could generate countless variations of these sorts of comparisons to express similar messages through different representations. Most metaphorms enable us to think inventively about the relationships between commonly known and unfamiliar things. Many make the familiar even more familiar and poignant—or curious. Others are less productive. In fact, some can be downright misleading and manipulative in ways that reveal the worst of our vices. Along with mastering our abilities to conceptualize and connect diverse forms of information, we need to

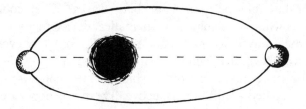

METAPHORM 5. The Arts and Sciences Symbolized as Apsides. *Two points in an eccentric orbit circle the Truth. One point (the higher apsis) is farthest from the center of attraction (the impulse behind the search for Truth); the other (lower apsis) is nearest to the center of attraction. The intimation: Neither art nor science is permanently situated in the upper or lower position. Neither one is The Truth, but both ride around the periphery of that unnameable netherworld of ambiguity we call matter. Their representations of the indisputable Truth about life and death in the cosmos and in ourselves are just that—representations. If we were to substitute for science-art the relationship of brain-universe, for example, the meanings of this metaphorm would instantly change.*

continue developing critical minds for evaluating our metaphoric creations. This includes learning how to pinch ourselves to examine what constitutes a good connection or analogy.

On a related issue, it's important to observe the types of transformations and consequences initiated by our acts of naming aspects of nature. A poet might call the stars "lightspheres" that warm our minds and illuminate our presence in space. What follows from this association is aesthetic pleasure. However, as soon as one assigns mathematical names to these imagined or real "spheres of light"—dimensioning them and detailing their mechanical properties—many consequences follow, some of which can transform our understanding of light itself.

Explorations in metaphorical thinking no doubt pre-date Hesiod in the eighth century B.C. Three special masters of this medium of communication (out of thousands over the millennia) include the Greek writer Aesop (in the sixth century B.C.), the Roman poet Phaedrus (in the first century A.D.), and Leonardo da Vinci—each of whom used the suggestive imagery of fables, for example, to convey insights into the world of human nature. In more recent times, writers, artists, poets, educators, and scientists have systematically explored the versatile "art of comparison and association" as a means of nurturing our impulse to connect and to discover.[3]

Physicists engage in metaphorming when creating words or images for describing novel relationships. If there is no existing word or expression to convey a concept or hypothesis, physicists simply invent one. Niels Bohr and his colleagues Hendrik A. Kramers and John Slater did just that when they invented the word "wavicle" to accommodate the descriptions of the behavior of particles as waves and of waves as particles, which they referred to as "probability waves." They coined these words in 1924 while developing a mathematical device that was designed to help particle physicists predict the probability of some event in the subatomic world ever happening or not happening.

The invention of these words, which helped convey the idea of probability, signaled a shift in perspective. There was now a new way of looking at the paradoxical relationship between the particle-like and wave-like properties of photons, the basic elements of light. As it turned out, nature surprised us; photons possess both properties. This point is central to the concerns of neurocosmology, not in the details of the theory of probability, but in the moving away from the notion of "predictability" which stood steadfast in classical physics. And, more central, is the realization that something as fundamental as a photon could have a complementary aspect or dimension to it. Immediately, the metaphysical and metaphorical overtones of complementarity had some concrete meaning, some material basis.

Most of these examples suggest how visual metaphorms, in particular, inspire hypotheses, and vice versa. In an informative essay on this subject Arthur I. Miller, a professor of physics at the University of Lowell, explains how essential visual thinking is in communicating a scientific theory. Discussing the use and disuse of visualizable models in development of the quantum theory, Miller relates: "There is a domain of thinking where distinctions between conceptions in art and science become meaningless. For here is manifest the efficacy of visual thinking, and a criterion for selection between alternatives that resists reduction to logic and is best referred to as aesthetics."[4]

This thought on aesthetics and visualizations in science is part of a trajectory of thinking in art and architecture that rises above the highest planes of analytic reasoning. Consider, for example, this inspired metaphorm—the ninth-century Buddhist stupa (shrine)-temple Borobudur in Java. This particular temple unites mind and cosmos through active meditation and metaphor. Its elaborate symbolic form, based on the mandala (a sacred, mystical symbol of the universe; see Metaphorms 6b and 6c), was designed to stimulate worshipers to swim and drift in the deepest waters of meditation. Devotees experience a sense of unity by physically moving through the mindful spaces. They would ascend from the five square-like galleries to each of the three ascending circular terraces. From there, they would travel upwards to the main stupa in the center of the temple where the spaces blossom like the artwork in each of these spaces. Here, galleries open or grow into terraces. As devotees journey upward, the storytelling changes tone.

The scenes of the Buddha's life, so clearly articulated at the outset, evolve into pure abstractions. Experiencing this evolution involves following with unbroken concentration the story as it evolves. Devotees engage this transformation to various degrees of enlightenment. They pass from the passive viewing of familiar objects and images (the known and the pictured) to the gradual, active letting-go of all recognizable forms (the act of embracing the unknown and the unpictured).

In this way a person experiences an altered state of existence—a rapturous moment that results from traveling inward in a focused and fused state while seeing outward toward the greater, less detailed expanse. This quiet way of enlightenment is suggested by the imagery and architectural *forms in transition* in the Borobudur temple, for example, where we go from representational to abstract images and back again. Experiencing this interchange of worlds between the purely abstract and the realistic, or naturalistic, is one aspect of the *art* of metaphorming.

The temple metaphorm flows between these complementary worlds in which material substance intermingles in the transmutable mind where everything imaginable seems possible. The early nonobjective

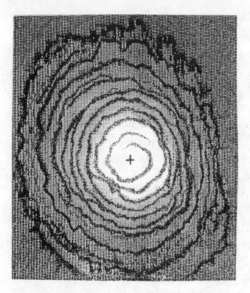

METAPHORM 6a. Betelgeuse, *a star 800 times the size of our Sun and 650 light-years distant in the constellation of Orion. This contour map is based on the potassium light intensities of this star. Like a visual echo, the "faceted," petal-like patterns of gas clouds emanate from the surface of Betelgeuse. The radiating cloud masses evoke T. S. Eliot's poetic image of the "fire and the rose," the eternally burning universe kindling life eternal—matter continually renewed and transformed by energy.* (From Michael Marten and John Chesterman, The Radiant Universe, 1980)

METAPHORM 6b. *The* mandala, *a sacred pattern of the universe, formed the conceptual foundation of the Buddhist temple Borobudur in Java (ninth century A.D.). The mandala emphasizes the process of spiritual growth, of the mind's evolution echoing the evolution of the universe. Upon entering the massive walls of the temple, devotees move through the square galleries towards the round terraces. Once they pass through these terraces, they approach mental paradise. This movement acts as a return to the primordial world some mystics call the cosmic consciousness.* (From Patrick Nuttgens, editor, The World's Great Architecture, 1980)

METAPHORM 6c. *An ancient tool for meditation, the* yantra *symbolizes the tension and integration of opposites through its nine interlocking triangles pointing upward and downward simultaneously. The triangles represent the male (Shiva) and female (Shakti) in the process of becoming united; the yantra invokes the union of the personal, ego-laden, time-conscious world we live in daily and the impersonal, egoless world without time we experience in our most liberated thoughts and in death. In Zen Buddhism, the circle symbolizes enlightenment; it embodies wholeness and "human perfection."* (From Carl C. Jung, Man and His Symbols, 1964)

painter Wassily Kandinsky called these worlds the poles of "great abstraction" and "great realism." As Kandinsky understood: "The poles open two paths, which both lead to *one* goal at the end"—unity. In describing this unity of abstract realism, our descriptions become diffuse and shadowy.

Metaphorms, then, imply relationships between things we cannot explicitly compare or literally equate. I mean, how would you depict Plato's notion of a "World-Soul" in his mythical dialogue *Timaeus*? As a free-floating ethereal body that has no distinct form or boundaries? As an elongated wavering figure—part human and part cosmos—painted in the style of the sixteenth-century Spanish painter El Greco? Neither one of these images is rich enough to represent the full meaning of this particular expression. And yet, we have a sense of what it means, just as we have some idea of what Plato meant when he wrote about our "purposively ordered rational universe." Some miraculous mechanism in the central nervous system, unbeknownst to all, allows us to take two things, which by themselves have distinct meanings (here, "World" and "Soul"), and relate them in a way that creates a new meaning which links these things.

The noted contemporary artist Arakawa and writer Madeline H. Gins identify this connective action as the "mechanism of meaning." As they have articulated: "Meaning might be thought of as the desire to think something—anything—through; the will to make sense out of the ever-present fog of not-quite-knowing; the recognition of nonsense. As such it may be associated with any human faculty."[5] Arakawa sees this process of relationship-making as the most essential part of our creative reach. It involves the transformation of words and images and ideas from one state of matter, or "being," to another—from one frame of meaning to another. In another interpretation, Jacob Bronowski writes: "The scientist or the artist takes two facts or experiences which we separate; he finds in them a likeness which had not been seen before: and he creates a unity by showing the likeness. . . . All science is the search for unity in hidden likenesses."[6] In the course of engaging this search, both artists and scientists produce discoveries whose range of expressions tests the authority of rational thought. (Some examples of contemporary art that support these general statements are presented in Branches #1 at the end of this book.)

Keeping these notes on metaphoric processes in the forefront of your thoughts, consider the museful question: In exchanging ideas about the brain-universe, must our ideas be amenable to all lines of inquiry and representations? Or can our metaphorms simply point and say "See here!"

Sit down before fact like a little child, and be prepared to give up every preconceived notion, follow humbly wherever and to whatever abysses Nature leads, or you shall learn nothing.—
THOMAS HENRY HUXLEY

True science teaches, above all, to doubt and to be ignorant.
—MIGUEL DE UNAMUNO Y JUGO, *The Tragic Sense of Life*

2

The Three Principles
of Neurocosmology

Neurocosmology adopts the attitude of the well-intentioned arbiter who tries to mediate between three antagonistic, though complementary, views of the world. The views are presented by three groups of explorers, some of whom vie for the glory of authority. One group swears that its logico-mathematical thinking is truly the only enlightened way to understand the whole world. The other group insists that only through aesthetics and poetic insight can we know our world. The third group is convinced that the two other groups need each other and other philosophies for representing the phenomena that *lie* before their eyes. After patiently listening to the complaints of one group about the incompleteness of the other two methods of inquiry, the arbiter decides that all three groups are basically "right"—that their world-views are self-consistent and correct. Immediately, someone protests: "But we *all* can't be 'right'!" To which the arbiter retorts, "You know, *you're right!*" The final blow to this fable was wielded by one who asked why we should believe this arbiter in the first place!

The attitudinal problem of this arbiter is not that he doesn't leave us with anything—leading us with a singular and profound viewpoint. The problem is he leaves us with *everything!* The statement that each of these complementary views is "right" implies that potentially *all* views are permitted (though not necessarily welcomed!). We might immediately conclude that with so broad a view we're left with *nothing*. On the contrary, this broadness implies that we can't rely on any one of the systems of thought created for discerning the merits of one person's theories, models, or ideas and another's. We can't completely count on any one of our methods for discovering nature. Instead, it intimates that we need most of them. One way of handling this relativism is by understanding that it is the *combined insights* from various views that determine the clearest picture of the brain-universe—a picture that is universal in scope. Meaning it is defined as "a whole collectively or distributively without limit," like the word *universal*.

THREE PONDERANCES THAT DRAW US TOGETHER

The brain is the grandmaster of propagating theories about itself. There are potentially an infinite number of theories and models of brain processes. This infinitude is echoed in models of the physical universe. There are no definitive models, since the source of what is being modeled —nature—is constantly changing in form through its "stable" of processes. Similarly, there are no bedrock sciences or mathematics that are the foundation of all other forms of knowledge, uniquely privileged in their representations of nature. The body of scientific knowledge is as transient as the human beings who create and feed it.

Each theory may have evolved from a unique angle of analysis, but ultimately our theories form from—and refer to—a sort of unified, collective mind. Some theories are, of course, more meaningful than others, stimulating new intellectual outlooks. The best are able to model many aspects of our world as they presently exist and as they may continue to exist indefinitely. A theory's value lies in its ability to predict or to explain the behavior of some studied phenomenon, providing symbolic models that lead to insight. However, various factors, such as contexts and values in the arts and sciences, influence our assessment of this ability. In neurocosmology, whether a hypothesis or theory proves to be valid is almost inconsequential. All that's important is that it direct us, constructively challenging our previous approaches or canons. In this way it is productive.

The three basic techniques of speculative inquiry used in neurocosmology employ both inductive and deductive strategies. Human and cosmic processes are explored from the data gathered and examined in both a narrow and broad sense.

FIRST PRINCIPLE

Study the brain vis-à-vis the universe and vice versa. Books about the universe may be interpreted as metaphorms representing the human brain (mind). Similarly, current studies in neuroscience that are used to determine cell sociology in groups, cell-neuron connectivity, and plasticity can be discussed in the context of the organization of the universe, which includes its rules of assemblage and behavior. As we read about stellar populations or the special features of a select cluster of galaxies, we can infer things about the populations of nerve cells and ganglia of which we are composed and which compose our worlds of mind.

The implication is that knowledge of the brain is not merely supplementary information for studying the cosmos—it is key. Knowledge of both things relates to two different aspects of one thing: energetic matter that is continually materializing as different forms of information. This knowledge is as *central* to our experience of nature as information is *essential* to our nature.

SECOND PRINCIPLE

Question everything about the *neurocosmos*, including your own questions (. . . says Nature.) In settling for well-packaged explanations, we rarely scrutinize the authorities responsible for producing these explanations and issuing them like patents. Just because people haven't considered asking questions in a rigorous way about the nature of the brain and cosmos doesn't mean they're not qualified to reflect on these subjects. In order to grow and survive we need to uncover some common ground to talk about what we're experiencing from our unique viewpoints. We need to develop a means of communicating with one another that is sensitive to—and even tolerant of—contrasting points of view. Our talking must "take into account," rather than talk in defense. Methods of refutation and analysis must be flexible enough and evenly tempered to balance rigor with a sort of disciplined permissiveness.

THIRD PRINCIPLE

Use every means to question everything about the brain-universe. To the visual artist, for instance, this principle might suggest that painting, sculpture, and graphics are appropriate media for exploring this relation —certainly as important as particle physics and cosmology, which apply mathematics as a medium for thinking poetically about the geometry of the universe. The "whole" idea of using *every means* welcomes us to think with the full spectrum of analysis and creative expression. The scientific method is only one means of formulating and investigating ideas.

The procedural style of the neurocosmologist is to consider the various theories and models as an ever-changing whole. This is necessary in order to construct a picture as large as this subject demands. These principles reflect our impulse to *question* how we know something and how we apply this knowledge. Or how we should best explore our actions and creations in relation to those of the cosmos.

By applying these principles, we can educate ourselves about the world—as we see it and as it is presented to us through the representations of others. We can also create a discipline that establishes broader guidelines, or criteria, for judging the quality of insightful explorations. Even though the concept of *quality* is as ethereal and tenuous as twilight, we seem to possess a sixth sense for distinguishing types of quality. Basically, disciplines help those who participate in the gamework of thought to maintain some means of evaluating and guiding their work.

One concern of neurocosmology is developing a common language and system of conveying, or sharing, ideas. The informal rules of communicating as neurocosmologists hinge on one axiom, which underlies the three principles of neurocosmology:

0. Suspend what you know or have been taught, in order to experience what you don't know. And if you don't know anything worth suspending, then simply listen or observe.

The three principles of neurocosmology are summarized here:

1. Through metaphorms, relate everything you do, see, feel, create, or experience in terms of brain processes. Also, relate the processes of every discipline or field of knowledge to the processes of the human brain. (This directive requires that you first learn something about the brain and then question what you've learned.)
2. Through metaphorms, relate everything you do, see, feel, create, or experience in terms of the universe's processes. (This requires that you learn about the universe and then question this knowledge.)
3. Interrelate the results and experiences from 1 and 2. Then, ques-

tion what these interrelations mean or suggest to you. Why is one metaphorm more relevant and meaningful than another?

Exploring these suggestions amounts to rethinking every aspect of nature. It means respectfully questioning the preconceptions and assumptions that underpin every model interpreting nature. What may result from these inquiries are more versatile theories and theorists— people thinking with more generous hypotheses about our nature. We need to continually see how the models we create re-create us in another form. Our models of nature (of matter and mind and their interactions), both release and entrap our imaginations in our unquestioning devotion to them.

3
Metaphorming Neural and Stellar Systems

There is an expression, and exercise, in figurative drawing that most people have either heard of or practiced. It's called the "figure-ground" relation. In drawing a still life or human model, you learn to see *the figure in relation to the ground*—the space or place that supports the object of your study. It is an invaluable analytical exercise that leads to glowing reward for those who are unaware that we are connected with our environment, that our environment supports us in a more generous way than simply gravity holding us in place. This gripping awareness can be conveyed by the gestural sweep of a graphite line under the feet of the subject of the picture. Or it can be communicated by scumbling paint around a hint of an arm or head. Otherwise the image floats disassociated and suspended, drifting in the spaceless environment of the empty paper. The message here is that the figure isn't studied independently of the ground. They are studied together, as each informs the other about its material form. It's so easy to forget that the figure has a ground and that the ground, without a figure in its midst, remains a blank abstraction.

Science studies nature in an analogous way. Some phenomenon is seen or identified in relation to its supporting environment. The interactions between the two things are often explored in the context of cause-and-effect relations. There are hundreds of examples, ranging from the hindsight sciences of pathology and etiology to the predictive science of meteorology.

THE FIGURE-GROUND RELATION: THE BRAIN-UNIVERSE METAPHORM

I introduce the concept of the figure-ground to suggest that we study the human brain *(the figure)* in the context of the physical universe *(the ground)*. Sounds simple enough. Instead of painting an apple on a table, we're painting the brain in the universe. But what does that entail? What are the details? Are we talking about liberally studying brain processes as though they were extensions of the universe? A sort of neurophysiology interpreted from the perspective of particle physics? Or are we talking about picking up where Aristotle left off in the seminal work *Metaphysica*, his treatise on the "ultimate causes and underlying nature of things"? A sort of neurophysiology wedded with metaphysics—or neurometaphysics or meta-neurophysics—in which we factor into our knowledge of things the influences of the human brain? I suppose the answer is *both*. As long as we build from observation and empirical work, there's little danger in these conceptual hybrids and proposals for disciplinary syntheses.

The scope of these suggestions may weary you. Admittedly, it's one thing to loosely ground some figure with a symbolic line or to casually recommend that we understand something in a broader context. It's quite another thing to cross-relate detailed information about the brain and cosmos (see Metaphorm 1). Both efforts—the one relating figure and ground and the one interrelating these different forms of knowledge —are similar in scope. Both are important to pursue. Both encourage us to experience our larger identity as human beings, to expand our outlook in being human, as we look out with a common vision. This changing vision allows us to see things in their changing contexts and to see all things as relationships.

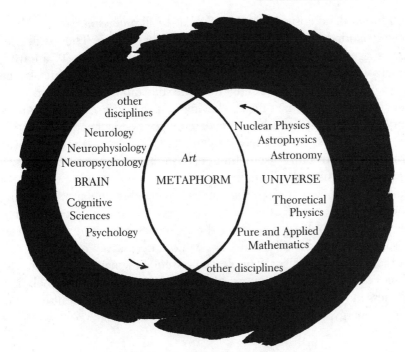

METAPHORM 1. Brain-Universe Relation—*combining neuroscience and cosmology through art, in metaphorming neural and stellar systems. In comparing these entities I proceed with the caution of a soldier crossing a minefield of speculations —where scientific method and factualism lie waiting to explode with one false step in concept formation. I intend to float above this field by metaphorically pointing out certain similarities in the dynamics of brains and stars. This is a far safer path to take, rather than generating theoretical neural-stellar models—whatever they would be. Such models would most likely remain in a state of suspended imagination.*

ENGAGING PROCESSES OF LIFE

In this chapter I will be examining the nature of relationships in terms of processes, emphasizing the likeness of processes in biological and physical systems. This examination will de-emphasize the issues of size, form, function, magnitude, and time. In the realm of processmorphs, everything may be "blown out of proportion" or re-sized, as there are no levels or hierarchies, or barriers and boundaries, separating one form of matter and activity from another. The *way* of something is more important than the *look* or *appearance* of something. Measurable quantities (of structures) are linked together through immeasurable qualities (of processes).

Neurocosmology is essentially a comparative study of processes in both human and cosmic creations—what I call *processmorphology*. The concatenation of *process* + *morph* creates *processmorph*. The word "process" suggests a distinct forward movement from one point to another, usually with a particular purpose and end result. "Process" may also be the transformation of something from one state or another without some specific goal. "Morphic" means having a specific form, as in "homomorphic" for crystalline forms between unlike chemical compounds.

But a process needn't have any visible form at all, as it does in the term "process structure" (sometimes used in thermodynamics). The concept of processmorphs intimates that two or more things can share a similar process even though they may not resemble one another in any other way. For example, the neural tissue we identify in microscopes and the veil-like nebulae we observe through telescopes may be processmorphs, possessing similar functions without necessarily resembling one another. The dynamics of nerve cell–assemblies may parallel the dynamics of clusters of galaxies or interstellar gases that make up nebulae. Likeness in *form* is only one of the likenesses between different forms of matter, energy, or information. (See Metaphorms 2 and 3.)

ART OF THOUGHT

The running commentaries that accompany the following metaphorms do not explain the sketches or drawings. They merely establish the sources of information from which the metaphorms are built. Here are a few metaphorical equations which explore the concept of proportion: brain ≈ star, brain ≈ galaxy, brain ≈ cluster, brain ≈ supercluster, brain ≈ universe, where the sign ≈ means that the concept on the left is implicitly compared to the concept on the right. Note that the two concepts are discussed in terms of relationships, not equivalencies or isomorphisms.* As we pluralize the words "star," "galaxy," "cluster," "supercluster," and "universe," the metaphorms change in complexity as well as in their meanings. The implications of each relationship seem to shift with the changes in *scale* (see Metaphorm 2), even though the processes may remain the same. Since we're comparing disproportionately sized objects, the concept of proportion—and, more important, *the lack of proportions*—weighs heavily in these comparisons. The lexical definition of *proportion* reads:[2]

---

* *Isomorphism* informs us about the relationships between *structures*, where the materiality, or form, of one thing maps onto the form of another. By contrast, *processmorphism* informs us about the relationships between *processes*, where form is not an issue—nor are time and space, which are both "structure-bound" concepts.

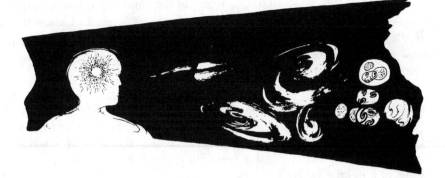

METAPHORM 2. Stars, Galaxies, Clusters, Superclusters. *Shifts in scale and form alter the scope of these comparisons. To borrow an analogy from set theory in mathematics: The mind is a subset and an emergent feature of a larger set, the states of brain, just as a star is a subset and emergent feature of a larger set, the states of cosmos. As we increase the number and size of sets, we expand our definitions of the forms of things. But we do not necessarily expand the number and size of sets of processes.*

1. A part considered in relation to another part or to the whole. 2. The relationship between things or parts of things with respect to comparative magnitude, quantity, or degree. 3. A relationship between quantities, such that if one varies, another varies as a multitude of the first; ratio. 4. Harmonious relation; balance. The verb form bears the following meaning: To adjust so that the proper relations between parts are allowed.

Each of these definitions extends the common view that *proportion* means *size* alone or some related concept of measurement. When we speak of the proportions of something, we rarely have in mind a thing's proportion *relative to* something else. Now imagine a world without proportion—where all *processes* defy measurable scale.[3]

By referring to the *brain as a collection of stars*, I'm implying either that the individual neurons which make up the brain are metaphorically like stars or that whole subsystems of the brain represent stars. Each "soma-star" (or cell neuron) would have its own energy output. No two somas would possess identical features or outputs.

Extending this reference, if the brain consisted of a galaxy or cluster or supercluster of stars (neurons), there would be some unspecified relationship with and influence over one another. The physiological behavior of the brain would be different than in the first case scenario. Or would it be? What really changes by these jumps in scale and in complexity? More complicated forms or structures? More processes? Increased order and organization?

This idea is in contrast to the notion of the *brain as a universe*—one containing billions of neural stars, where the thoughts and ideas the brain creates are analogous to the stars and galaxies the cosmos creates and where a concept consists of a constellation of stars held together gravitationally by a central theme.

As I stated at the outset of this chapter, the aim of this exercise in comparative study of processes is to observe how our interpretations of both brain and cosmos change with respect to these singular and plural case scenarios. As you would expect, discussing the lives of a star is orders of magnitude easier than describing the life events and interactions of an entire galaxy or cluster of galaxies. The same may be said in discussing the human brain. Reporting on a single living neuron is relatively simple compared to representing the constellations of neurons and functional anatomy of the whole nervous system.

In Lewis Thomas's important book on creative seeing, titled *The Lives of a Cell*, the author mused about the same issue concerning proportion. He writes:

> I have been trying to think of the earth as a kind of organism, but it is no go. I cannot think of it this way. It is too big, too complex, with too many working parts lacking visible connections. The other night, driving through a hilly, wooded part of southern New England, I wondered about this. If it is not an organism, what is it like, what is it *most* like? Then, satisfactorily for that moment, it came to me: it is *most* like a single cell.[4]

Although the discussion in this chapter principally considers the scenario of the *brain as a star*—as opposed to the *brain as a universe*—I don't mean to suggest that the latter scenario isn't the case. Each neuron in the brain may well be equivalent in its energy-production process to a single star—that is to say, on some unthinkable, relativistic scale. This would leave every human being with literally a universe on their shoulders! This ever-expanding, open universe grows in all directions within the compound of one's relatively fixed cranium. This thought may empower those who, in moments of intense self-doubt, see only their bleak smallness and unspoken insignificance, those whose egos are inversely proportional to the expanse of their neurospace.

What lifts this metaphorm out of the dull drums of truism is that there may be some physical connection here that's worth exploring further. By applying the constraints of the scientific method in teasing apart the literal and metaphorical aspects of this comparison, we may arrive at some new insights.

It's irrelevant whether the literalness of this metaphorm is truthful or not. What's important is that it binds our small expendable lives to the vast unconsumable life of the universe. It makes every human being (and perhaps, every thinking life form) as "big as life" or as "large as the whole of nature." The metaphorm is also positive in that it leads us away from the myopic thinking that sees every facet of human behavior only in terms of the tangible deeds of neurons. Our lives are more than the summed actions of the smallest functional units of the brain. Our thoughts are more than the interactions of these units, as our societies, cultures, and civilizations attest to. Similarly, the life of the cosmos is more than the summed actions of its largest functional units.

The metaphorm hints that our present views of the brain may actually limit the properties of neurons. What about the reverse? Do our limited views about the dynamics of neurons reflect the full capabilities of neural systems? We may never know. The only sound thought worth building on is this: Most of what we now rely on as bedrock information concerning neural and stellar systems will probably be supplanted in fewer than 50 years.

By connecting the processes of these diverse systems, we can transform the way we think about neurons, human behavior, and the integrated workings of our societies and civilizations. They can awaken our minds to the possible—though possibly unlikely—attributes of nature.

A HEATED POINT: RELATING THE UNRELATED

Nowhere in my comparisons do I intimate that there are hot plasmas (16 to 100 million degrees Kelvin) in the human head! Amusingly, I was once asked by an eminent physicist whether my expression "intuition: the plasma state of mind" implied this absurdity. No, the interior of the brain doesn't appear to be *literally* a high-temperature, high-density region where atomic nuclei violently and randomly collide to form the ionized gases of stars. No, the physiology of the brain is not identical to the physics of the Sun. The products of neurons are not like photons which disseminate from a star's core to its radiative zone to the convection zone to the photosphere where they light up the solar atmosphere in spectacular colors churning in turbulent seas.

The one point I suggest we coolly explore is how a star's primary mechanisms for transporting energy—convection and radiation—are similar to the ways in which the human brain "transports" energy. Consider the implications of Metaphorm 3.

BRAIN   STAR

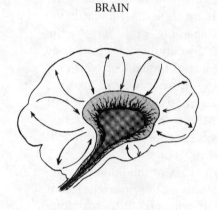

Cerebral Cortex   Convection Zone
(thinking)   *(transporting heat energy)*

Limbic System   Radiative Zone
(feeling)   *(release of nuclear energy,*
                                      high-energy photons)

Brain Stem   Core
(acting)   *(thermonuclear reactions)*

METAPHORM 3. The Brain and Star as Processmorphs—Alike in Process but
Unlike in Form or Appearance. *Although the Sun doesn't "think" or "feel," at
least not as these words are defined, the processes of thinking-feeling-acting (creating) may be similar to those of its biological extension. One similarity is that
both systems fuse and split different forms of matter. (In Chapter 8, I discuss these
processes in some detail.) Both brain and star are open systems that operate at far
from equilibrium. Both are self-organizing. Both consume fuel—matter and energy (or information)—and experience the process "feedback."* Thus cerebreosyn-
thesis *is the complement of* nucleosynthesis; *the latter is the means by which stars
undergo nuclear fusion in sustaining their lives. The general principles of energy
conversion are compared here—not their details.*

MINDFUL ILLUMINATIONS

In metaphorming neural and stellar processes, I'm suggesting that there's
a consistency in the way energy is transported and converted in both
systems. One way of transporting information (or energy) in the brain is

via synapses; in Greek, this word means "connection."* "Is there an 'integrative action' of the star system?" a neurocosmologist might ask, while pointing to the three interpenetrating regions of the Sun.

By comparison, one way of transporting energy (or information) in the star is via conduction, radiation, and convection (note Metaphorms 3 and 4). Conduction involves the collision of energetic atoms in the core. In radiative transport, the energy from these collisions flows outward into sections of the star. The energy subsequently diffuses as it is scattered and absorbed in the plasma of the radiative zone. And in convection, heat energy is transported by means of mass thermal motions which make their appearance on the Sun's surface.

On a related topic, the astrophysicists Michael Zeilik and Elske van Panhuys Smith write: "Because stellar luminosity represents energy loss, no star is perfectly static . . . stars *must* evolve because they lose energy to space."[5] This statement reiterates the point that stars are open systems which struggle to resist entropy like most life forms. Astrophysical research supports the following scenario in the life and death of a normal, healthy star. It starts out feeding itself the lightest of elements, hydrogen. For billions of years a star burns hydrogen, transmuting countless tons of this element into helium and other elements. As the hydrogen cycle begins to tire, the helium cycle takes over, transmuting numberless tons of helium into successively heavier elements.

There are always boundless exceptions to these evolutionary scenarios for stars. Astronomers have recorded novas and supernovas whose abnormal behavior and energies seem to violate cosmic laws. These aberrant astronomical bodies practically rip space apart as they explode with tumultuous force. One parenthetical note: In 1885, a supernova was observed in the Andromeda galaxy (M31); it was said to be one-tenth as radiant as the entire galaxy. You should shield your eyes in just *imagining* the brilliance of this luminous supernova!

---

* The term *synapse* first appeared in Sir Michael Foster and Sir Charles S. Sherrington's *Textbook of Physiology* (1873). Sherrington referred to the synaptic elements as the key to internal communication—or neurons conversing with neurons. They occupied the central theme of his classic *The Integrative Action of the Nervous System* (1907), which helped transform an abstract concept, synapse, into a concrete and testable mechanism that can be studied on both molecular and cellular levels.

ILLUMINATING THE NATURE OF MIND

Why suggest that something as gigantic and luminous, as fiercely energetic and seething, as a supernova resembles our relatively fragile and quiet brains? Do all processmorphs share the same ancestry of process? I'm inclined to believe that the more substantial connections in nature have more to do with processmorphism than they have to do with isomorphism or homology, which deal with likenesses between seemingly diverse structures. As open systems, the brain and cosmos "live" in the way we know all life-giving processes live—"experiencing" their environments through flowing exchanges of information, matter, and energy.

What do we really know about our physical connections with our stellar cousins? For the sake of exploring an oblique angle of perspective, let's say, "We don't know very much." In fact, we understand so little about these connections, in part, because we're just beginning to understand these two systems of nature. We're dumbfounded anytime someone questions *basic*, *scientifically grounded facts*, such as: The temperature at the core of the Sun ranges from 8×10^6 Kelvin to 20×10^3 Kelvin. How do we *know for certain* that the interior of the Sun has this temperature? Have we ever visited this core technologically? No. Have we ever even penetrated the photosphere, a region some $1\ R_\odot$, or 6.96×10^5 kilometers, from the core, with our remote sensing, photorecording devices? No, we haven't.

WHAT ARE WE REALLY OBSERVING

The closest we've come to inspecting the interior of a star is through our mathematical constructs guided by observational data, extrapolations, and educated guesses. But what are these guesses actually based on— empirical data gathered from the depths of the source itself? No. Data assumed from some ingenious speculative inquiries? Yes. We could press that a poet can see the inner life of a star with the same degree of precision and truthfulness as an astrophysicist. Good astrophysicists are poets in this sense. They're able to describe, for example, the interior of a supernova as a "cosmic onion." However, they're also able to show symbolically (mathematically) how the concentric, ring-like organization of matter at the core of a supernova burns from hydrogen (over 10 million years) to helium (over 1 million years) to carbon to neon to oxygen to silicon to sulfur and so forth. It's the calculations of temperatures and densities that punch the poetics into light-speed, as we learn how the core can heat up from 40 million degrees Kelvin to 3.4 billion degrees—all within 11 million years.[6]

At the core of these exploding stars, the elements are presumably "layered in order of increasing atomic weight."[7] At least this is the report of two principal investigators of supernova phenomena—professor of astronomy and astrophysics Stan Woosley of the University of California at Santa Cruz and physicist Tom Weaver of the Lawrence Livermore National Laboratory.

"Are there comparable rises in 'heat' in our imaginings?" you might wonder. "Is there a similar layering of the 'atomic weights' of thought?" probes a neurocosmologist. This question may be meaningless in terms of our current understanding of matter or energy or information. However, it may prove to be meaningful as we grow to understand the broader properties of matter which the ancients intimated as being related to the properties of mind and life.

To be clear: There is a big difference between the poet's perspective and the astrophysicist's perception. The latter is intent on describing and quantifying various aspects of these physical phenomena. The poet, on the other hand, is content with embracing their mystery and, at best, providing insights into their ontology.

HOW DO WE INTERPRET WHAT WE OBSERVE

This comparison between observers and types of observations prompts another point—one concerning methods of observing these phenomena. In any methodology, there are inherent flaws in the ways things are observed and described. For example, in attempting to observe the structure of the Sun, astrophysicists have relied on two groups of observations for judging the depths of this solar mass. One set is based on measurements of the neutrino flux from the Sun. (Neutrinos are massless, chargeless, elementary particles that are extremely hard to detect.) It turns out that the neutrino flux is one-third the force predicted by solar models. And so these theoretical models are inaccurate (but are currently being revised), as are the theories that outline the properties of neutrinos based on these models.

The second group of observations for determining solar depths is predicated on analyzing the motions of gases on the surface of the Sun. The gases have spectral lines that reveal these motions. The magnitudes of the shifts in frequencies of waves in these spectral lines (called "Doppler shifts") have been interpreted as oscillations from sound waves that are believed to emanate from the interior of the Sun. Apparently, the

Sun oscillates (or pulses) with distinct periods and patterns. Its heartbeat-like pulse is read by such sensitive devices as helioseismographs. These operate on the same principle as a seismograph, which listens to the acoustic waves rolling through Earth's interior landscape. Helioseismology is most helpful in sketching the Sun's layered inner world. However, there are many unknowns that need to be clarified concerning the origin(s) of these acoustic waves before the accuracy of this technique can be assessed. Where geology shows the Earth as having conformable and unconformable strata (resulting from crustal movements), it's not so clear that helioseismology can show how these sorts of strata, or a similar model of stratified structures, are representative of the Sun's internal anatomy.

Returning to this issue of observation: The way we observe something influences our description of the thing we observe—and, ultimately, our experience of it. Insofar as our descriptions are dependent on ordinary language, they are intrinsically ambiguous. The combination of these inadequacies in our observations, descriptions, and languages creates the Achilles' heel of facts, knowledge, and communication. Translating our observations into mathematical expressions helps to filter out some of these ambiguities, but certainly not all.

WHAT DO OUR INTERPRETATIONS MEAN

One final arrow of a note aimed at this heel: Even if we could probe directly the secrets of the Sun's core (or the interior of any star body for that matter)—even if we could provide a direct account of its physical actions in living detail—we *still* wouldn't be able to corroborate our general theories about the internal dynamics of celestial objects, no matter how much evidence we amassed. As the philosopher of science Sir Karl Popper expressed, no matter how much evidence one provides in support of a general theory, one can never prove that theory completely as there may be some evidence one's omitted. In the process of verification—or falsification—to the degree that one cannot find any evidence that contradicts one's hypothesis, then the probability of its likelihood rises accordingly.

TOWARDS INCREASING THE LIKELIHOOD
OF PRODUCTIVE THEORIES

Only recently have we learned of a star's interior dynamics. Early in this century, astronomers got their first conceptual glimpse of a star's thermonuclear reactions through the work of the American physicist Hans Bethe. In 1930 Bethe theorized that the stars shine by means of a fusion process. He referred to this process as *nucleosynthesis*, whereby light elements merge to form heavier elements.[8] His theory, which was awarded a Nobel Prize nearly 40 years later, was informed by Einstein's monumental concept equating matter and energy.*

Before Bethe's theory of thermonuclear reactions, all sorts of theories were sported about the Sun's energy source. The most commonly held view was that the Sun was a coal-like burning furnace. Another, more sophisticated concept describes the radiant energy of a star in terms of increasing pressures and temperatures. This increase causes compaction. As particles collide more violently, their collisions produce more energy. Although stars undergo something similar in their formative stages, if they sustained such activities, they would shrink to oblivion in the breadth of a million years or less, rather than many thousands of times this number.

AT THE CORE OF OUR BEING

To date, we understand that a living star, such as our Sun, is a gaseous object with an imposing mass of 1.99×10^{30} kilograms (equaling 333,000 Earths) and a radius of 6.96×10^5 kilometers (about 109 radii of Earths), making it appear to be as immovable as our terrestrial globe.[9] The gaseous matter that composes this light sphere is tethered at its center by immense gravitational attraction. Theoretically, no foreign object can perforate the core because of its density, unless of course an enormous projectile, as dense as a neutron star and traveling close to the speed of light, were to plunge into it dead center. As one might expect, the projectile would punch through the Sun's core as cleanly as a 22-caliber bullet passes through a ball of steam unnoticed. Even though we can simulate this perforation by first calculating the density of the Sun and the density and momentum of the projectile, we still would never be able to experimentally test this thought for whatever reason. There are just too many unknown variables involved that shadow our knowledge of the interior of stars. Similar thoughts can be posed regarding the effects of stirring up the well-defined, concentric layers of the Sun depicted in most diagrams. In reality, these virtual layers of gaseous rock may not be so immaculately stratefied and visibly distinct (Metaphorm 4).

---

* Around 1904 Einstein posited that any material, irrespective of its composition, could be converted into pure energy—releasing almost inconceivable amounts of energy.

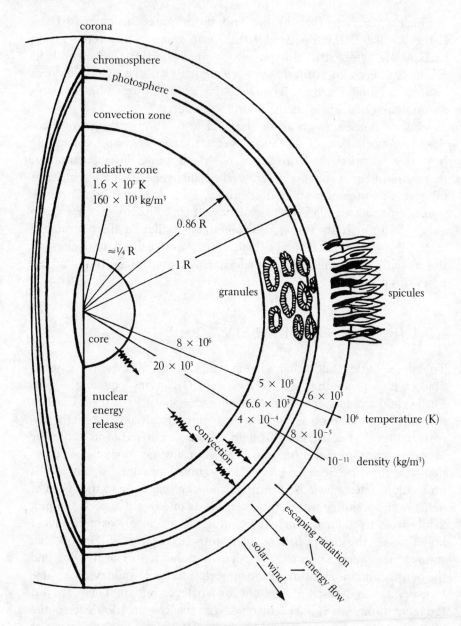

corona

chromosphere

photosphere

convection zone

radiative zone
1.6 × 10⁷ K
160 × 10³ kg/m³

0.86 R

≈¼ R

1 R

core

granules

spicules

8 × 10⁶

20 × 10³

nuclear
energy
release

5 × 10⁵

6.6 × 10³

4 × 10⁻⁴

6 × 10³

10⁶ temperature (K)

8 × 10⁻⁵

10⁻¹¹ density (kg/m³)

convection

escaping radiation

solar wind

energy flow

METAPHORM 4. The Structure of the Sun. *(From Michael Zeilik and Elske van Pan-huys Smith,* Introductory Astronomy and Astrophysics, *1987. In the original caption it is stated that the granules and spicules are not drawn to proportion.)*

One remarkable illusion of the Sun is its virtual surface, called the *photosphere,* or "light sphere," which shimmers like ageless coals glowing at a temperature of about 6,000 Kelvin. This temperature range is close to freezing, compared to the temperature at the core (20,000,000 K) or in the corona (1,000,000 K). The oceans of photospheres of other stars have temperatures ranging from a cool 2,000 K to a searing 500,000 K. Suspended above this "surface" hovers the *chromosphere*, or "color sphere," which is a loosely layered veil of gases between 1.5 and 15 megameters thick; that's many times the diameter of the Earth. Within this region *sunspots* (whirlpools of plasmas stirred by strong electromagnetic forces) and *flares* (whose energies equal millions of hydrogen bombs exploding simultaneously) and *prominences* (which shoot skyward, some at 500 kilometers a second, arching at heights greater than 100,000 kilometers) continuously form. These riptides of gases burst into the expanse of the corona as if to gasp for air, only to be dragged downward into plasma oceans by the gravitational hands of the solar body. Every so often these bulging masses of plasmas slip beyond the Sun's reach, roaring silently into deep space. The corona, more restless than the energetic clouds of billions of hurricanes combined, makes up the outer atmosphere of this spherical furnace. In the shadow of the corona floats the Earth. Some 150 million kilometers away, the Earth faces these solar eruptions as it survives on the surplus of the Sun's life-supporting light.

BEING PART OF A GRAND ILLUSION

Perhaps the greatest illusion of the brain is its basic solidity in structure. This feature is in contrast to its unfirm forms of mind, which cognitive scientists call *mental architectures*. The "architectonics of thought"— meaning the structure and organic order of thoughts—are as illusory as the Sun's light sphere. On the one hand, the brain's cytoarchitecture (the structure and organization of nerve cells) has very specific geometries. In fact, there are seven layers of nervous tissue that connect the neocortex to the more diffuse geometries of cells that join one another in the hindbrain. On the other hand, the types of "geometries of thought" these cells create are unlike anything we have ever seen or have ever touched in physical reality. Either that or our mental architectures are like *everything we see around us* as tangible entities. No one knows which is the case.

Anyone who has had the pleasure of studying human neuroanatomy often zeroes in on the ponderous question: How does the functional architecture of the neocortex—with its 7,000 million or more neurons of various types coordinated together by about 9,000 miles of fibers for each cubic inch—"think"! How does this web of energy contribute to the interrelated processes of our thoughts, feelings, and creative actions? Why do we refer to the neocortex as the "cerebral" cortex only, instead of the "sensual" cortex, which it also happens to be in that our sensuality and intellect are both established and continually manipulated by this brain mass? As you look at this biological star, or "biogalaxy"—our cerebral-sensual universe—this question is amplified. One might respond to such thoughts by trying to depict the actions of the mind. The interpretive images and metaphorms we create would not only represent some aspect of human mental activity, they might also re-phrase the question such that we have new ways of exploring the phenomena in question.

⦿ THE FIGURE

The brain doesn't appear to be some sort of "outgrowth," "projection," or "protuberance" of the universe. Rather, both systems seem to undergo a metamorphosis of matter, energy, and information. In effect, they each evolve out of the other. Neither neuroscientists, physicists, nor metaphysicians can agree on their accounts of this evolutionary anomaly. Nor can they come to terms with the brain's powers of virtual transformation. It creates its own categories with which it classifies itself; and then, almost predictably, it challenges these abstract categories and the criteria it used to establish the categories in the first place! It continually pulls the rug out from under itself every time it gets too comfortable with the idea that it knows itself. Evolution does the rug pulling.

Perhaps, with its will-o'-the-wisp production of nuclear weapons, the human brain extends or mimics the dynamics of stars, which are perpetual hydrogen bombs. It's a curious thought, albeit troublesome, to consider that the brutal energies stirring overhead are as much a part of us as the flesh covering our bodies. When I think about the ordered chaos of the cosmos, I'm left wondering whether we are in fact some evolving form of the heavenly infernal machine—a biophysical machine that is infinitely more compact than its celestial counterpart, disguising its explosive and destructive energies within its gentle facade.

⬤ THE GROUND

Normal stars, like brains, are hydrostatically and thermally well adjusted. By this I mean that they burn at a balanced rate—neither too rapidly nor too much—consuming their nuclear fuel at the rate of millions of millennia or more. (This, of course, does not apply to the brain except perhaps in a strictly spiritual sense!) In spite of their immense size, nature has taught them to pace themselves. Our Sun is no exception. Smaller than average and fairly unathletic—emitting about 5.4×10^{33}

joules of energy each year (compared to 8.5×10^{43} J of other stars)—the Sun exerts itself like a long-distance swimmer who's mastered her breathing so as to avoid tiring. After a star exhausts its hydrogen supply and begins its helium-burning cycle, its longevity is fairly ensured. Both hydrogen- and helium-burning cycles are processes of nuclear fusion. They are also metaphorms and processmorphs of the human brain.

Like human beings, there are normal and abnormal stars—meaning, main-sequence stars that evolve and burn with almost predictable regularity. Upon hearing this statement, one mental image you might form is cast in a social context: For example, the image of a "normal" law-abiding star versus an "abnormal" anarchistic star may come to mind. These terms are generally based on genetic, anatomical, physiological, or other factors influencing the organisms' patterns of behavior: from simple, multicellular organisms to complex creatures. By applying these terms to stellar bodies and "behavior," I mean to evoke an image of stars as being human-like and vice versa.

This gesture is in the tradition of the ancients, who were able to intuitively represent the smaller entities, such as humans, by the larger one, the universe. The nature of the universe, then, was human nature in disguise. The ways of humans were the ways of the cosmos. This mythic anthropomorphism was one means of making visually concrete this peculiar relationship linking our minds and bodies to the stars. It's hardly surprising to hear, some 2,000 years later, astrophysicists remark that stars and "galaxies are like people: when you get to know them they're never normal."[10]

We trivialize the word *star* each time we ascribe this word to a celebrity as opposed to the whole of humankind. (At the very least we could be a little more inventive about this ascription or label.) There's a star on every shoulder and not just on some shoulders. Unfortunately, we'll never understand this—how we are all stars—as long as we practice idolatry rather than exercise our respect for all life. To quote one of the world's most influential authorities on the development of science, Francis Bacon: "There is a great difference between the idols of the human mind and the ideas of the divine. That is to say, between certain empty dogmas, and *the true signatures and marks set upon the works of creation as they are found in nature*" (my italics).[11] So spoke the essayist, philosopher, statesman, and parent of the Scientific Revolution.

I imagine the combination of our descriptions of this brain-universe metaphorm will provide a broader view. Whether or not this view is plagued by paradoxes is beside the point. Paradoxes are synonyms for reality. For instance, we cannot possibly see the whole of what we are, if we persist in seeing the parts of our nature and nature's parts as disassociated entities. Defining things without considering their relationships with other things is a fatal flaw in our descriptive analysis. It's as fatal as searching for unsuspected likenesses between patterns of forms—to the exclusion of processes.[12] If we persist in thinking with our *fragmentary* specializations about the structures alone of all these disassociated parts, the most we can expect to understand are the apparent connections between things, not the unapparent, deeper ones that relate one process to another.

Consider our current situation. We use billions of words and images to relate potentially billions of perspectives about the brain and the universe as seen separately. Each perspective refers to one and the same reality, although there's little agreement about what this singular reality is. Isn't it possible that the brain-universe processes are as integrated as the sides of a Möbius strip? And that it is only our reflections on the nature of these processes which are divided and "dis-integrated"?

To paraphrase a Chinese proverb: Are we not riding on the back of the ox and looking for the ox at the same time? I think so. But then what is the character of this ox and the meaning of our ride?

If our explanations or our understanding of the universe is in some sense to match the universe, or model it, and if the universe is recursive, then our explanations and our logics must also be fundamentally recursive.—GREGORY BATESON, *Steps to an Ecology of Mind*

Universe is an evolutionary-process scenario without beginning or end, because the shown part is continually transformed chemically into fresh film and reexposed to the ever self-reorganizing process of latest thought realizations which must continually introduce new significance into the freshly written description of the ever-transforming events before splicing the film in again for its next projection phase.
—R. BUCKMINSTER FULLER, *Operating Manual for Spaceship Earth*

4

The Evolving Brain in the Context of the Evolving Cosmos

ateson's and Fuller's refreshing views (above) apply to the brain as well. Like vibrating prongs of a tuning fork, their views tune our thoughts to the concordant ways of nature and its continuity. We can choose to think of the human brain's recursive nature as an "evolutionary-process scenario" without points of commencement. However, once we take this thought route or listen closely to the harmonics of this concept, we're soon stymied by the reality of birth and death. Undisputedly, things are born to die—human beings, stars, and all the things they create between them. And yet, as many mythologies convey, this reflection on beginning and end points is as fragmented as our picture of reality. Disputedly, things also die to live. Or so our imaginations confirm with the support of theosophy.

This ageless thought of a "world without end" implies that the brain, like its celestial godmother, is a continuum—a timeless world in which matter and energy are the same throughout. The *microtime* (10^1), with which we measure the birth and death of our bodies, and the *macrotime* (10^9), which we use to determine the birthdays and deathdates of celestial bodies, are conceptual illusions. Matter continually reorganizes itself in one form or another without the discretion of time.

EVOLVING WITHOUT WARNING

Could there be some imperceptible, "timeless" process at work outside (or inside) of visible reality? A process that affects evolution the way neurochemistry affects the human mind? Some *superprocess* that distorts (or heightens the realness of) time-space and matter? This unknown process could be larger than all of our concepts of nature assembled together. It could also be so subtle as to exist at the cliff of our know*ledge*, at the *ledge* of the corporeal world, at the *edge* of physical laws. (I sound like the science fiction writer Rod Sterling introducing some supernatural delight!)

We sense this extended life-death process like animals sense the presence of an intruder or guest. But what is this "other world" that billions of people have sensed for eons? (Who can say?) Is it tangent to yet part of the world we wake up to daily? (How can we know?) Are our imaginations overshooting or falling short of reality when we declare: The universe never dies, and we, as physical reflections of the universe, never die.

TIMELESS EVIDENCE

Many may regard this as a pretty steep assumption lacking conclusive circumstantial evidence—that is, hard-core testable evidence. But then, what is eternally tangible "evidence" *anyway*? Considering the paucity of human life experiences over a handful of millions of years, how definitive can our evaluations of life be? The volumes of recorded history barely occupy the last 10 seconds of one day in the life of the universe's 15-billion-year history—a history that can be compacted into one year on the cosmic calendar—as the scientist-educator Carl Sagan cleverly demonstrated. Maybe we're only authorities *after* a few billion years—when we've been around some and have witnessed a pattern in the evolution of life forms and knowledge; when we've seen in detail and in aerial-like overview the projections of our ideas. Perhaps we'll never attain full authority. It's as if we have to literally live beyond our means to understand the means by which we live. And the means by which the universe lives.

Since we possess neither preeminent experience nor supreme technological sophistication that equals nature, we have to rely on at least two things in discerning our brainways and in discovering the ways of the cosmos: our reflective arts and speculative sciences. This chapter proposes that as we learn about the evolutionary processes of the universe, we should bear in mind brain processes. This study first entails knowing something about their anatomy and physiology. Where anatomy equips you with an understanding of the interdependent elements of the brain (or cosmos), physiology informs you about how these elements function in part and in whole. *I urge you not to rely on what I've singled out as points of similarities.* The sets of relations I deem important may not be as important or interesting as those you will discover in your readings and experiments.

DESCRIBING EVOLUTION
WITHOUT EXPLAINING EVOLUTION

In relating the evolution of the brain to the cosmos, I find it useful to conceptualize this relation through a metaphorm. A pantograph seems appropriate. This instrument, which is employed in mapmaking, can enlarge or reduce geometric figures and motions (see Metaphorm 1). The "motions" of the "geometric figure" I wish to metaphorically map refer to brain processes. Instead of literally enlarging the brain four times its original scale and ending up with only an outline of a bigger brain, I'm suggesting that you consider a different view, one that only your imagination can provide. Imagine that on one sheet of paper are the most widely accepted scientific descriptions of brain processes and on the other sheet is information on the evolving universe. The mechanics of the pantograph are such that as you physically draw your conclusions about the processes of the brain, you are simultaneously extrapolating about its extension, the universe. The converse is also true.

I have to speak metaphorically here because we cannot literally, or physically, use a pantograph to relate these entities in the way I've just described. In reality, this rigid technical drafting device is designed to mechanically reproduce a scaled map outline of some figure without transforming this figure (other than enlarging or reducing it). There's no conceptual transformation as suggested here. I find this artistic, idealized

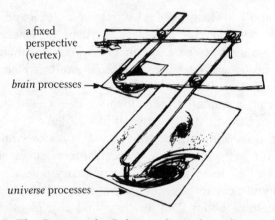

a fixed
perspective
(vertex)

brain processes

universe processes

METAPHORM 1. The Pantograph. *Relating the evolving brain and cosmos, the hinged bars of a pantograph represent the complementary disciplines of art and science. The instrument itself represents technology, the dynamic union of the arts and sciences involved in this mapmaking process.*

version of the pantograph useful for discussing the "expanding nervous system scenario" as seen in relation to the "expanding universe scenario." Of course if we were to talk about realistically mapping their evolutionary processes—and relating what we mapped—we would need a number of Cray computers and inventive programmers to animate this comparison. For the purpose of conveying the gist of this relationship, the pantograph is a good deal simpler (and more economical).

PARALLELING INFORMATION

Another path in the journey of neurocosmology is created by reading "parallel chapters," or twin topics, in books on the brain and cosmos. There are chapters that discuss the genesis and development of these entities in an evolutionary context. They may start with star and galaxy formation (in astrophysics) and finish discussing the formation of the entire universe (in cosmology). This progression is echoed in the studies of neuroscience and psychology in which we learn about the growth of neurons and nervous systems up through the development of our social organizations for which neurons are wholly responsible.

Instead of tracing the evolution of the human brain over the course of millions of years—comparing ancestry and relatives (as in phylogeny)—we'll be looking at aspects of changes spanning an average lifetime. Further on, we will relate the long-distance study of the human brain and behavior to the long-distance study of the phenomena of the universe.

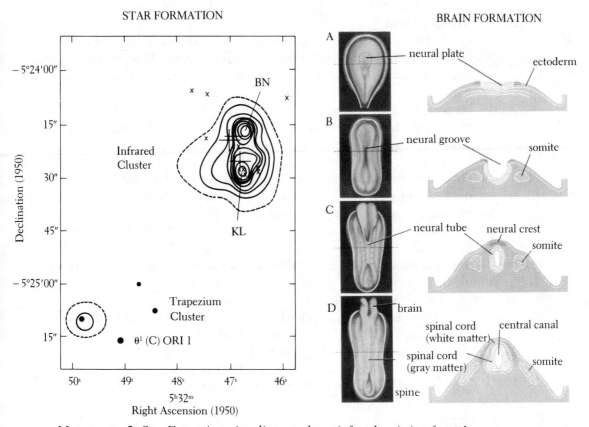

STAR FORMATION

BRAIN FORMATION

METAPHORM 2. Star Formation. *A radio map shows infrared emission from the Orion nebula. There are two distinct sources of the clusters pictured here. One source is the Kleinmann-Low nebula (KL) and the second source is the Becklin-Neugebauer (BN) object which is entrenched in a molecular cloud. This object is described by some astrophysicists as the "embryonic cloud" that is part of the star-forming medium. The Trapezium cluster is the most evolved part of Orion. The dot here represents a newly formed family of several hundred stars that are no older than a million years, living within earshot of each other—a mere 1 light-year distance. (From Michael Zeilik and Elske van Panhuys Smith,* Introductory Astronomy and Astrophysics, *1987)*

METAPHORM 3. Formation of the Human Nervous System. *The images show the third and fourth weeks of development after conception. Left: external view of the embryo; right: cross-sectional view. (Adapted from W. M. Cowan, "The Development of the Brain," Scientific American, Vol. 241, No. 3, 1979, pp. 112–133.)*

In Michael Zeilik and Elske van Panhuys Smith's excellent textbook *Introductory Astronomy and Astrophysics* there is a chapter entitled "The Evolution of Stars." To me, this information begs to be placed in tandem with a parallel chapter on "Development" (of the nervous system) by the neurophysiologist Samuel Schacher, included in another superbly written textbook by Eric R. Kandel and James H. Schwartz, *Principles of Neural Science*.

In bridging this material, the first thing that struck me is how blatantly different neural and stellar structures appear. (They would definitely be rejected from a look-alike contest! At least, any contest whose standard of judgment is based solely on appearances.) I found that by allowing my imagination to do what it does best—imagine—I soon distinguished a few semblances (see Metaphorms 2 and 3) that were vaguely intriguing. I say *vaguely* because at first glance they appear to be similar. However, on closer inspection, these similarities seemed to be as vaporous as naphtha. In opening their "container" (scrutinizing the diagrams), their meaning vanishes. As I read about what I was looking at, the picture-statements became significant again. The similarities returned to their liquid state. I could see and touch them and feel refreshed by them. Once again, they became suggestive and meaningful. The images led me to the words which led me back to the images, in a cycle of understanding that renewed my curiosity.

There's an important thought here. In identifying a likeness of process, visual inspection is limited literally to your depth of field (as in physics, philosophy, art, etc.), not just your field of vision. We need words and mathematics—as unsighted guides—relying on other senses. Together the images, words, and mathematics direct our awareness, inspiring us to search more thoroughly the surfaces of the visible.

ALL THAT YOU ARE . . . IN BEING AND BODY

In venturing into the following comparisons, observe one thing: What you're really learning about are *your* origins, *your* formation, *your* evolution and demise. You're not reading about some abstract fictionalized creation that has nothing to do with your life. You're not contemplating some obscure material that is separate from you.

This material is you. Just as these comparisons are *about* you. The neural plate and ectoderm (the place of origin of all cells of the nervous system), the neural tube (from which the three principal sections of the brain emerge—the forebrain, midbrain, and hindbrain), and the somite (the series of segments into which the central nervous system is divided) in Metaphorm 3 immediately describe *your* anatomy as you were forming. These images document your history. Likewise, the map created by a radio telescope (Metaphorm 2) shows the infrared emission of ionized gases of the Orion nebula* as it forms from the "embryonic cloud"—an

---

* *Nebula* is a general term for different classes of gas clouds that are charged by stars which they cloak. Some of these clouds of gas are internally illuminated by hot or radiant stars;

expression warmly used by astrophysicists. The map also presents an overview of our origins from space. As a model, it offers us a rare symbolic look at our galactic home and parents from which humankind itself formed. Imagine, then, the details of star and galaxy formation representing our genesis and vice versa. These imaginings are what neurocosmology is all about—the imaginable mapping onto both the possible and the improbable.

GENETIC AND EPIGENETIC INFLUENCES

In discussing the role of genetics in the developing nervous system, we need to examine how "intrinsic and extrinsic factors" affect the developmental process. The *intrinsic factors* refer to processes within the cell that direct the actions of the genes, turning them on and off. The *extrinsic* or *epigenetic factors* help shape the genetic blueprint, or the *phenotype* as it is called in biology. The extrinsic factors originate in the cell's environment.[1] They influence the cell in the same way our external environment influences the well-being of our internal one. The interactions of genetic and epigenetic signals are what gives the nervous system its form. They also influence the way neurons develop and interconnect with untold confidence.

These interactions occur in two stages of our nervous system's growth known as *determination* and *differentiation*. The first term means that specific cell populations are guaranteed their place in the nervous system. Like ticketholders to a play, they have reserved seating that is generally honored. The *proto-neurons* and *glial cells* that hold neurons together like glue (from which the latter got their name) make up these populations. Both groups of cells pretty much sit in their designated seats for a lifetime—barring disease, which is no welcomed intermission. The second term, "differentiation," refers to the process whereby the cells that drift downward from these "determined" populations become neurons. These young neurons are born sophisticated. They form very precise connections with one another, not to mention their target cells. Without cluttering my report with too many specifics, I need to add a few other details before challenging you with some provocative questions and images that relate this information to the genesis of stars and galaxies.

---

these are called *bright nebulae*. There are also *planetary nebulae*, or *gaseous nebulae* as they are called, which consist of the remnants of ordinary stars or supernovas that are in the throes of death. Then there are interstellar clouds that rely on reflected light for their illumination. These are called *reflection* or *diffuse nebulae*. Finally, there are moody cloud masses that look as though they shut all their operations down like a retired power plant. These are called *dark nebulae*.

Determination results form the middle layer of the embryo, or mesoderm, interacting with the outer layer, or ectoderm (see Metaphorm 3). This concept suggests that the embryonic cells are constrained in their ability to become something other than what they were programmed to become—perhaps growing into some other tissue. In effect, once a neuron always a neuron, like "a rose is a rose is a rose." Determination gives rise to *neuroectoderm*—which is the nervous system by any other name. The neuroectoderm owes its existence to the process of *neural induction*. This term doesn't mean "reasoning inductively" with or about neurons—certainly not at this stage of development anyway! The expression refers to the process by which the neuroectoderm is prevailed upon by the embryonic mesoderm.[2] As you might expect from a science that names every process it studies, *gastrulation* is no exception. This process involves the invagination, or inward folding, of cellular material. Interestingly enough, the expanded definition of this process smoothly folds into physicist David Bohm's metaphor of the universe as an infolding-enfolding structure. One is a double of the other.

INFLUENCING OUR DESIGN

These processes are responsible for the design of the whole nervous system. In particular, a process called *regional specification* (which makes sure that the main sections of the neuroectoderm are clearly specified) establishes the regional organization of our nervous system. There's much evidence indicating that the structure of the neuroectoderm is always open for modulation. In modulating any one of these design processes, you affect populations of neurons and their interconnections.[3] In the absence of regional specification, the cells of the nervous system would be hopelessly confused, as would all bodily functions. The effect is comparable to modulating a musical instrument in such a way that all you produce is noise rather than music—"noise" that even the composer John Cage would find objectionable!

Without burdening you with lengthy notes on the three phases of differentiation (which, by the way, include neurons proliferating and growing, migrating, and maturing), ask some basic questions. How are these processes (determination and differentiation) represented in the universe? How do we connect these cellular changes with stellar changes? What metaphorms would best relate these things? Without having to be fully articulate, and without having to pinpoint the details of this relationship, how do we discuss the ideas these metaphorms convey? It seems the more ambiguously they're rendered, the more room for flexing our imaginations.

OUR LITHE MINDS AND COSMOS

You may find that, in manipulating these comparisons, it's helpful to think as a conceptual contortionist. The more supple you are, the more you can twist and reshape your body of logic—without experiencing the tightness of double-talk. As we search for likenesses between the brain and cosmos, we may do well to borrow Paul Klee's advice "to think not of form but of the act of forming." Expanding on this thought, Klee relates: "The artist does not ascribe to the natural form of appearance the same convincing significance as the realists who are his critics. He does not feel so intimately bound to that reality, because he cannot see in the formal products of nature the essence of the creative process. He is more concerned with formative powers than with formal products."[4]

The genesis of the nervous system, which I've just briefly described, involves the mesoderm interacting with a special region of ectoderm. You're now left to discover how this process is akin to the formation of stars—a process requiring the tender loving touch of interstellar medium in the form of dust and gas. (I could spend the remainder of this book just describing the different types of molecules that make up this interstellar medium aside from its principal components, molecular hydrogen $[H_2]$, carbon monoxide $[CO]$, silicate, and carbon among other inorganic and organic molecules.) Neural and stellar systems are related insofar as their formation involves taking at least two media and joining them through intense and sustained interactions.

GROWING LIKE "MAIN-SEQUENCE" STARS

In *Introductory Astronomy and Astrophysics*, Zeilik and Smith inform us that stars evolve in a similar sequence: from protostar and pre-main sequence to main sequence and post-main sequence. Roughly speaking, a star's mass determines its rate of evolution.[5] There are at least two types of star births which include massive stars (with 10 or more solar masses) and solar-mass stars. The luminosity of massive protostars is substantially greater than solar-mass ones. Upon maturing to the main-sequence phase, massive stars convert the gas around them into ions (electrically charged particles or clusters of atoms). This ionizing action is obscured by dust. So the only way to peer into the stellar womb is by infrared and radio observations.[6]

The process of formation consists of approximately six steps, as determined by radio telescope studies, which detect the activities of ionized gas. In the case of massive stars, first, they condense from molecular clouds. Next, this condensation event, which is called a "free-fall collapse," warms up the interstellar dust* to low temperatures that are estimated to be around 30 to 50 K. At this point, the earliest form of the star emerges. This *protostar*, as it is called, bakes in the higher temperatures of the interior dust, which are about 1,000 K. (The exterior dust is plenty hot at 100 K. But compared to the interior temperatures, it seems cool.) Once the protostar enters into the main-sequence phase the hydrogen gas surrounding it is ionized. ("Main sequence" refers to the band range of spectrum luminosity of about 90 percent of the stars.) This results in the creation of what is called the *H II region*, an area that may be described as white hot—glowing with thermal energy that generates young stars (spectral types O and B).

The H II region is the most mature part of the molecular star-forming cloud. In the final two phases, the ionized gas mushrooms into a spherical form, taking the dust with it as it expands like a placenta. The dust eventually disperses as this region continues to expand, its volume increasing all the while. The protostar becomes visible once its thermal energy has created a large enough window for its light to radiate through the thick, opaque atmosphere of interstellar medium.[7]

DYING LIKE "POST–MAIN-SEQUENCE" MINDS

Some of these formative processes may be found in Metaphorm 2. It is up to you to decide exactly how each of these phases of development corresponds with my description of our evolving nervous system. Here you must stretch your imagination and shape—through reason—all that you've stretched. Although the infrared and radio maps shown here are not as concrete as "hands-on" anatomical dissections, they're nevertheless equally accurate. The point is that these two complexes of matter are connected, even though we have not observed this connection directly—let alone measured it. And yet its presence is very much sensed. Perhaps in time we'll be able to confidently discuss what it is we're sensing.

One belated commentary on Metaphorms 2 and 3: In buttressing facts on stellar formation and neural evolution, I'm not inferring that the early stages of both systems are identical on some relativistic scale. Nor am I implying that the Orion nebula eventually elongates and subdivides further like a newborn nervous system. This comparison serves only to

---

* Much of this dust consists of modest remnants of supernovas blasted away at thought-defying energies. Supernovas are known to throw about 1 to 50 solar masses' worth of material into space, which is a staggering amount.

establish that both systems mysteriously share certain processes in their evolution.

The presence of these similarities—even though they may be superficial in the final analysis—teases our imaginations in a positive way. Perhaps in the same way that a powerful optical illusion (such as the famous Penrose "impossible triangle") teases our perceptual apparatus, all illusions lead us to wonder about the reality of this biological apparatus we respectfully call our visual system. It's worth noting that this system is intimately connected with the association cortex from which our more imaginal thoughts are often tendered by logical and dream-like reasoning.

One feature that appears startlingly similar is the form of the neural plate and the pattern of the radio maps showing the expansion of the Orion nebula. It's interesting how the neural plate, which is the cornerstone of the nervous system, undergoes a process of invagination in becoming the elongated structure of the neuroectoderm. This structure is later subdivided into the main regions of the brain: the spinal cord and hindbrain (or *rhombencephalon*, which includes the pons and medulla oblongata), the midbrain (or *mesencephalon*), the between brain (or *diencephalon*), and the neocortex (or *telencephalon*). If you were to touch your head this very instant, you would undoubtedly be activating the cells in each of these regions. And if you were to wonder how this process of elongation and expansion is played out in the births of stars and galaxies, each of these brain centers would join you in this wonderment.

MINDFUL LUMINOSITY AND STELLAR INTELLIGENCE

On a related topic, there is another unusual correlation you may appreciate. It concerns the relation between the different masses (or energies) of stars and their luminosities. A similar relationship exists between the *virtual energies* and *luminosities* of the human nervous system—specifically, those indescribable energies we associate with intelligence.

Consider this: In the creation of protostars, their rate of development is contingent upon their total mass. The bigger (more massive) and brighter (more energetic and luminous) the star, the shorter its life. Massive stars retire earlier from the main sequence than do the smaller-mass stars. In comparing rates of energy output (or luminosities), we can see this relationship. Hot stars burn at far higher rates relative to their masses than does the Sun. The amount of mass, or energy, a star has—and its consumption of energy—determines its lifetime. (Note Metaphorm 4.)

BRIGHTNESS = LUMINOSITY-MASS RELATION

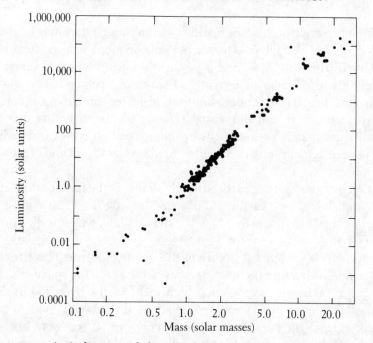

METAPHORM 4. A *diagram of the mass-luminosity relationship for binary star systems. The few stars that are off the beaten trail have low luminosities. They include white dwarfs, which are made of deteriorated gases rather than ordinary gases. (From Michael Zeilik and Elske van Panhuys Smith,* Introductory Astronomy and Astrophysics, *1987)*

LUMINOUS MINDS

What does the preceding paragraph suggest to you about human stars? To me it intimates that mass and luminosity—meaning brain energy and the rate at which this energy is used—are somehow connected to intelligence. This thought is reminiscent of the correlation between brain mass and intelligence discussed by Carl Sagan in his insightful book, *The Dragons of Eden: Speculations on the Evolution of Human Intelligence.* Sagan examines the statistical correlation between brain size and intelligence. He concludes that there is no one-to-one correspondence. In short, people with larger brains are not necessarily smarter people. If a few hundred grams of thinking tissue were absent from the cerebrum there would be little or no functional impairment affecting one's cognitive abilities.[8] He proposed that "a better measure of intelligence than the absolute value of the mass of a brain is the *ratio* of the mass of the brain to the total mass of the organism" (see Metaphorm 5). As you might imagine, there is a fairly generous range in the mass of a normal adult

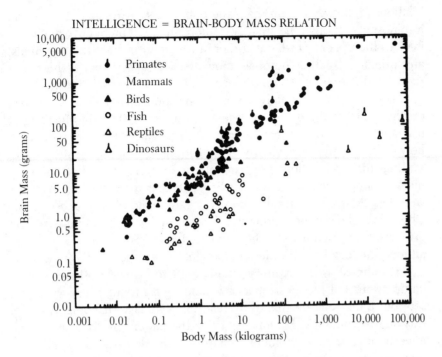

METAPHORM 5. *A scatter diagram of brain mass versus body mass for primates, mammals, birds, fish, reptiles, and dinosaurs. (From Carl Sagan,* The Dragons of Eden: Speculations on the Evolution of Human Intelligence, *1977)*

human brain (which weighs approximately 1,375 grams, or 3 pounds) and the body. Although there are many researchers who have trouble with this equation—intelligence = the ratio of brain mass to the organism's whole mass (or energy)—it is nonetheless an intriguing idea with growing experimental support.

If we were to creatively relate this discussion of brain mass versus body mass to the mass-luminosity relations in stars, we would shortly find ourselves without an analogy to stand on. I mean a star is not a carefully partitioned life form complete with a head, torso, limbs, and appendages. Nature designed stars to be much less compact than we (or so we assume from our small perspective; then again, given the immensity of space, even the red giants seem petite and compact). It seems nature brilliantly packages its goods, fitting them to the needs of their primary consumer, space.

Concerning this reflection on the nature of intelligence: You'll notice that Metaphorms 4 and 5 are strangely similar in their design and even more alike in their content. Both convey information about the relationship between brain mass and body mass, between star mass and luminosity. Although the analogy is interesting, is it productive and credible? In the upper right-hand corner of Metaphorm 5 are the mammals, and primates (such as humans, monkeys, and apes), and in the upper right-hand corner of Metaphorm 4 are stars with very high luminosities.

You might immediately conclude that our imaginations can travel on the backs of these sorts of visual comparisons. They're engaging at first and even curiously convincing, until you peer further to discover how dissimilar they are when you change the logarithmic scales in presenting this information. For example, if we were to change the size-mass volume scale (along the X-axis) such that the more massive stars were less dense, then the luminosity scales would change accordingly. The higher the number (along the Y-axis), the lower the value. We soon discover that anything can be compared vis-à-vis scatter diagrams by altering the logarithmic scales or changing some aspect of proportion.

It's almost as if our imaginations want to believe in these sorts of comparisons but then a stronger awareness—regarding the way we present our data—claps its hands like clashing cymbals, snapping us out of this quasi-hypnotic seduction. Isn't this the pattern of discovery in science as in art? One moment something stimulates an association or memory and in the next moment you question not only the association and its implications but also the source of the stimulus!

ILLUMINATING THE MASSES AND ACTIONS OF OUR MINDS

In the end, our responses to the value of the images and concepts in this chapter are limited to the degree our imaginations are constrained. The more we confine our intuitions, the more these presuppositions and propositions seem to be neither here nor there. The more we liberate our reasoning, the more enthralled we are with the possibility that we share some direct tie with our celestial family. The similarities are indeed compelling. The dynamics of human beings (including their relationships) seem to resonate with the dynamics of the heavenly "beings" (and their "relationships"). The implication is that our energies, for example, represent another form of starlike energy. They can be as compressed, or dense, as a neutron star, quasar, or pulsar.

In terms of energy production, our minds are like stars—though we're unevenly matched. In terms of matter, we're unlike stars—though we're evenly matched—in our complex and diverse forms. We obviously don't have the musculature, energy, or luminosity of stars (a rather lame understatement). But I'd venture that the *virtual* musculature, energies, and luminosities of our creative minds are fair competition for the boundless output of stars. We're just beginning to understand how impressive our creative output is. And with understanding comes appreciation.

Most stars viewed through optical telescopes are either binary or multiple systems. In the latter, stars are physically related by means of their "mutual gravitational attraction." Newton's law of gravitational attraction was confirmed by the eminent eighteenth-century astronomer Sir William Herschel, who observed the orbital motion of the Castor binary stars. Since this important discovery in 1804, astronomers have learned much about the dynamic relations of stellar binary systems (two stars that are bound together by gravitational fields).

SEVEN TRAITS

Astronomical observations have established that there are seven different classes of binary systems. Each class seems curiously similar to patterns of human social interactions. And I will speculate that the dynamics of binaries resemble patterns of interactions between cerebral hemispheres.

Depending on the plane of physical reality in which we're entertaining these comparisons, we're going to interpret them differently. For instance, if we regard the whole brain as one solar-mass star, the concept of stellar binary systems as related to *human star systems* would have one connotation. It would imply that the interactions of two people resemble the actions of different types of binary stars. It would also imply that human interaction isn't completely dependent on what we commonly regard as brain functions. If, on the other hand, we regard the cerebral hemispheres as two binary stars, then this concept would change dramatically. It would then mean that the actions between brain hemispheres would resemble the interactions of the seven binaries. I feel more comfortable with the analogy that the cerebral hemispheres are best represented as galaxies, clusters, and superclusters rather than as single stars.

As you may have already anticipated, in the nomenclature of astronomy and astrophysics there are *binary galaxies*, too—that is, two galaxies under the influence of each other gravitationally. There's no reason why this relationship I've already described cannot be extended further—and copied. The concept of binary galaxies would imply that a group of galaxies (or people) would interact with another group of binaries. Nature seems to enjoy this sort of repetition of process, or what some sardonic writers might call "redundancies" or iterations. It repeats the same process like an eternal chant. This is to keep life simple and coherent. But life without its varieties of form would be certifiably boring. So nature surprises us, itself, with its constantly changing forms of matter, each form feigning a unique process when, in reality, there may only be as few as three processes. This thought on *process* returns us to the pre-Socratics, who searched for the "thought through which all things are steered through all things." As we grow to understand our union with nature, we will discover that we are connected in process—though unconnected in form—to all events and elements of nature we systematically study.

As part of our discussion on the evolving universe, consider the following facts about seven unique relationships—between and within brains and binary stars. It's hard to resist relating one or more of these descriptions of binaries to your personal relationships (by which I mean with yourself and with others). Perhaps you may never look at stars the same way again.

Apparent (or *optical*) *binary stars* are not a "true couple," meaning they're not married to one another by gravity. They only appear to live together because they exist on the same plane of sight. On closer inspection, their motions in space show that they are not related as a binary system. (Which implies that they may "think alike" but they don't act alike; by "thinking" I'm referring to their energy output or luminosity, not actual biological "thinking.")

Visual binaries, on the other hand, are a truly bound couple. The mutual orbital motions of these stars show there's a fairly stable and healthy relationship. Their orbital periods range from approximately one year to thousands of years. This period, when related to the human brain, may refer to periods of major change in the relationship between brains as whole systems or between cerebral hemispheres as parts of whole systems.

Astrometric binaries are less obvious about their relationship. When observed through a telescope, only one star is visible. However, its oscillatory motion manifests the unexpected. It turns out that this star has a "hidden companion," as though one star's actions outweigh the other's. Applied to the human brain, this reference implies that one hemisphere is more forcefully dominant than the other—like the actions of "major" and "minor" cerebral hemispheres.

Spectroscopic binaries represent an "unresolved system." Zeilik and Smith explain: "In some cases, two sets of spectral features are seen (one for each star) oscillating with opposite phases; in other cases, one of the stars is too dim to be seen, so that only one set of oscillating spectral lines is recorded. Typical orbital periods here range from hours to a few months."[9] Here the expression "[stars] oscillating with opposite phases" might be interpreted to mean two people with complementary personalities or perspectives. The expression also reminds me of the complementary actions of the cerebral hemispheres, as described later on in Chapter 8.

Spectrum binary systems, like spectroscopic binaries, also represent an unresolved system. Their stars do not appear to have orbital motions, but instead have two distinct spectra (spectral features) that overlay each other. Astronomers look to these stars as the producers of this composite spectrum.

In *eclipsing binaries*, two stars take turns occluding each other. These periodical eclipses affect the brightness of this binary system. I'll be speaking more about their dynamics momentarily.

And then there are single stars—"oddities" such as our Sun. (Quite symbolic: our Lone Star of space, our Lone Brain Star.) Astronomical observations indicate that solar-type stars have a ratio of single:double:triple:quadruple systems represented as 1:8:46:45, respectively.[10] Perhaps this ratio tells us something about the rarity of biological stars and human beings in the cosmos.

These thoughts may make no sense neuroanatomically or astronomically. It's what they *imply* that makes so much sense to me. Their different forms turn us away from looking deeper—in the direction of process. To emboss the point in your mind, these relationships are simply ideas presented, without pretense, for meditation.

"TRADING PLACES"

While all these cautionary notes and ideas mix in mid-air, let me close this chapter with one other special comparison that is pertinent to our discussion. Since we're on the subject of binary stars, let me extend to you a few additional notes on *detached, semidetached,* and *contact binaries* that come under the category of the *eclipsing binary system.* In imparting this information, I would prefer that you personalize it. Think of the described processes as taking place inside *your head* and around it (meaning your relationship with others and your environment.) This way the following notations on binaries will feel especially relevant to you. They'll have a special "ring of truth," to borrow a phrase from the renowned physicist Philip Morrison. You're asked to introspect on your brain processes in the context of Metaphorm 6 and its accompanying statement-picture. Zeilik and Smith write:

> We picture the interactions of these [binary] systems by considering the effective gravity at many points locally. The effective gravity results from the combination of the real gravitational attractions and the centripetal force from orbital motions. If you explore the space

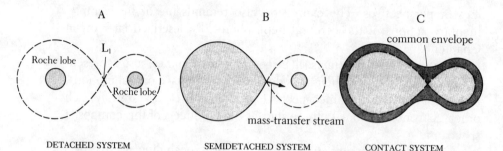

A   B   C

Roche lobe   common envelope

L₁

Roche lobe

mass-transfer stream

DETACHED SYSTEM   SEMIDETACHED SYSTEM   CONTACT SYSTEM

METAPHORM 6. *Diagram of a* contact binary star system, *such as an eclipsing binary. A: Figure represents a* detached system. *In this system two stars are observed to be smaller than their Roche lobes (that is, the effective gravity of each star). B: Figure shows a* semidetached system, *in which one star fills its Roche lobe—after which its mass flows to fill its companion star. C: Figure indicates that the Roche lobes of both stars are filled as a* contact system. *They are, metaphorically speaking, fulfilling the complementary aspects of each other in their union. This union is symbolized by a common envelope thought to be of uniform temperature. L₁ signifies the region where the gravities of two stars touch. In this region, at this point, the effective gravity is zero—like the path of least resistance.*

In neurocosmology, this information is related to human brain processes, where the interactions of the Roche lobes represent the actions between two groups of people (as superclusters), between two people (as clusters), or between cerebral hemispheres (as galaxies or as stars). (From Michael Zeilik and Elske van Panhuys Smith, Introductory Astronomy and Astrophysics, *1987)*

around the stars, you will find a certain region, shaped like a figure 8, where the effective gravities of the two stars touch (L₁). Here the effective gravity is zero. Each half of the figure 8 indicates the regions controlled by the effective gravity of each star; these are called *Roche lobes.*

We can now classify close binaries on the basis of how large each star is relative to its Roche lobe. If both stars are smaller than their Roche lobes, the system is detached. If one fills its Roche lobe, the system is semidetached; matter can flow through the contact point L₁ to the other star. One star eats up the mass of the other. Finally, if both stars fill their Roche lobes, they are in contact and a common envelope of material enshrouds them both.[11]

CLUES TO THE NATURE OF OUR BEHAVIOR

These notations on binary stars are like rough clues to the internal dynamics of our nervous systems. They also clue us to our relations with our fellow human beings and environment. We can interpret these notations the way an architect reads a blueprint of a building, translating symbols into realities. In Metaphorm 6, *Roche lobes* represent either neurospheres or people or the collective lobes of brain hemispheres. The point L_1, marking the intersection of the figure eight, represents the region of information exchange between and within the brain. The process of "a star filling its Roche lobes" represents the temporary dominance of one cerebral hemisphere over the other—or of one person's personality overpowering another's. And the act of completely filling the Roche lobes represents the mutual fulfillment of the brain's "complementary hemispheric functions"; this action is a testament to the unity of brain and star processes alike—conjoined in their acts of creation.

One concluding image which relates to this description of contact binary stars is the *Lorenz attractor* (Metaphorm 7)—an emblematic drawing I've borrowed from chaos theory. I thought this was an appropriate metaphorm to include here as it implies that within the orderly ex-

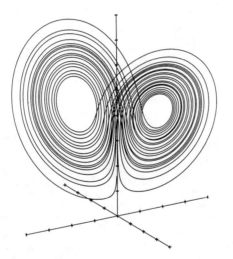

METAPHORM 7. The Lorenz Attractor—*representing the interactions between contact binary stars and "contact binary brains," or, more precisely, between stars and galaxies/cerebral hemispheres and human relations. The drawing is named after the mathematician and meteorologist Edward Lorenz, who was one of the first interpreters of the chaotic behavior of matter. In James Gleick's superb book* Chaos, *the writer describes the implications of this image: "At any instant in time, the three variables fix the location of a point in three-dimensional space; as the system changes, the motion of the point represents the continuously changing variables. Because the system never exactly repeats itself, the trajectory never intersects itself. Instead it loops around and around forever. Motion on the attractor is abstract, but it conveys the flavor of the motion of the real system." (From* James Gleick, Chaos: Making a New Science, 1987)

changes of energies in contact binaries exist disorderly events such as those associated with the phenomenon of chaos. Investigating the chaotic aspects of nature—aspects which test our patience with such untidy things as disorder, fuzziness, unpredictability, and irregularity—was only recently recognized as a legitimate issue of science.[12] As you will read, the science of chaos reaches into the very pulse of life from nervous tissue to interstellar star-forming clouds. I think cognitive scientist Douglas Hofstadter's perceptive remark best summarizes the mystery of this subject: "It turns out that an eerie type of chaos can lurk just behind a facade of order—and yet, deep inside the chaos lurks an even eerier type of order."[13] *Chaos* describes the *process of instability* in neural and stellar growth alike, as both systems evolve by counterbalancing their tendencies toward instability in stage after stage. It also describes the *process of stability* towards order. Such is the creative process.

Thought is only a flash between two long nights, but this flash is everything.—HENRI POINCARÉ

Intuition flashes are the lightning bolts and solar flares of the mind . . . As processmorphs, they represent the "plasma fusion state of mind."

5
Comparing the Brain's "Apples" to the Universe's "Oranges"

With only seconds left to the close of this millennium, the technological sciences, aboard their satellites and electron microscopes, have looked long and far over the orchards of life. Through their detailed views we now see the brain and cosmos as creative, open systems. As all living organisms grow things, I like to think that the mind and universe are no different. Both are fruit-growing trees of sorts, where the brain's *apples* represent all of its concepts and physical creations and where the cosmos's *oranges* represent all of its phenomena (which indirectly include those of the brain's).

To compare these two fruits of nature is to invite an early frost. Everyone knows that comparing apples and oranges is "pointless." We're taught that it's futile to relate *any* opposite things. And yet, there is a surplus of solid examples that show the contrary. Most opposites, when stretched far enough apart, paradoxically come full circle only to be rejoined at some point of confluence, some moment or area of tenseless harmony. We know from color theory, for example, that the extreme colors of pure black and ultra white represent two different aspects of the same thing—the absence or presence of all the colors in the solar spectrum combined in one color, either black or white, respectively. We've also witnessed how the most abstract, nonobjective paintings resemble the coherent, energetic patterns of a material viewed under magnification or from aerial perspective. In distinguishing between abstraction and realism, there is *a point* where these worlds blur like the two ends of *a line* that have been physically and mentally brought together. Along this "line of intuition" Wassily Kandinsky's aphorism holds: "Form is the outer expression of the inner content." At this "point of reasoning," we sense that *the form of the cosmos is an expression of our minds' content.*

DREAMING OF BRAIN—UNIVERSE METAPHORMS

A number of metaphorms relating physical, biological, and social systems are introduced here. They act as liaisons between these seemingly opposite systems. Like dreams that work associatively to connect our inner and outer world experiences, metaphorms fuse the abstract and the real. The captions for these images intertwine neurological and astronomical terms. For instance, you will be reading my comments about the large-scale grouping of neural bodies in the cerebral hemispheres (Metaphorm 1), but you'll be looking at an image of the "Local Supercluster"—the *cosmopolis* of our galaxy (Metaphorm 2). In discussing the various forms of star clusters and superclusters, I mean to direct your attention to the human brain's *clusters and superclusters*—its large complexes of nuclei (as shown in Metaphorm 3b).

As I emphasized in Chapters 3 and 4, we're searching for likenesses in how these entities work, not how they look. Although I present images of both systems, I invite you to see beyond any superficial similarities of their forms (as in Metaphorms 4 and 5)—to search the more unapparent connections.* Investigating the discrepancy between process and struc-

---

* Any diagram offered as a description of a process is misleading graphically because the drawn image has a specific form or shape. In a word, it's "suggestive" by nature. So there's little we can do about our graphic representations of processes—which include flow charts, schematics, interpretive drawings, etc. These visual notations, whether drawn by hand or by computer, are all we have to conceptually render these ideas.

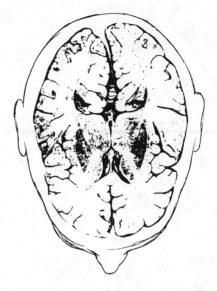

METAPHORM 1. The Local Supercluster of the Brain. *Shown is a three-dimensional view of the human brain and its densely packed nuclei (horizontal section). The only relatively open space is the third ventricle in the middle of the brain at the center of this image. The section reveals the diencephalon, or the "between-brain centers," which is subdivided into four major nuclear complexes. Many neuroscientists regard this region as one of the keys to deciphering the integrated world of the central nervous system.* (Adapted from Gray's Anatomy, 36th edition, 1980)

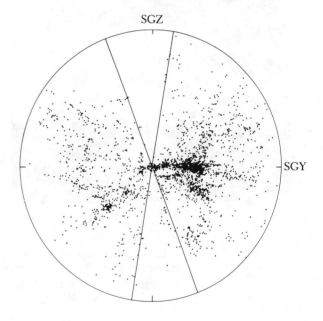

METAPHORM 2. Local Supercluster. *Shown is a three-dimensional map (top view) of the large-scale structure of stars called the Local Supercluster. Each dot signifies a galaxy. The X-shaped wedges, denoted SGZ and SGY, represent the areas that are blocked from our view by the dust clouds in our galaxy. Zeilik and Smith state that "most of the Supercluster is empty space: 98 percent of the visible galaxies are contained in just 11 clouds that fill a mere 5 percent of the overall volume. Yet, the clouds do delineate a disc structure."* (From Michael Zeilik and Elske van Panhuys Smith, Introductory Astronomy and Astrophysics, 1987)

ture, you indirectly question an ancient canon in architecture which volleys that "form follows function." In the comparisons ahead, the functions of both systems do not follow any prescribed form. In fact, you may delight in how their processes are more universal than the particular mechanics of their forms.

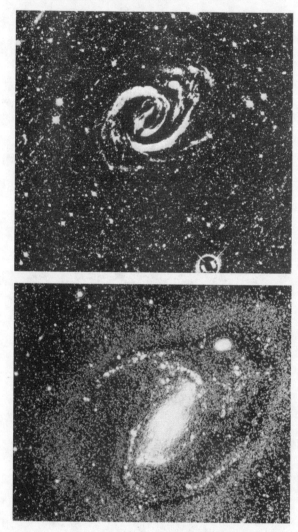

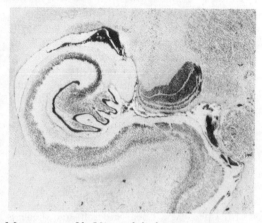

METAPHORM 3a. *Two views of the* spiral galaxy, NGC 1097, *constructed from computer-enhanced, image-processed photographic plates which were exposed over 7 days using the 4-meter Cerro Tololo telescope in Chile (1975). Note the increase in the star field with this magnification, revealing a gossamer, halo-like image surrounding the galaxy—a halo composed of superheated gas and dust from which new stars form. Note also that this galaxy is only one of the dots represented in Metaphorm 2.*

METAPHORM 3b. *View of the* hippocampus, *a section of neural tissue from the temporal lobe of a human brain. This section shows a river of cortical fibers flowing into the hippocampus. At the bottom right is an area called the entorhinal cortex, which links the neocortex to the hippocampus like an estuary connects one body of water to another.*

METAPHORM 4. *The veil-like* shock waves *forming from a supernova's explosion—the massive waves traveling outward in all directions for hundreds of thousands of years. In death and life, supernovas seed the living universe. They provide the machinery and raw materials (such as carbon, uranium, gold, silver, platinum, mercury, and iodine, among other heavy elements) for star-forming factories. (Metaphorms 3a and 4: from Michael Marten and John Chesterman,* The Radiant Universe, *1980)*

METAPHORM 5. Neural tissue *stained with HRP, a peroxidase enzyme extracted from various plants. The enzyme is used in the retrograde tracing of axon terminals in neurons. This technique involves tracing the axon backward—meaning in the direction of signal conduction. The destruction of neural tissue in this technique provides invaluable knowledge about living neurons. The retrograde technique was introduced in the 1880s and is still applied today. The photomicrograph shows the subthalamic nucleus (an area near the hippocampus) of a rat. (Metaphorms 3b and 5: Walle J. H. Nauta and Michael Feirtag,* Fundamental Neuroanatomy, *1986)*

HOW PROCESSES CONNECT DISCONNECTED FORMS

One process, then, can be related to many dissimilar forms. For instance, a neuron has at least one integrated process that is connected with several distinct structures (see Metaphorm 6): The dendrites receive inputs; the cell body integrates these inputs; the axon performs a conductive function in passing this information along; and finally, the terminal fibers relay the news to neighboring neurons.

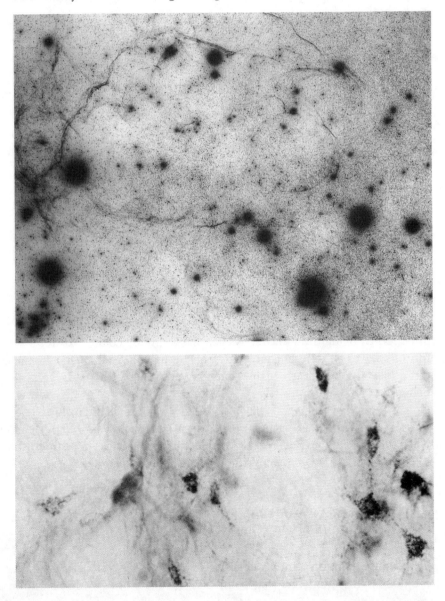

These multiple processes relate to all neurons, including sensory, motor, local interneurons, and neuroendocrine cell-neurons. Each shares this same process of communication, even though they are structurally different.[1] What is more significant is that a neuron in the neocortex, when combined with its cortical cousins, can yield a potentially infinite number of *virtual structures*. These imaginary structures are called "mental architectures" by cognitive scientists and "forms of thought," or "thought forms," by philosophers and poets. You can imagine how these virtual structures extend the dimensions of the mind (that is, the brain).

My point is that where the "real" (or physical) structures of neurons appear to be limited, the "virtual" (or nonphysical and symbolic) processes do not appear to be affected by time-space-form constraints. With this consideration in mind, I'm going to relate the processes of one structure to those of unlike structures.[2] In grasping this distinction of real versus virtual, you may want to look ahead to my concept of the virtual world of *reflectionism* (Chapter 6).

"REAL" NEURONS CREATE "VIRTUAL" FORMS OF THOUGHT

One caveat: The more literal you are in relating the evolution of the brain to the evolving cosmos, the farther you stray from recognizing any likeness between them. Oddly enough, we do not crash headlong into contradiction here. In suggesting that we can search for literal likenesses between processes—but not forms—I mean this: The forms of these entities are like different masks worn by the same actor, nature. It's easy to be fooled by nature's masks, as each has a unique color, texture, and form. They're all different looking, even though they all co-star in the same play (or processes) of life—an ongoing play with many acts of being.

In the following pages we will consider the more distinguishing features of the brain that—when viewed in a larger context—might lead you to rethink the features of the cosmos that we've come to distinguish separately. In the *neurocosmos*, the real *and* the virtual are integral.

THE NEUROCOSMOPOLIS: OUR "COSMOPOLITAN" NERVOUS SYSTEM

The "contemporary" human nervous system can be conceived of as a cosmopolitan center, one that connects with the whole world of information (including matter and energy). This cosmopolis is an internationally important city inhabited by many different citizens who call themselves cell-neurons.

Although these neurons occupy a specific space, just as a city takes up a certain amount of acreage, the neurons actually exist within a much larger, abstract space which I imagine to be a *neurospace*. This infinite space of the mind complements the more finite space of the *neurosphere*, which refers to the virtual globe of the body. The neurosphere is to the human body what the biosphere is to the Earth's body.

Irrespective of what region these citizens live in demographically, neurons are free from local, national, and international "prejudices." (Perhaps there's a straightforward reason for this: They're blind, so they can't harbor the sight of racial or ethnic tensions.) They act in an integrative way towards news that pertains to any region in the neurosphere —irrespective of how remote that news is from the more immediate affairs affecting the cells in their vicinity. All news affects their mainstream actions in thought, dreaming, and collective creation.

Although all neurons share in this centralized process of internal communication, the inhabitants of the neurosphere are nevertheless divided into two distinct groups. There are those that live inside the central nervous system (CNS); these include the *secondary neurons* that make up the brain and spinal cord. And there are others that live outside this central system, occupying the territory of the peripheral nervous system (PNS); these include the *primary neurons* that receive and send all sensory information to the brain. A sprightlier way of putting it: The primary neurons of the limbs inhabit the rural areas; the secondary neurons of the head and torso inhabit the urban and suburban areas of the neurosphere. Even with their different living arrangements, they live humbled by a larger conception of themselves.

The CNS and PNS, then, consist of many interrelated metropolises (*nuclei* and *ganglia**), comprising numerous areas of nervous activity. All the messages, which are transferred from one city or town to another, are expressed in a single dialect with complementary parts—synaptic and action potentials. Another means of communication which seems to integrate these complementary potentials (and fulfill their complementarity) are electromagnetic waves. To keep my description as simple as possible, I will speak only of synaptic and action potentials. These potentials form the language of neuronal signaling, the medium of our "talking

---

* A *nucleus* is a group of central neurons (sensory or motor) that live in close proximity to one another. The term implies a "nucleus of neural activity" and is only loosely associated with the nucleus of a cell. The two words share the same name as homonyms. The term *ganglia* refers to the clustering of uniform peripheral neurons. The neurons in a brain nucleus, by contrast, have many shapes, sizes, and functional assignments.

heads." Without neurons talking, we couldn't talk. Without neurons acting, we couldn't act. And without neurons caring, we couldn't care less!

The neurophysiologists Carl Cotman and James McGaugh describe concisely this communicative process in their beautifully illustrated introductory textbook *Behavioral Neuroscience:* They relate how the nervous system deals with its most fundamental problems—"relaying information and integrating it." The receptor organs must transfer the signals they receive to the centers of the brain that process this flux of information. The signals travel over long and short distances through the vehicle of *action potentials.* Once these potentials arrive at their temporary destination, they are unloaded like cargo. The contents of this neural cargo are then translated and integrated in that particular stop. The integration involves converting the action potentials into smaller messages called *graded potentials.*

According to Cotman and McGaugh's interpretation of this process, the action potentials deliver the messages and the graded potentials gather and sort the messages. Depending on how far the information has to travel, the procedure varies. Action potentials suit the demands of "interneuronal communication over long distances," whereas "graded potentials, usually in the form of synaptic potentials, are the signals generally used over short distances." Moreover, graded potentials "weigh" the information they carry. The weighing process is called neuronal integration.[3]

I associate the *relaying* of information through action potentials with the process of fission. It involves the splitting or separating of signals towards the ends of branching, axonal structures (Metaphorm 6). In a complementary way, *integrating* information through synaptic (graded) potentials involves a fusion-like process. Here the term *fusion* refers to synaptic vesicles attaching themselves to the release sites, merging with the plasma membrane, and expelling their transmitters.

CELLULAR FUSION-FISSION/ COGNITIVE FUSION-FISSION

There is extensive evidence that these two basic processes occur in all nerve cells. Different sources of input converge and numerous outputs diverge. How are these processes represented in thinking and creating? Could these processes, which occur on a cellular level, also occur on a mental level, as in cognition? Are they present, for example, in our acts of cooperation and competition—between and within nervous systems?

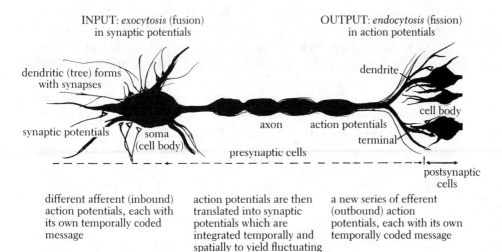

INPUT: *exocytosis* (fusion) in synaptic potentials

OUTPUT: *endocytosis* (fission) in action potentials

dendritic (tree) forms with synapses

dendrite

cell body

synaptic potentials   soma (cell body)

axon   action potentials

terminal

presynaptic cells

postsynaptic cells

different afferent (inbound) action potentials, each with its own temporally coded message	action potentials are then translated into synaptic potentials which are integrated temporally and spatially to yield fluctuating graded transmembrane potentials	a new series of efferent (outbound) action potentials, each with its own temporally coded message

METAPHORM 6. Fusion and Fission Processes in Neuronal Signaling? *Here the process of fusion refers to* exocytosis, *or the releasing and transforming of chemical "transmitter substances" at nerve endings. The fission process refers to* endocytosis, *or the splitting of the signals into graded potentials at neighboring cells. (Diagram adapted from C. F. Stevens, "The Neuron" in* Scientific American, *Vol. 241, No. 3, 1979, pp. 54–65)*

Are there moments of "cognitive fusion" (integration) followed by "cognitive fission" (separation) in the central nervous system? Accordingly, vast streams of different sources of information would be integrated, separated, and relayed in the neocortex and its associated subsystems. An even bigger question: If these fusion-fission processes do in fact occur inside us, how are they unlike the processes that occur naturally in the cosmos? I believe that these processes are indeed present in every "way" of nature and in all its manifestations. (I discuss my reasons in depth in Chapter 8.)

THE "TOUCHING" LANGUAGE OF NEURONS

To continue with this description of the neurocosmopolis: The electrochemical language of neurons translates into K^+ (potassium) and Na^+ (sodium) fluctuations in cell membranes. Neuronal signaling and electromagnetic waveforms are the means by which a single neuron can talk to billions of neurons through the voices of one of many neurotransmitters.[4] These "voices" are defined as electrochemical substances that are released synaptically by neurons and that affect other neurons or effector organs in one manner or another.* They're "touching" in a persuasively moving way.

If a family of neurons in the hippocampus (the cluster of central neurons in the *archicortex*, or *primitive cortex*—shown in Metaphorm 3b) participate in the creation of an idea, everyone living in the neurocosmopolis not only knows who said what and when, they remember aspects of the idea, having contributed to its creation. Nothing's kept secret. No information is proprietary. Maintaining a "continuous dialogue" is the rule of the "mainstream." Somehow the mind converts these clear conversations into cryptic parables and paradoxes. It seems to complicate these messages that emerge from spontaneous assemblies of neurons. It's as though the brain invented the mind to camouflage its actions, to cover its tracks, by often *confusing everything* with its turbid interpretations. (Just a muddled thought.)

MOMENTS OF EXCITATION AND INHIBITION

When neurotransmitters make contact and begin to speak through synapses, the conversation can be either excitatory, inhibitory, or modulatory. Depending on the nature of the message, the postsynaptic responses range from rapid local changes in the postsynaptic membrane to slow, global changes. The messages are apparently as specific as barcodes—those black and white, pin-striped identification numbers on all "consumer" products. I'd best restrain myself here from relating the mechanisms of thought to the contents of thought—a relationship that is perhaps larger than the topic of this book! More on the neurocosmopolis.

Communication is as follows. Primary neurons speak to secondary neurons in the CNS. These secondary neurons in turn evaluate this incoming information at the same time that they take care of old business. The flurry of news is integrated on the cells' surface membranes. Once this information is processed, the details are passed along by way of their axons to other smart cells in the central nervous system whose intelligent relatives voice their opinions. Soon, billions of neurons, representing various provinces of brain nuclei, let us in on the details of this internal public conversation.[5]

It may amaze you that the neural highways and backroads, on which

---

* The British neurophysiologist Sir John Eccles and pharmacologist Sir Henry Dale observed: "A mature neuron makes use of the same transmitter substance at all its synapses." There are approximately 10,000 synapses per neuron of this type, etc. (Quoted in James H. Schwartz, "Chemical Basis of Synaptic Transmission," in Eric R. Kandel and James H. Schwartz, eds., *Principles of Neural Science*. New York: Elsevier/North-Holland, 1981, p. 108.)

this news travels, ramify into many hundreds of thousands of routes. Each route terminates in practically every minuscule square acre of our bodies. Remarkably, no region is left unconnected with another region. In a normal brain no neuron is an island. The only poverty in communication occurs when the neuro-environment is afflicted with a pathology or is wounded. Suddenly (and sometimes not so suddenly), the processes of neuronal signaling are temporarily disturbed. In many instances, major roadways are swept away by disease—their transportation systems permanently destroyed. More than 120 years of clinical neurology, which studies brain lesions and disorders, supports this view.[6]

THE WAY OF COMMUNICATION

Think back on what you have just read. Do these descriptions of neurons or nuclei or assemblies of nuclei suggest anything about the nature of the cosmos? Anything that you didn't already know? Does our network communications system resemble that of the universe? These thoughts share a kinship with more pointed questions posed by Zeilik and Smith: "Does the Universe have a higher level of organization than clusters of galaxies? Are there clusters of clusters—superclusters?"* Their response: "For some years, astronomers were extremely skeptical about the reality of superclusters. Before observations drove home that reality, the standard picture was that of more or less spherical clusters embedded in a very uniform distribution of background, noncluster galaxies. That view has recently and rapidly changed to one in which superclusters have a strung-out, filamentary structure some 100 Mpc [100 million parsecs] long. Between them lie vast voids, empty luminous matter such as galaxies. *The superclusters may be interconnected—the fundamental network of the Universe*" (my italics).[7]

With the idea of this "fundamental network" in mind, you could say that all stars, galaxies, and clusters "communicate" via gravity. That is, they speak to one another through the commanding "voice" and force of *gravitons*, as they are called by theoretical physicists.[8] Their pulling and pushing motions, over literally astronomical distances, represent the process of "dialogue." You might even relate these motions to the receiving (or pulling) and sending (or pushing) process of synaptic and action potentials in neuronal "conversation." It is through such conversation that neurons extend their influences over one another. Although no one knows exactly what the properties of the forces are that underlie these

* Supposedly there are billions of stars within galaxies and millions of galaxies within clusters and, perhaps, thousands of clusters within superclusters, all packed into one living room.

influences, I'll seed your imagination by saying that the range in meters of the electromagnetic force is infinite. This force, which is part of the neuron like skin is part of the body, is difficult to detect with its range of 10^{-12}. This is also true of the force of gravity, whose range is infinite and whose relative strength is 6×10^{-39}—a number that whispers like a remarkably faint force.

Neurons are the smallest functional units of the brain. Galaxies are the grandest units of the universe. They appear to be the largest assemblages of stars, gas, and dust. The Milky Way galaxy, for example, contains 10^{11} stars—a figure close to the number of neurons in the mammalian nervous system. If galaxies were physically related to brains, there would be billions of them—as there are. Within our galaxy (and others) there are assemblies of stars that comprise what may be called the "nervous system of the cosmos." Within the human brain, there's another kind of assemblage of matter: "cell-assemblies," as the Canadian neurophysiologist Donald Hebb called them. Hebb used this term to describe the dynamics of neurons forming connections—assemblies—between one another and between groups. He theorized that "repeated exposure to a given sensory stimulation will organize an assembly (a number of neurons in the cortex that become interconnected)." Basing his hypotheses on experimental evidence, Hebb conjectured that "if thought is a series of cell-assembly activities, these must ordinarily be excited both sensorily (PNS) and centrally (CNS)."[9]

LIKE GALAXIES COMMUNICATING
BY MEANS OF GRAVITY
WE COMMUNICATE
BY MEANS OF CONSCIOUSNESS

The implication here is not that stars and galaxies "think" (whatever we regard "thought" to be or not to be). It is that these celestial bodies organize similar "assemblies" through which they communicate gravitationally with one another and within the enormous celestial network of the whole universe. *Gravitation is to galaxies what consciousness is to human beings.* To say that a galaxy is "conscious" of its actions is to speak metaphysically about how a galaxy *acts* through gravity. Its "thoughts" are its actions or movements. As mystics have expressed, often to the annoyance of scientists, for all we know *we* may be a thought of the galaxy—or a dream as it sleeps, an episodic dream that will disappear once it awakens. Perhaps the intervening time between the Big Bang that marks our beginning and the "big crunch" or "big squeeze" in years to come will be a mere three-second dream sequence for the universe.

Billions of years to us may be milliseconds to the cosmos having dreamt about itself—in nightmare and repose.

HYPOTHETICAL ANATOMIES OF COMMUNICATION

A related thought, though smaller in scope, conceives of the anatomy of a galaxy as a sort of nervous system, where the central bulge, or nucleus, represents the central nervous system, and where the star-forming regions in the spiraling arms represent the peripheral nervous system (Metaphorm 7). Here the bright nebulae, which are known to form new

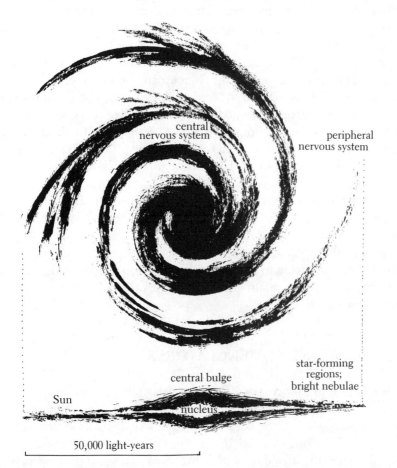

central nervous system

peripheral nervous system

star-forming regions; bright nebulae

central bulge

Sun

nucleus

50,000 light-years

METAPHORM 7. Milky Way—*galaxies as virtual nervous systems of communication. Here the "language" of this galaxy is gravitation. Both stellar and galactic structures "communicate" via gravity (or gravitons, one of the interactive forces of nature). The range of "dialogue," or conversation between stars and between galaxies, is measured by the degree of space–time curvature. View is of orthographic projection. (Drawing adapted from Timothy Ferris,* Galaxies, *1982)* stars, act as neural-like ganglia. They create or receive and "process information" (in the form of gravitons). This information is sent to the central nervous system of the galaxy. Like the ganglia in the peripheral nervous system of the human body, these *galactic ganglia* are repeatedly exposed to "sensory stimulations." To apply Donald Hebb's concept: This "exposure" is the mechanism by which "star-assemblies" are organized. In this regard, they are similar to the brain's "cell-assemblies."

EXPLORING AD ABSURDUM

Hopefully, you've heard these sorts of analogies and references so often that you're now prepared to map the nomenclature of astronomy, astrophysics, and cosmology onto the nomenclature of neuroanatomy and neurophysiology. To some small degree this has already been initiated. For instance, in the past century neuroanatomists have assigned the name *corona radiata* (Latin for "radiating sun") the massive, fan-shaped neural fibers that grow from the base of the brain to the top of the neocortex—widening along the way as they combine three other major fiber tracts. Also, note that there are *astrocytes* ("star-shaped" cells) in the central nervous system. These cells are one of the five types of *glial* cells (cells without axons) that bind neurons together.

Without going into too much detail here, either about the types of astrocytes or about the corona radiata, it's important to note that the origins of their names were not inspired by the same kinds of comparisons that this book advocates. By all accounts, neuroanatomists didn't think about the relation between solar radiation and the radiating internal structure of the brain. And yet, some other intuition guided their impulse to name.

EXCHANGING NOMENCLATURES

The names and processes of stars, galaxies, clusters, and superclusters can be—consciously—applied to neurons, nuclei, and large-scale neuronal bodies such as the basal ganglia, etc. I can envision, for instance, applying the terms *H I and H II regions** to the activities of the entorhinal cortex, the area deep within the brain that's part of the hippocampus (note Metaphorm 3b). The hippocampal galaxy-like subsystem is known to be involved in initiating the production of ideas; it's regarded

---

* These terms, *H I region* and *H II region*, are used in astrophysics to describe the areas where star birth is occurring. (See Metaphorm 2 in Chapter 4.)

as one of the roots of *ideation*. In this region ideas form trillions of times faster than stars form. Note that stars are created over millions of years by both mammoth and small molecular clouds of interstellar medium. Although my imaginary calculations may not match those of Shakespeare, who said, "Thoughts, which ten times faster glide than the sun's beams," they express the same thought (though, perhaps, at a different speed).

"WHY?" (ASKS INTUITION) . . .
"BECAUSE" (ANSWERS REASON) . . .

Why speak anthropomorphically? What are the real benefits of these excursions from pre-rational science and rational science fiction? I can think of three reasons, hierarchy not intended. One reason is that it acknowledges our connection with our celestial progenitor. The events and processes that occur above our heads also occur inside them. Between and within stars there are connections. Between and within our brains are connections. The mind serves as the *connector* in comparing the patterns of these connections along with its own patterns. And as Douglas Hofstadter remarked, "Mind is a pattern perceived by a mind."[10] (Mind is also a pattern *conceived* by a mind, insofar as the mind created the concept of "pattern.")

A second reason for speaking metaphorically is that it establishes our oneness with nature. If biological life and human social structures are not basically different from the physical systems of nature—a concept I believe many people subscribe to—exactly how much alike *are* humankind and nature anyway? Why do we think our thoughts and actions resemble only *some aspects* of nature and not *all aspects*? Although nature's forms are varied, the vast differences of form are contrasted by a simplicity of likeness in process. *Our internal processes mirror nature's internal processes and vice versa.* The third reason I offer isn't a reason at all. It's an impulse—the impulse to connect and create meaning, to convert the improbable to the possible through dream.

PATHOLOGIES OF MINDS AND UNIVERSES

But we all need to dream together and discuss our dreams like the Somoan Indians when asking: What would be analogous to the brain's communications system in the cosmos? Is gravity its only "language"? Does this system ever break down, as though it were struck by some neocortical dysfunction such as *apraxia* or *aphasia*? (Apraxia is a neurological condition which affects a person's capacity to act purposefully. Aphasia is defined in the next paragraph.) Do galaxies suffer similar "speech, motor, and comprehension pathologies" which prevent them from fully communicating ("listening," "comprehending," "speaking," and "acting") in accordance with other galaxies—like human beings? Are there syndromes of behavior in galaxies that affect certain regions of space and matter yet do not affect others—like the linguistic disorders in humans called *Broca's aphasia* and *Wernicke's aphasia*? Are there such things as Broca and Wernicke "aphasic galaxies"?

Aphasia is a condition affecting the speech and comprehension centers of two distinct regions in the left hemisphere of the human brain. The area is at the level of the superior temporal lobe, which is just a little above the middle of your ear. These regions are separated on the surface by a continental divide called the *Sylvian sulcus* (or fissure). The area in front of this fissure is called Broca's area, or the *precentral gyrus*. In this region motor neurons flourish.

The area on the other side of this great divide is called Wernicke's area, or the *postcentral gyrus*, a region inhabited by sensory neurons. Lesions in either area affect the coordinated actions of a human being in comprehending the world and in expressing this comprehension.

What would be the correlate of these afflictions in celestial bodies? I can imagine stars or galaxies or clusters whose motions could be interpreted as "impairments of motor activity and comprehension." A "Broca galaxy" can understand the language of gravitons but it cannot speak normally (meaning "act" or "respond") due to its "motor" dysfunction. These galaxies move or speak disjunctively. A "Wernicke galaxy," by contrast, can speak but it does not completely understand what it's saying (what its actions are). It moves and functions fluently but this movement lacks meaning. It can't "comprehend" the messages of gravitons and so it acts erratically, or rather speaks nonsensically, implying that the graviton force is not fully comprehended. This doesn't mean that a Wernicke galaxy is necessarily violent. It's just confused. Perhaps the aberrant motions of interacting galaxies that are measured in radio waves reveal similar types of impairments. (There's also a greater possibility that this primitive comparison is an absurdity raised to the highest exponent conceivable! My intuitions inform me that it isn't *that* absurd, no more than the concept of "constants" in physics, and I trust my intuitions.)

EXCHANGES WITH AND WITHOUT EMOTIONS

When you think about the different types and intensities of galactic exchanges (or "intergalactic encounters," as author Timothy Ferris calls

METAPHORM 8. *A section of cat neural tissue. Are we looking at a biological version of "colliding stars and galaxies"? By colliding I mean that they're interlocking networks that slowly move (not necessarily over a single lifetime) over thousands or millions of years through evolution, paralleling galaxies that appear to move over millions of years. This movement changes both the neural architecture and the behavior of the organism, just as it changes the "behavior" of galaxies. The neurons' axons and dendrites shown here are distinguished by Golgi and Nissl staining techniques, which are two of the most important techniques in neuroanatomy, developed about 100 years ago. The two techniques illuminate different aspects of neural structures.* (From Walle J. H. Nauta and Michael Feirtag, Fundamental Neuroanatomy, 1986)

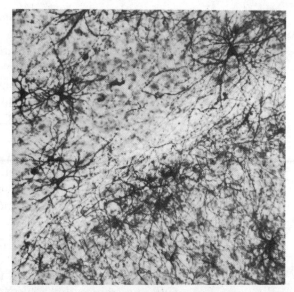

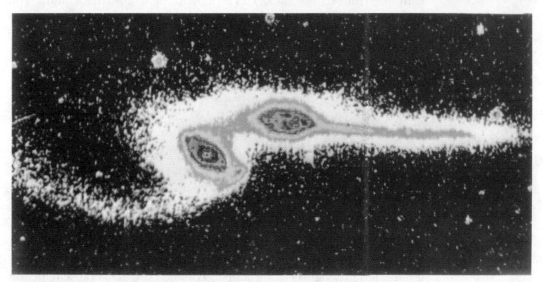

METAPHORM 9. *A photograph of colliding galaxies. Although the distances between stars are enormous (approximately 30 million times their own size), the distances between galaxies are considerably less—hence more collisions. In fact, the ratio of collisions to solitude is estimated at 25:1. This photograph was created by the 4-meter telescope at Kitt Peak, near Tucson, Arizona. The bridge, or conduit-like structure, that seems to connect these two galaxies spans some 5 million light-years! Note: There's absolutely no pun intended by the fact that the neural tissue in Metaphorm 8 is from a cat's brain and these galaxies are called "The Mice" by Kitt Peak scientists. Humor aside: In a strange way, it is a sort of "cat-and-mouse game" we play with nature insofar as the getting of scientific wisdom involves the* hide-and-seek of information *or the* chase of knowledge. (From Michael Marten and John Chesterman, The Radiant Universe, 1980)

them), the thought of "afflictions" seems quite natural. There are healthy humans and ill ones. Why would we believe differently about celestial bodies? Some interacting galaxies appear to be fairly gentle, like our spiral galaxy making waves with the Magellanic Cloud with which it presently interacts. Others can evoke horror at the magnitude of their vicious collisions (contrast Metaphorm 7 with Metaphorm 9); the giant elliptical galaxies, such as M87, and the exploding "Ring" galaxy are two exemplars. In the case of the latter, the whole nucleus has been swept away. Still other galaxies seem to behave like Broca and Wernicke aphasics as they try to communicate. You can almost feel their torment.

Eventually, as you read about normal brains, as well as the tragedy of those who suffer deficits and lesions as studied in neurology, you will experience a double-think. That is, you'll accept as valid all inconsistent versions of facts regarding these two phenomena, having trained your mind to ignore the conflict or unresolved points between these facts. At that time, you'll be able to think about the possibility of stellar bodies suffering from pathologies. You'll begin to factualize this fiction. You may read about the birth of massive stars and follow their development to their traumatic conclusion as supernovas, like a neurologist observing the aberrant behavior of a patient suffering from some brain pathology. It's not as if we're ever going to prove any of these ideas, like *astrophysicians* autopsying galaxies with the same intent as forensic medical practitioners.

THOUGHTS WITH NARROW CONTEXTS BUT BROAD CONSEQUENCES

I'd like to see some of the more important researches in twentieth-century astronomy, such as Edwin Hubble's classification of galaxies in his classic *Atlas of Galaxies*, fueling the reflections of neuroscientists. If the insights from neurocosmology do nothing more than provoke us to entertain these thoughts—for example, Hubble's classification scheme * in a broader context (such as brain processes)—then neurocosmology will have served its invention.

As an almost trivial example in comparison to the broadness of the changes proposed, think of how the classical concept of interneuronal communication was subverted when neurophysiologists discovered that

---

* Hubble's classifications of galaxies are not cast in some fixed evolutionary sequence. Their forms depend upon one thing in particular, *angular momentum*; the more momentum, the flatter the galaxy. The *Atlas of Galaxies* classifies galactic structures according to their principal characteristics, which include their color spectra, masses, neutral-hydrogen content, sizes, luminosities, mass-luminosity ratios, etc.

dendrites (the receiving end of a neuron) could also transmit information. It had been widely held that this key structure in all neurons was simply a passive receptacle for collecting inputs. It passed along information, as opposed to participating in some dialogue—actively contributing to the scope of this dialogue through its dendrodendritic connections (that is, the tips of one dendrite conversing with its own or with another neuron's dendrites).

EXPANDING THE CONCEPT OF CHANGE

It could be that a similar "exchange" of information or energies was taking place in the substrate of space surrounding planetary bodies or stellar bodies. Using an example close to home, let's say that in an area referred to as the transition region around the Earth's magnetosphere (see Metaphorm 10) there are interactions occurring that are analogous to the electromagnetic field around the soma of a neuron or the entire neuronal body. Of course, the scale of events and the energies and distances involved are vastly different. Imagine that at the "shock front," which appears to be a buffer zone between the solar wind and the edge of the outer belt of the magnetosphere, the protons and electrons that are contained inside the magnetosphere are sort of conversing with the incoming (afferent) particles of the solar wind. As their "conversation" is limited to brutally intense collisions somewhere on the order of 100 MeV (million electron volts), you wouldn't expect to hear any sweet talk or pleasantries. Their conversant actions would more likely resemble a brawl or street fight than a formal gathering. And yet, nature constantly surprises us in all its brute impulses. When we dispassionately expect to find nothing but more violence behind the edifice of our violent world, we're gently touched by other seemingly organized and harmonious interactions of nature. These interactions are not without their lessons. It's how we observe phenomena and interpret our observations or data that "makes all the difference" in the world.

Think of how orchestrated the actions of dendrites are in talking to other dendrites and of neurons conversing with their peer neurons. Contrast this with the jaded actions of charged particles "trapped in the radiation belts"* and spiraling "along lines of magnetic force while

---

* According to satellite data, the particles that are bound inside the concentric, inner belt (which measures between 1 $R_\oplus$ and 2 $R_\oplus$) and outer belt (measuring 3 $R_\oplus$ to 4 $R_\oplus$) of the magnetosphere are derived from solar winds and formed from cosmic-ray interactions in the upper atmosphere of the Earth (a region about 50 to 500 kilometers above the Earth). Our planet's strong dipolar magnetic field holds these energetic particles like a mother cradling her baby. Note that the symbol "$R_\oplus$" refers to the radius (R) of Earth ($\oplus$) at the equator, $R_\oplus = 6,378$ kilometers.

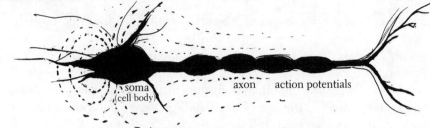

Weathering the Neural Winds of the Nervous System

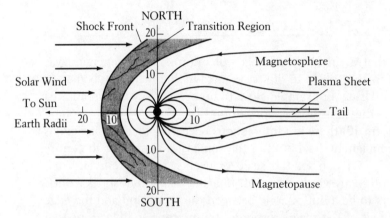

METAPHORM 10. Magnetosphere. *In interpreting the interactions of very low fields in the Earth's magnetosphere and in the human nervous system, the solar wind represents information incoming to the neuron. The spiraling motion of charged particles that arrive from the solar wind interacts with the dipolar magnetic field of the Van Allen belts. These high-energy particles are thrown back and forth between the "mirror points," which are formed from the North and South magnetic poles of the Earth. The protons (positively charged particles) and electrons (negatively charged particles) that drift in both the large and small Van Allen belts travel in opposite directions. How does the meaning change as you substitute the Earth for the brain? How does it further change as you substitute the brain for the dendrites and cell body of a single neuron in the brain? Instead of charged particles from the Sun, consider the flow of electromagnetic energy around the bodies of neurons. Then consider the types of information exchanges. (Diagram at bottom from Michael Zeilik and Elske Van Panhuys Smith,* Introductory Astronomy and Astrophysics, *1987)*

bouncing between the northern and southern mirror points."[11] The particle actions seem to have little organization and lots of anxious energy.

Has anything meaningful emerged from contrasting these different sources of information on brains and solar-mass stars? Have we come any closer to understanding the virtual energies of cognition (those invisible energies which we associate with the forming of ideas)? Or have we merely frightened all driven individuals who burn brilliantly in their industrious thoughts—reminding them to either conserve fuel or die early from rapid exhaustion?

I KNOW WHAT YOU MEAN
BUT I CAN'T EXPLAIN HOW I KNOW

How does this information help tie together our discrepancies between biological and physical "organisms"? My feeling is: If we can't bridge the two forms of matter through either dialectics or physics, at least we're encouraged to think of these organisms, brain and cosmos, in a different —broader—context. Instead of conceptualizing them as distinct forms

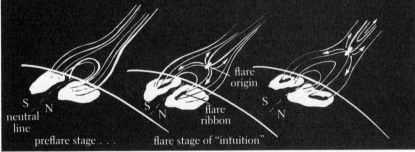

METAPHORM 11. *Lightning flashes as nature's "neurons" . . . remind us of our nature. (Photograph copyright © 1972 by Gary Ladd)*

METAPHORM 12. Solar Flare—*the "plasma fusion state of mind." Diagram of a solar flare showing the stages in its creation: pre-flare stage, where the magnetic field lines are unconnected as they swell beyond the coronal loop; flare stage, where the field lines connect, sending one current downward towards the base of the star and another upward into the corona. (Diagram of solar flare adapted from Michael Zeilik and Elske Van Panhuys Smith,* Introductory Astronomy and Astrophysics, *1987)* with fixed anatomies, you're asked to see their *dynamic morphology* (their "processmorphic" properties).

In comparing intuition flashes to lightning bolts and solar flares (Metaphorms 11 and 12, respectively), I am speaking of *processmorphs*. The implication is that these forms of matter are forms of plasma fusion. Plasma represents the fourth state of matter—a state that is neither solid, liquid, nor gas but extremely hot, *ionized gas.** The expression "plasma state of mind" then may represent a state of heightened consciousness (as in intuition), but it also insinuates something else. It suggests that the mind undergoes the same process in producing this so-called "plasma state" as a star undergoes in its production of a solar flare (Metaphorm 12).

LOST IN FORM, FOUND IN PROCESS

Although lightning, solar flares, and related phenomena appear to be unlikely candidates for describing brain phenomena, they're not as alienable as their differences in size, energy, and media make them out to be. These differences in proportion are all relative. I support this statement with a recent finding by the Swedish Nobel laureate and father of plasma physics, Hannes Alfvén. Dr. Alfvén and his colleagues have discovered that microplasma fusion structures created in the laboratory resemble cosmic plasma fusion structures such as solar prominences on the sun and galaxies. Alfvén claims that the universe "has always existed pretty much as it is now, will go on forever, and is dominated by the same electromagnetic forces that shape nearby plasmas," which he defines as "gases that conduct electrical currents." [12] Plasmas include the undulating auroras such as those Alfvén had observed while living in Sweden. Dr. Alfvén suggests that the universe is electrical in nature, a view that continues to meet resistance in the scientific community.

Question: How are *we* not "dominated by the same electromagnetic forces that shape nearby plasmas" (as defined here)? Although I don't mean to suggest there are smaller types of "microplasma fusion structures" created in the human brain—plasmas that are a billionth of the heat intensities of the microplasma structures Dr. Alfvén and his colleagues synthesized in the lab—I do think there may be a similar phe-

---

* On this large scale, ionized gas is electrically neutral. At temperatures above 10,000 K, the electrons are stripped away from the atoms. This action results in the creation of a flowing plasma which is composed of electrons and protons in equal number. The heat is enormous because plasma has a high thermal conductivity. If the arts represented protons and the sciences represented electrons, their mutual presence and equal mixture in a plasma is an interesting metaphorm to contemplate with regard to the making of an intuition.

nomenon that occurs on an infinitesimal level in the nervous system, involving the conduction of electrical currents.[13]

MAKING THE FAMILIAR MORE FAMILIAR

To return to the orchard: We need to mix our theories of cosmology (or oranges) with our theories of brain processes (or apples). This way, we stand to discover the peculiar mixture of nature's fruits and the "jams" we create with them. The new combined theories of the neurocosmos will no doubt either be sweet or sour, fully cooked or half-baked. As I've pointed out throughout this book, these value judgments are based on the integrity of both the astronomical and the neurobiological data. Although one theory may be shown to be more probable than another, in the largest picture conceivable or in the broadest frame of reference, we can never be certain.

This uncertainty is understandable when you weigh the facts of our labor. We're only able to see 10 billion galaxies out of another 10 billion, or 20 billion collectively—perhaps even these numbers are too low. Understand that I'm including the galaxies within clusters—not just the clusters of galaxies. That's a lot of matter in a space that seems to expand beyond the instrumentation we invent to measure that space. In short, we can't possibly expect to see or listen to the whole of the universe without deceiving ourselves. To make such claims would be as ludicrous as a microbe thinking it has "seen" not just the human body it inhabits but every human body ever created. It's hard enough to fathom "cosmic dimensions" that we presently assess to be on the order of 10^{26} times our size. After you play with the idea of more *super*cosmic scales, you're ready for the outdoors or some immediate distraction as the figures are absolutely vertiginous! Moreover, the concept of scale becomes unthinkable altogether.

MIND, AT THE VANISHING POINT OF MATTER

Some examples: In cosmology, we can choose to look at the origins of the universe from two broad perspectives with equal detail. We can conceive of mind and cosmos as biological organisms or treat them as strictly mechanistic constructions. (Many theoretical physicists are uneasy about granting the first option. For them, there's no *choice in the matter*, so to speak.) Frankly, I think both perspectives—the biological and the physical—lead to a similar conclusion or "vanishing point." Only the details obscure our vision, like dense, intergalactic clouds occlude our vision of certain regions of space. One view which seems most consistent with the ways of nature conceives of it as being both mechanistic and organic, implying that they share the same processes of creation *(chaos)*, preservation *(equilibrium* or *homeostasis)*, and destruction or transformation *(nonequilibrium)*. Here the two end points on the line of evolution—creation and destruction—come full circle. Only the details differ about the nature of these points and how they're joined.

These interrelated processes have been described for centuries by ancient mythologies and more recently by modern "mythologies." As Joseph Campbell and many other scholars have pointed out, these processes are intimated in the sculptures, paintings, writings, and prayers of traditional Indian philosophies as well as in the contemplative artworks of Chinese culture. However, their views on creation-preservation-destruction are encoded in the language of artistic symbols and nuances —the very things that are deemed unapproachable by scientific investigation (as it is presently practiced). As the sciences expand their horizons, the arts will enter their field of view—changing the shape of the horizon altogether.

A REVERSED PERSPECTIVE
AT THE POINT OF ORDER

We now recognize, for instance, that the random actions of disorderly particles are coherent patterns of a larger, orderly system experiencing some transformation. We now believe that to transform *something* is to move some *thing* from one state of energy to another; this includes passing from one form of matter to another. A rapid shift in thermodynamic equilibrium triggers the kinds of transformations we observe in seemingly chaotic patterns of energy, with the patterns of chaos reflecting the processes of creation.

Transformation in all systems is a universal creative process. I define "creativity" as any *unconditioned response* by one system towards itself or its environment. "Conditioned responses" represent systems that operate near equilibrium (as in *linear systems*), whereas "unconditioned responses" relate to those systems that radically depart from these harmonious conditions. Instead they can operate within the fuzzy state of nonequilibrium (as in *nonlinear systems*). Human beings are about as nonlinear a system as nature makes. A creative system can accommodate abrupt changes and catastrophes with its agility. All life forms are creative. The cosmos is a form of life—one that is as agile as the life forms it has created.

With these considerations in mind, I believe we can choose to see the birth and evolution of the cosmos in the same light of understanding as we see a living organism. Accompanying the concept of *organism* are the "seed" and the "blueprint"—metaphors for a somewhat predetermined form. I say "somewhat" because there are no guarantees. Although all biological forms are believed to possess a fair amount of programmed knowledge—growing by instruction, as it were—biologists know that the environment also acts on the organism, affecting its evolution. (Recall my notes on the notion of *phenotype* discussed in Chapter 4.) In the case of the universe, it supposedly created its own environment as it brought itself into existence *by itself*.

PLANNED GROWTH

Few cosmologists would concede that the model for the genesis of the cosmos is a gene or a seed—as in the "seed of singularity." The reason for this is that the metaphor implies some genetic program of growth in which all of matter and space unfold with determination—like the neuroectoderm. And yet, the universe possesses the equivalent of genes and chromosomes in its *growth process*. It seems to grow like its progenies— the "gardens" and "orchards" and "organisms" it has created on this planet. Taking a cue from our evolutionary sciences: Nature has a propensity for producing *variations on a theme*, or what some writers might call "redundancies." There are redundant features of neurons, for example, which are thought to serve as backup systems in the event of emergency. There are no doubt redundancies in stellar bodies. (Some astronomers will surely protest, pointing out the idiosyncrasies of some 2,000 of them that have been catalogued. With a few hundred billion others as a remainder, I'm certain there's room for repetition.)

UNPLANNED ACTIONS

Consider another perspective: Why wouldn't nature repeat itself—with some variation—in its creation of "pluriverses" (more than one universe)? After one universe has "experienced life," from the singular moment of its singularity to its hot or cold demise, a new one would form. This thought, too, has been intimated through the artistic works of the ancients. One modern artwork which can be interpreted along these lines is the Romanian sculptor Constantin Brancusi's *Endless Column*. The convolutions of his monumental free-standing sculpture—with its concave and convex "motions"—suggest to me the continuity of an endless stream of creation. Each movement, from convex to concave, might symbolize the ascent and growth or descent and decay of the cosmos, the outward and inward movement of matter and thought. The movement is repeated in such a way that the details of life in between these planes of creation and destruction are always varying, while the overall pattern of growth remains intact. Brancusi's work of thought, reflecting the divergence and convergence of matter, resonates with about 4,000 years of Western and Eastern philosophies. This art can also speak for the complementary concepts of an inflationary and deflationary universe. All of these concepts in modern cosmology may be applied to the workings of the human brain, with imagination.

Although nature seems to have its "reasons" for this replication of similar processes attached to the many faces of its diverse structures, it seems reluctant to share this wisdom with us. Or perhaps it does, but our assumptions regarding its complexity obscure our vision and congest our interpretations.

IN ALL LIVING ORGANISMS

Accepting the idea of an underlying, biological order to the cosmos is less troublesome than it seems. Although this organic order implies that the human gene cycle is analogous to the hydrogen- and helium-burning cycle of a star, you don't have to believe this implication. To believe that the growth of the universe reflects the growth of our bodies is to sense the gaping space beyond all physical evidence. To think that in the compressed moment of singularity, at the onset of the Big Bang, there was some virtual blueprint that presently guides the growth of matter is to reason by feeling, not logic. And yet, modern cosmology is constantly moved by "deeper feelings" embodying intuitions. So often prescient thoughts have hit the bull's-eye of physical law and pinned the tail on our stubborn assumptions.

When I read in a technical journal that the universe began from randomly composed rudimentary elements—elements that somehow evolved into the beautiful world that is inside and that exists before our eyes—I hear myself ask: What universe are we talking about? The one all clergymen describe as being organic and soulful, or the one theoretical physicists describe as being mechanistic and secular? Are we speaking about the universe that rises to complexity by means of the *logos* (the rational principle governing its growth) or by means of chaos? "Out of all things there comes a unity, and out of a unity all things," wrote Heraclitus. This reflection is now transmuted by the thought: Out of all things comes chaos, and out of chaos all complex things.

How could matter *and space* evolve from a pin-size dot into this staggeringly complex universe without some plan? (Recall the plan of "determination and differentiation" in the genesis of the nervous system.) How could the cosmos come about without laws of order and disorder, growth and entropy? (Recall the insights of traditional science that describe "closed linear systems."* Also, recall the nontraditional science that relates "nonlinear open systems" to all forms of matter.)

HUMANS AS ONE OF MANY MANIFESTATIONS OF THE COSMOS

How could the universe establish itself without some intrinsic knowledge? Not knowledge as Plato, Aristotle, or Kant defined this word with all its limitations, but knowledge as it *includes* the concept of our unity with nature. A concept that is as large and full as we dare to imagine and literally realize, *or to think literally through metaphorms*—modeling what we think. This "unity" suggests that *we are* a manifestation of the universe—one of its many forms of matter—as numerous ancient cultures remind us and as contemporary science observes. We share the properties and processes of the universe—in all its manifestations. This includes its *virtual* processes, for which we have no confirmed image or immaculate description—only symbolisms. And our symbolisms no more represent the caprices of human invention than nature's invention of human beings represents caprices.

THE TABULA RASA METAPHORM

The idea that everything grows from nothing is reminiscent of the concept of *tabula rasa* (Latin for "blank slate") as it was debated by neurobiologists some years back. The big, cumbersome question was: Are we programmed, or "wired," from birth by means of our genetic constitution? Or are we born "blank"—gifted with only the basic components that allow us to acquire knowledge through our experiences of the world, growing from experiences as we go? This notion seems to parallel the idea that the universe grew out of practically nothing—primordial matter making space for itself in what was, possibly, a virtual spaceless environment.

The *tabula rasa* concept has been contested by the fact that neurons show a great deal of specialization from the very start. We're born informed. When a chance intuition befalls, our brains are prepared to deal

---

* *Closed (linear) systems* are defined as the sum of their parts and nothing more. *Open (nonlinear) systems* are more than the sum of their parts: Examples include stellar bodies and human bodies. The term "open" refers to the system's capacity to interact with other matter and systems, exchanging energy and information.

with it and whatever else comes our way. From infancy to the instant before expiration, we're naturally equipped to handle the throes of a beautiful and bellicose nature. The question remains: How informed are neurons at birth and what is the source of this information? Was matter "born" with similar "information" about its path or purpose? Did the initial "start-up" conditions of the cosmos include some genetic map of sorts? (Like *egg*, like *chicken*? Like all organisms?) Perhaps one of the requirements of any creative system is that it possess some plan to which it might adhere or deviate from unwittingly. One positive conclusion from looking at this scenario—a rather narcissistic view—is that neurons are nature's way of recording the acts of nature. Our brains are one of nature's inventions for interpreting its (nature's) actions and creations. Like the handsome youth Narcissus in Greek mythology, nature's in love with its own image—as it changes face, as it experiences its own eternal presence.

AS TRUE AS HEAT

Whether these statements are as true to nature as the laws of thermodynamics or Einstein's theory of general relativity is a moot point. The fact that we *think these thoughts* is proof enough that nature has thought of them *and perhaps lives them* as we think—perhaps.

In the thick of our diverse views on these issues exists a common pattern of perception, but we have to search hard to recognize this pattern. With this common perception we can redefine the human nervous system and the cosmos's system in a way that allows us to map the processes (and general principles) of one onto the other. The wider our definitions, the more the conceptual boundaries between the beholder (the brain) and the beheld (the universe) overlap. My critics will immediately intercept this suggestion, inquiring: What's wrong with our present definitions? They seem accurate enough. Are you suggesting that we rewrite our textbooks on neurophysics and astrophysics when so much progress has been made with this knowledge? Anyone who has some understanding about the subjects of brain and cosmos will readily agree that many of the newly discovered laws and principles of these systems are precise and productive, so what is to be gained by this revision other than unnecessary mayhem?

Everything is to be gained by expanding our perceptions, gained without intellectual havoc and without falling into the mindless black hole of nihilism. Most immediately, we stand to gain an understanding of ourselves that is unequaled in its wide implications and scope. This understanding hinges on the idea that our current definitions of these two most complex entities are, in fact, interlocking and interrelated. Moreover, they're capable of being unified by our only tool for unification, the mind. Our leviathan minds are the true *hidden* or *missing matter* which links us to the universe (Metaphorm 13).

The only way of expressing a unified field theory for matter and mind may be through poetic realism. This form of realism is frequently referred to as abstraction. It is by means of this abstraction and symbolism that nature "describes" itself.

As a postscript: If the comparisons explored here seem heavy-handed, I'll remind you with bounteous examples how *life is a heavy-handed master*, how *nature is graceless*, and how *reality is oppressive*. We also possess opposite qualities—the ones that make us effervesce with periodic enlightenment. Everything stated here, then—however awkward—reflects nature's awkwardness. And beauty.

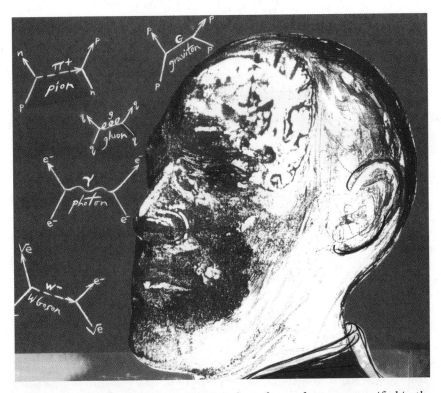

METAPHORM 13. Mind As Nature—*all the basic forces of nature are unified in the human brain.*

6

A World View— Encyclopedic Imagination

This chapter provides a break point for summarizing the previous cosmological ideas and for preparing you to consider in the next chapters the nature of mind and its relations to the things it creates.

COMBINING OTHER VIEWS—FOCUSED ECLECTICISM

The gathering of ideas in neurocosmology—as well as its philosophy—I call *reflectionism*. Imbedded in the concept of reflectionism is the dialectic (as in *dialogue*). The dialectic represents a process of dealing with contradictory arguments and perspectives as a means of arriving at truth.

Since neurocosmology is more concerned with inventing new forms of dialogue for communication than it is in discovering truths, I have modified the original concept to suit my needs.

Dialectics are present in nearly every phase of Western philosophy, from pre-Socratic to Hegelian philosophy. According to Hegel, things are constantly coming into being and ceasing to be. Everything is "permanently" changing and being superseded. This view finds its precedence in Eastern dialectics which speak poetically of the processes through which our minds and nature change. Tension, or strife, and fluctuation are two principal means of change.

Through *reflectionism*, we see the world as a whole entity with reducible properties. The dialectic steers us towards a more complete understanding of nature. Its view intimates that "all is flux, nothing is stationary," as the Greek philosopher Heraclitus thought in the sixth century B.C., and elaborated:

> Things taken together are whole and not whole,
> something being brought together and apart,
> something which is in tune and out of tune;
> out of all things there comes a unity,
> and out of a unity all things. (Fragment 10)[1]

IN BECOMING ONE

Reflectionism considers three associated pairs of relationships: matter *And* nonmatter,* brain *And* mind, science *And* art. The word *And* is the mirror conjunction that simultaneously integrates and separates these relationships which comprise the whole of reality (or nature). (Note Metaphorm 1.) Paradoxically, the integration of these relationships can only occur *without* this conceptual mirror which separates the object from its virtual image. The presence of the mirror reveals nature's diverse and dissimilar forms, whereas its absence conceals nature's unified and similar processes. A definition of nature (of which the physical universe is part) is "complete" only when it can describe the connections between all the elements of these three relationships.

On the subject of symmetry, the renowned theoretical physicist C. N. Yang writes: "Nature seems to take advantage of the simple mathematical

---

* "Nonmatter" is not to be confused with anti-matter, which is material in nature and which has been observed in cosmic-ray showers and particle accelerator experiments. Anti-matter is composed of anti-particles as opposed to particles. An example of an anti-particle is the positron (or anti-electron), which was discovered in 1932. Anti-particles differ from particles in that either they're oppositely charged (for example, anti-proton and proton) or their magnetic momentum is opposite their spin (for example, anti-neutrino and neutrino). Nonmatter, on the other hand, can never be "seen" and quantified, or measured, except in an abstract or symbolic sense.

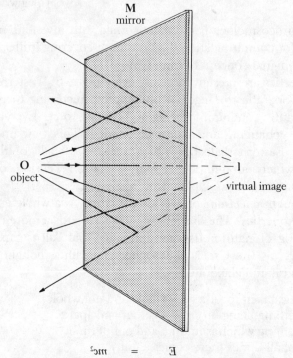

M
mirror

O
object

l
virtual image

$$E = mc_2$$

METAPHORM 1. *Everything is influenced either conceptually or physically by mirror reflection—one of the symmetries of nature. Whether we're discussing the material structures of molecules or the symbolic processes of thought, both matter and mind are affected by reflection symmetry.*

> Between matter *And* nonmatter is the *Mirror, Reality.*
> Between brain *And* mind is the *Mirror, Reality.*
> Between science *And* art is the *Mirror, Reality.*

The Mirror is reality's "ideal" mechanism for grasping the nature of reality itself. Paradoxically, the moment we see the integrated nature of nature, the mirror removes itself from our consciousness or mind's eye in this moment of synthesis.

representations of the symmetry laws. When one pauses to consider the elegance and the beautiful perfection of the mathematical reasoning involved and contrast it with the complex and far-reaching physical consequences, a deep sense of respect for the power of the symmetry laws never fails to develop."[2]

THE ILLUSION OF TWO
CREATED BY THE NON-ILLUSIVE ONE

The plane mirror is analogous to the reasoning process of the dialectic which involves comparing a thesis (object) with its antithesis (mirror image). The dialectic searches for the truth between opposing ideas, seeking the synthesis of ideas. It suggests that there is more than one way to observe and interpret reality. There are ways to look "beyond" all transitory judgments. The mirror exemplifies this. It represents one means of dividing and synthesizing ideas. Or more accurately, *the act of removing the plane mirror is an act of synthesis*. Simply, the antithesis disappears into the thesis—the virtual image vanishes into the object, like a shadow erased from the form creating the shadow. Reconciliation or gathering occurs. Two things are seen as one thing, bearing the properties of its opposite.

THE SEDUCTION OF MIRROR REFLECTION

Mirror reflection has inspired, seduced, provoked, and perplexed humankind throughout recorded history. In learning about this concept, one is bound to inquire: What aspect of "reflection" (thesis, antithesis?) attracts and repels us so (towards and away from synthesis)? Is it the physical characteristics of light radiation and specular reflection—like speculative philosophy—that "touch" us? Is it the metaphysical and metaphorical characteristics that "move us"—as we reflect on intangible things such as thoughts, ideas, and memories? Or could it be that the simple act of one thing mirroring or replicating itself evokes the dynamics of the mind in reflecting on something?

The history of mathematics illustrates brilliantly how one principle of mirroring—namely, the equality sign—has been used forcefully towards this end in weighing unknown variables (the mirror image) alongside known variables (the object). The expression $x = 1$ is the same as saying x *is* 1. It can also imply x *mirrors* 1.

Apparently, the idea of mirror reflection lodged itself in our imaginations long before we had any clear understanding of optics. Perhaps the mirroring process is a physical facet of our minds in the same way it is a facet of the atoms, molecules, and proteins that make up the macromolecule DNA in our bodies. In his book *Life Itself: Its Origins and Nature*, the Nobel Prize physiologist Francis Crick discusses the role of mirror imagery in the "handedness" of molecules. He concludes: "It is an experimental fact that the asymmetrical molecules on one side of your body have exactly the same hand as those on the other side. But could we not have two distinct types of organisms, one the mirror image of the other, at least as far as its components are concerned? This is what is never found. There are not two separate kingdoms of nature, one having molecules of one hand and the other their mirror images. Glucose has the same hand everywhere. More significantly, the small molecules that are strung together to make proteins—the amino acids—are all L-amino acids (their mirror images are called D-amino acids: L = Levo, D = Dextro) and the sugars in the nucleic acids are all of one hand. The first great unifying principle of biochemistry is that the key molecules have the same hand in all organisms."[3]

METAPHORM 2. The Mirror Plane as a Möbius Strip—*the human brain's mechanisms for transforming a perceptual illusion into a conceptual illusion (and vice versa). Consider the implications of these influences as the mind contemplates the brain and universe.*

HANDLING OUR SEDUCTION

On a more poetic note: Perhaps the concept of reflection enraptured our minds when we first saw ourselves and the world's reflection in still water. Whatever the case, certainly a mirror's illusory properties have intrigued humankind since this discovery. I suspect that our awareness of the reality of illusions prompted insights into the illusions of reality. Where the former relies primarily on *perceptual illusions*, the latter relies on *conceptual illusions*. At some point we twisted and then connected the ends of this awareness, thus creating a Möbius strip—a topological melding of illusions (Metaphorm 2). The ancient Indian concept of reality echoes this thought by saying: "Reality is one thing but the learned call it many things."

Once you accept the world-view of reflectionism, where everything is connected, no aspect of reality is seen as being separate and unrelated. Neither the universals nor the particulars of matter *And* nonmatter, brain *And* mind—nor the languages we use to describe these things, science *And* art—are seen to be in conflict with one another. There is only confluence.

UNCONFUSING REALITY

One brief interjection for those versed in the history of philosophy: The words "mirror" and "reflection" are not used here in the same way as the seventeenth-century systematic philosophers used these words to describe the mind as a mirror—a mirror reflecting reality.[4] Reflectionism is *not* based on the analytical philosophers' concepts insisting, as Kant did, that "our chief task is to mirror accurately . . . the universe around us,"[5] as though we had the equivalent of a glassy substance in our brain's mind that reflected the outer world; as though all we have to do is clean this mirror to see more clearly the whole world. Also, note that reflectionism is not founded on the ideas of Democritus and Descartes, who assumed "that the universe is made up of very simple, clearly and distinctly knowable things, knowledge of whose essences provides the master-vocabulary which permits commensuration of all discourses,"[6] as though the purpose of the mirror was to "reflect" these "simple things," and in doing so, reveal the "foundations of knowledge." The philosophical basis of reflectionism is closest to the Eastern dialectic, where the concepts of complementarity, parity, and unification are emphasized (Metaphorm 3).

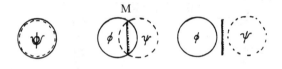

METAPHORM 3. *Tracing the processes of union, intersection, and separation— from monism to dualism to pluralism—in the philosophies of brain and mind. The plane mirror (in the middle diagram) is both a model and a metaphor for describing this relationship.*

AS REALITY BECOMES IMAGINATION

Reflectionism directs your attention to the fact that we understand the reality of the brain and cosmos by peering from different sides of the mirror. Like Alice (in Lewis Carroll's book *Alice in Wonderland*) and Black Orpheus (in Marcel Camus' film about the legend of *Orpheus and Eurydice*), both of whom journeyed through the looking glass into the realm of pure imagination, we tend to look and speak from one side of the *Mirror, Reality*; rarely do we consider the perspectives of both sides. We view the world either from the side of the physical object (monism) or from the side of the object's virtual image (dualism)—that is, the world of reflection. Reflectionism suggests that both monistic and dualistic perspectives represent reality. One view needs—literally, kneads—the other. Each view is more or less valid, although this issue is still heatedly contested.

AS MATTER BECOMES MIND

Many verbal battles between monists and dualists have scarred our world. In the context of philosophies of the ancient Greeks, monism conceives of reality as a whole with all its parts being interdependent. It is based upon a materialist philosophy which states that what you see and experience in the physical world is all there is to see. There's nothing more behind the face of nature. By contrast, dualism conceives of reality as consisting of corporeal and noncorporeal substances that aren't necessarily dependent on one another. It is based upon an idealist philosophy which states that what you see and experience in the physical world is only half of reality. Understand, these philosophies have played a crucial role in shaping the scope of modern thought about the universe

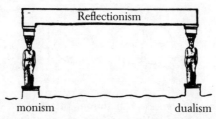

(and reality). Don't think for an instant that such philosophies are merely scholastic exercises for the unathletic. They're about as esoteric as a boxing match! I'll remind you that monistic and dualistic philosophies have directed the consciousnesses of whole cultures. They currently influence our views of life, death, and moral laws and our concepts of physical laws. If you follow the arguments over millennia regarding the right-to-truth of these philosophical perspectives, you might be surprised how profoundly they've affected the way we think about one another, the way we treat the human body and our planet—either with reverence, insolence, or indifference. They have divided Eastern world-views from Western, the monk and sage from the physical scientist. The drama of our spirits, as perceived by an audience of strict monists, is little more than a play performed by invisible actors on an imaginary stage. Sadly, the skepticism of this audience deprives them of imagining the existence of this play.

THE MINDLESS ONGOING BATTLE
FOR THE MATTERLESS

To balance my criticism, the imagination of dualists drives wedges into reality, splitting or shattering our sensibilities. Dualistic perspectives separate human beings—specifically, the "human" (body) from the "being" (self)—into neat halves. They follow the lead of Descartes, who conceptually divided the material body, the *res extensa* (or "extended thing"), from the immaterial mind, the *res cogitans* (or "thinking thing"). Descartes's dualism has wrestled our minds into a headlock for 400 years. This dualism found its way into the vocabularies and perspectives of contemporary science. For instance, the Canadian neurophysiologist Wilder Penfield's *The Mystery of the Mind*, an important study of consciousness and the human brain, discusses the subject of mind in noncorporeal terms. Similarly, the neurophysiologist Sir John Eccles' sensitive descriptions of brain processes in his books *The Neurophysiological Basis of Mind* and *The Human Mystery* clearly distinguish between the physical aspects of the brain and the nonphysical (mysterious) aspects of the mind that, in his view, cannot be correlated in some balanced one-to-one correspondence. By contrast, the Swiss psychologist Jean Piaget's *Biology and Knowledge*, which investigates the relations between biological and cognitive processes, is entrenched in materialist monism. "What mind?" Piaget might ask. "All I see are 'emergent brain functions' that we call mind," he might say, waving his hand in open space as he invites you to show him "a thought."

A WORLD UNDER THE INFLUENCE OF MATTER

The point is that we're still under the influence of these divisive philosophies. And out of necessity, we've invented ways of dismissing or ignoring the whole issue of relating matter (the physical, knowable world of brain) *And* nonmatter (the nonphysical, perhaps unknowable, world of mind). In the pit of our knowledge we sense that ultimately there may be no resolve—no universal fact—with which we can determine the validity of either monistic or dualistic perspectives and related concepts. Note that the concept of the dialectic, as espoused by Plato, Aristotle, Kant, and Hegel, is dualistic. It separates the physical world into "two distinct worlds"—a distinction that many scholars believe is partly conceptual and illusory in nature. But who can say with patent certainty?

HOW MIND IS THE INTEGRAL OF MATTER

Reflectionism implies that the terms *body* and *mind* refer to two different aspects of one reality. Any *prima facie* evidence for the biological basis of mind will also be evidence for the brain's mysterious nature. It may be that the covert nature of mind will continue to evade scientific explanation, if this explanation depends entirely upon quantification. Clearly, we can quantify (or measure) all that we see, but we still can't explain what these measurements really "mean"—either in whole or in relation to what cannot be seen and measured. It seems our descriptions of both biological and mental systems need to be stretched or shrunk some in order to fit one comprehensive description.

If the mind *is* the brain, then all the abstract, ambiguous, and undefinable properties of the mind must be accountable as brain processes. If our monistic theories cannot relate such things as extrasensory perception and aesthetic experience to the brain's physiology, then either our descriptions of brain processes are too narrow (and need to be broadened) or they grossly overestimate the nature of these mental processes (and thus need to be limited). Either way, our views and descriptions are due for revision.

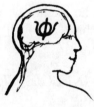

METAPHORM 4a. Reflectionist Monism—*perceiving the brain as one entity. A materialist philosophy of reality without mirror reflection.*

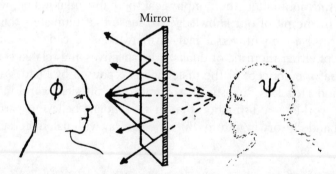

METAPHORM 4b. Reflectionist Dualism—*perceiving the brain and mind as two separate but interdependent things. An idealist philosophy of reality with mirror reflection. By introducing either a physical or an imaginary mirror we immediately create two separate worlds from one reality. All dualistic descriptions of brain and mind processes set up this conceptual illusion. The illusion can be explained away once the mirror vanishes from sight (as implied by Metaphorm 4a).*

EXPERIENCING "THE ONE"

Reflectionism is a philosophy that paradoxically accepts both monistic and dualistic views. This amounts to "having your cake and eating it too." The monistic view of reflectionism (Metaphorm 4a) considers, for example, how the brain and mind are the same thing; they are so many manifestations of a single physical entity. It is a view *without* mirror reflection. Without the mirror, the elements of reality are naturally fused. The worlds of the subject (you as observer) and its mirror image the object (what you observe) are integrated by nature.

EXPRESSING "THE TWO"

The dualistic view of reflectionism (Metaphorm 4b), on the other hand, is *with* mirror reflection. It suggests that the properties of mind are opposite and reverse the properties of the brain or body to which they correspond. This means that where the body is bound to the laws of physical reality, the mind is perhaps unbounded by these same laws. Figuratively speaking, the mind may live outside the world of matter, beyond the boundary of physics. Reflectionism claims that both viewpoints are necessary for comprehending the nebulous nature of the brain-universe relation.

THE FINITE(?) BRAIN AND COSMOS REFLECTING THE INFINITE(?) MIND

Since the subject of this book considers the brain and cosmos together, I think it's valuable to air these ideas which are relevant to both entities. As I've stressed, these ideas address the foundation of our knowledge of the world. As any serious inquiry into the study of brain processes reveals, at some point this inquiry entails entanglement in philosophical discussions on the nature of its counterpart—experience, the mind, spirit, soul, or whatever you wish to call the ineffable phenomena of human mental life studied by philosophers rather than neurophysiologists. If you speak with many experts in the field of brain science, they will tell you that this "brain/mind" issue is essentially a non-issue. What they mean is the idea of a mind being detached from its neural substrate, to which it belongs, is an idea which drifts in metaphysics and theosophy, not neuroscience. Contemporary brain science studies only the paths of neurons as opposed to the paths of consciousness. Most neuroscientists will gently point out that the latter idea is reserved for philosophical inquiry and meditation, not for physical studies of the material mind. They may describe to you the "neural identity" theory, which was created to clean up the messy mind-body problem.[7] This "problem" has befuddled some of the best researchers who have devoted their full intellectual powers towards understanding how the nervous system informs the mind.

THE PROBLEM OF IDENTITY

The way most neuroscientists resolved this issue was to adopt a materialist philosophy in proposing the neural identity theory. This theory maintains that our mental life is essentially the same as that of the central nervous system, that the physical matter and nonphysical matter of mind are in reality one and the same thing. Although we employ different expressions with insulated terms to describe this matter, the expressions essentially refer to the same tangible entity—the brain. Phenomena such as states of mind, moods, thoughts, etc., are all studied empirically according to this theory. They're treated as states of brain, or *brainways*, and are discussed as electrochemical activities involving neurons, neurotransmitters, hormones, etc. The implication is: As the states of brain change, so change the states of mind simultaneously, since the brain *is* the mind like an object *is* its mirror image.

MANY LANGUAGES WITH ONE REFERENCE

Because the brain *And* mind are *one* entity, the disciplines we use to analyze and represent this single entity must be united. If you accept the monistic idea that there is no distinction, then it is only our ideals which divide this entity. Our perceptions of this division continue to influence the very languages we use to express our perceptions.

Here it becomes clear why the physical body, symbolizing science, is often considered separately from the mind and spirit, symbolizing art. It's as if we "split" reality into two worlds. Reflectionism attempts to heal this split. But before it completely heals another relationship confounds us—the brain *And* universe relation. Our divisive tendency to dichotomize the world immediately snaps into action and we begin to think of these two entities in divided terms. In trying to synthetically represent animated biological matter and unanimated physical matter as one class of things, we're left puzzling, "One what?! What could this entity physically be?" Adding to this puzzlement is the problem of describing this "one entity." How can we capture the truth about its reality, when reality is many truths. As the theoretical physicist David Bohm points out: " 'Reality' means 'everything you can think about.' This is not 'that-which-is.' No idea can capture 'truth' in the sense of 'that-which-is.' " Is it any wonder why the sciences are troubled by concepts of reality and truth?

And so we confront one steep question after another which we must mountaineer to see over the peaks of our old ways of seeing.

TURNING WHEELS OF MYSTERY

The brain isn't the only mysterious material body that is swathed in cloths of monism and dualism. Current studies of the cosmos (its laws, forces, and principles or *ways of being*) often venture into the shadows of nonmaterial reality. Descriptions of its properties are frequently cast in metaphysical terms, as though half of the universe exists unseen and cannot be quantified. The other half conveniently moves and evolves according to mathematical physics. Like cogwheels, in which one *visible* rimmed wheel (the solid-state world) meshes with an *invisible* rimmed wheel (the quantum world), the universe transmits and receives motion.

DRIVING US TO DISCOVERY

Reflectionism combines Western and Eastern sciences of mind and philosophies. It is a compound concept that permits the physical world to coexist with the nonphysical—with physics fulfilling metaphysics and the reverse. It concludes that brain *And* mind are two complementary aspects of the same thing and process. A "complete" description of the nature of the human brain and body—if such a description is attainable —must implicate everything: from nucleons to molecules to nervous tissue to consciousness to cultures and societies. All of these elements of nature must somehow be factored into this Promethean description of ourselves which includes our mental processes, actions, and creations. With this description we will have some whole view of what we are. We will also have a broader understanding of our relation to nature. Faced with this staggering ambition, you can appreciate why the scientific method prefers to work with smaller, more manageable descriptions of nature to keep its clarity and confidence high.

DESCRIBING MIND THROUGH MATTER

The same statement regarding the summing up of knowledge about the brain is relevant for arriving at a whole description of the properties of the universe—its actions, creations, and physical processes. I believe that a truly comprehensive description must implicate every detail of nature, if it is to present the total picture. Any mathematical equation that assuredly describes the workings of the universe in its entirety—having left the human brain out of this portrait—is at best a beautifully conceived but incomplete *representation* of the total cosmos. It is no more accurate in its representational powers than the most meticulously rendered landscape painting, where every detail appears lifeful and truthfully realistic, where even the enigmatic atmosphere of light sources, which illuminate the painted passages, resists identification. An equation is one picture of reality that correlates as much to its perception as does any painted abstraction to its perception. But often the equation plainly neglects the presence of its creator, a presence that, I would argue, is more than the sum of mind + touch. In reflecting on this view, one of the great cosmologists of our time, Stephen Hawking, notes: "It turns out to be very difficult to devise a theory to describe the universe all in one go. . . . If everything in the universe depends on everything else in a fundamental way, it might be impossible to get close to a full solution by investigating parts of the problem in isolation."[8]

WHERE THE METAPHORICAL
FLOWS INTO THE LITERAL?

The mirror concept in reflectionism can be comprehended either as a physical analogy in which tangible objects are literally reflected in a physical plane mirror or as an "amorphous relationship" and philosophy, like the concept of Brahman in Indian philosophy. The scholar Betty Heimann in her book *Facets of Indian Thought* describes the Brahman in terms of transition, transformation, and reflection. In Heimann's words: "The Brahman is not bound to such or such definition and arbitration. . . . It defies all prediction and discrimination. It is constant, and yet dynamically changing in visible existence. . . . It is in-divisible in its unity—even when manifesting itself in the Universe through particles of its essence. . . . Thus the sum of all [its] manifest forms comprises only a small, or even negligible, part of the Whole."*[9]

* There's an interesting section in this book on the concept of "zero" as it relates to the Brahman. Heimann informs us that "the zero-concept is not only a mathematical discov-

A REAL WORLD TOGETHER

There's an inherent paradox in reflectionism introduced by the words "And" and "Mirror." The paradox is this: *Without the mirror,* reality is one thing or entity. It is whole; however, it appears to be only physical. There's no hint that there could be a parallel reality or some complementary reality that is not as physical as our bodies. And so, what we cannot see or detect in any way we assume is just not there. Neither mind nor spirit is physically present, according to this view, because they can't be counted, measured, touched, or dissected. These intangible things have remained on the casualty lists of philosophical battles since we first saw the world and debated what it is we're really seeing. Descriptions of mind and thought have always referred to intangible objects. A thought is something invisible to our touch on any physical plane of reality—except in its secondary forms and effects, as in artifacts of human creation.

This said, I have a problem with philosophies without mirrors. They tend to overlook the other half of reality. The half that doesn't appear to exist in physical form, and yet we feel its presence. Materialist philosophies bypass the views of those individuals who sense that there's a hidden reality inside and outside us that can't be explained in physical terms. In short, all idealist views are left without a world in which to discuss these views. Or worse, the sanctity of materialism (or monism) blinds us from seeing that other worlds may exist, such as those proposed by idealists. Accordingly, the following statement is incomplete: The laws of physics govern all physical systems. These same laws govern all living systems as they, too, are physical in nature. What's missing here is the possibility that (*a*) there may be nonmaterial systems and (*b*) these systems may not be governed by the laws of physics. Consider, for example, that high-energy probability physics is about things that can be detected but are not material.

AN IDEAL WORLD APART?

The flip side of the paradox is equally troublesome: *With the mirror,* we think we live in a divided world. The brain is investigated by neurophysiology and the mind is studied by psychology and philosophy. Thought

---

ery, but was originally conceived as a symbol of Brahman. Zero is not a single cipher, positive or negative (growth and decay), but the unifying point of indifference and the matrix of the All and the None. Zero produces all figures, but it is itself not limited to certain value. . . . Zero is the transition-point between opposites, it symbolizes the true balance within divergent tendencies" (p. 97).

processes are probed by cognitive scientists (and our souls are explored by theologians). Thus follows the tug-of-war between disciplinary studies in searching for truth, in fact-finding, in pursuing knowledge of reality. Just the thought of this division leads us to fragment our perceptions of reality. And these fragmentary perceptions lead us to group our societies into those who investigate one world or another and those who investigate their beliefs and preconceptions about the world. We find ourselves constantly choosing between these philosophies rather than accepting each for its unique perspective. As a result we will continue to live in tense confusion rather than clear the smog of all our logical analyses created from our conflicting perceptions of the world. On one hand, we want to preserve the powerful analytic tools the scientific method has evolved over centuries in its analyses of nature. On the other hand, we want to integrate the best insights and theories emerging from observation and intuition rather than relying completely on empirical studies.

The upshot is, our world-views seem to need the mirror—if only to serve as a reminder, or metaphorm. Reflectionism helps us conceptualize how one world may also consist of many worlds. As reality is both and more.

It's appropriate that I summarize this section by citing a theory that indirectly addresses the paradoxical nature of reflectionism and that employs the principles of symmetry. The theory of a "three-universe cosmology" (also called the Tachyon Universe) has been proposed by the Princeton University cosmologist J. Richard Gott III. Pursuing an idea that grew out of Albert Einstein's theory of relativity, Gott's theory considers a world of matter where "nothing travels as slow as light" (as opposed to traveling as fast as light) (Metaphorm 5).

In the final chapter of *Galaxies*, Timothy Ferris describes J. Richard Gott III's model: "He [Gott] suggests that the big bang generated not only our universe, but also a second universe composed of anti-matter and evolving in reverse time, as well as a third universe made up exclusively of particles that travel faster than light. The fleet particles of this ghostly third universe, called tachyons, are permissible under relativity theory, which requires only that nothing in our universe can be accelerated to the velocity of light; tachyons need not worry about this provision, for they have always been going faster than the speed of light. They occupy a mirror universe where everything travels faster than light and nothing can be reined to a velocity as slow as that of light." [10]

Gott's concept of tachyons hints at the relationship between the interactive force that holds all matter together and the "X" force (or unknown properties) of the human mind that seems to fit his description of tachyons. I'm speaking here of those metaphysical interpretations of mind and thought that have "no boundaries" or feelings that are "im-

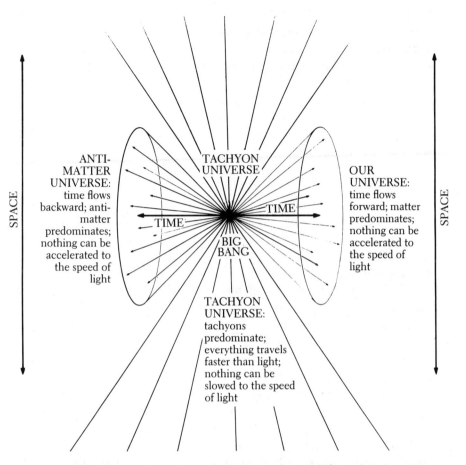

METAPHORM 5. The Tachyon Universe. *J. Richard Gott III's model of a "three-universe cosmology." For a synoptic description and finer illustration of Gott's "Tachyon Universe," see Timothy Ferris*, Galaxies, *1982.*

measurable" because they are so fleeting—like virtual (transitory) high-energy particles. In the context of reflectionism, you might contemplate whether the physical elements of our thoughts are like tachyons. Do these hypothetical particles make up the *Mirror, Reality?* Although Gott hasn't discussed this connection within the framework of cosmology, his model prompts this association.

Instinctively we sense that the mind surpasses matter in both its degrees of free will and its freedom, motion, form, and substance. Our intuitions tantalize us as we envision the mind and imagination without boundaries in terms of time-space. Unlike the speed of light there may be no boundary to "mindspeed." Even though electrophysiologists can measure in milliseconds the transmission of electrical signals from nerve to nerve and subsystem to subsystem, it is possible that the measurements are only representing one dimension or facet of the human nervous system. The imagination seems to form ideas at the speed of insight —a speed which seems faster than electrical transmissions or light itself. There may also be no limit to "mindsize," especially since the mind involves our capacity to learn a potentially infinite amount of information; this we conclude in discussing such abstractions as the size and shape of our imaginations. Unlike the dimensioning of matter, the mind can only be dimensioned in an abstract and symbolic manner. There may be no ceiling to lower its substance or walls to contain its essence. In contrast to celestial bodies, which endure measurement, the mind (brain)—with its capacity to imagine—is dimensionless. It defies definitive description, precisely because of its immeasurable qualities. How can one quantify the height, width, and depth of the mind? How can one confidently measure the volume, density, or luminous intensity of one's imaginings?

These metaphysical statements aside, many hurried steps are being taken to quantify aspects of human mental activity. In one area of cognitive science, for example, researchers are examining how we think spatially with mental imagery. As the science of spatial cognition stresses, mental images are not literally pictures conjured up in the brain. Rather they're representations in the mind that somehow evoke the "experience of seeing" without being prompted directly by visual input from the eyes. Some cognitive scientists are intent on numerically describing the full nature of this experience, without necessarily detailing the physiological and psychological events involved in this act of creative seeing. Whatever the intention, perhaps only the fragments of imagination—like the artifacts of cosmic creation—may lend themselves to descriptive analyses. As for defining or measuring the processes and properties of imagination itself, we may find ourselves re-evaluating the concept of measuring things before we try to capture the imagination with this concept.

We're still quite a ways from achieving a theory that not only includes *but integrates* the functions of the human brain—the grand mechanism that is *doing* the unifying, and not just perceiving the unification. In acknowledging the possibility that our theories of the universe are theories about ourselves, one question immediately springs into view: Do we have to study everything about ourselves—in conjunction with studying everything about the universe—to actually *know* the universe and to know ourselves?

ANSWERING OUR QUESTIONS IN PART
WITHOUT QUESTIONING OUR ANSWERS IN WHOLE

Whether or not this study should proceed by splitting the problems into small bits and creating thousands of theories to interpret our observations is another important subject of debate. This system of investigation may have worked for science in the past, but there are no guarantees for what lies ahead of us. In the course of interrelating diverse sources of observations and theoretical models, all the partial theories created might lead to an increasingly fragmented perspective. Or these theories may not overlap enough to form a consistent and coherent view of nature. Some argue that this overlay is unnecessary, as nature is demonstrably inconsistent.

Perhaps we will discover that our elegant abstractions, such as concepts of Big Bang and singularity theories—which describe how the universe first formed from a point of zero-matter that was infinitely dense and hot—equally describe the "Big Bang of Consciousness," a consciousness that occurred or formed in the human mind millions of years ago, introducing the phenomenon of learning. One thing this phenomenon has yielded is the idea that one day we may see how our poetic intelligence arose from the stars' energies and ashes.

Until that distant moment when mathematics and the rest of our abstractions can speak for "everything"—where *everything* includes the organization, behavior, and processes of nature—no definitive theory or explanatory model will reveal the whole story about the brain's relation to the universe without missing some key details. Each theory, with its systematized assumptions, tests, and experimental results, might be looked at as a piece of an ever-changing puzzle. And this "piece" might even be studied one day as an extension or projection of ourselves. Ironically, the human brain is currently acquainting us—through metaphorms—with its whole puzzling self while our intellects are busy engaging only with what is recognized as tangible evidence. Meanwhile, we may be missing the message of nature as we mistrust the beauty of its wholeness.

Ignotum perignotius—*The unknown explained by the still more unknown.*—AUTHOR UNKNOWN

7

Cerebreactors

Cerebreactors are artworks of neurocosmology that interpret the connection between the brain and its creations. They are metaphorical models of imaginary particle accelerators, nuclear fusion and fission reactors designed after the human nervous system. Where these devices were invented to observe and utilize the nature of matter, *Cerebreactors* were created for studying the nature of human brain matter, processes, and energies. These models offer ideas and images on how the human brain works, inspired by the analogy of how reactors work. The drawn models are not intellectual contrivances. Rather they're intuitive constructions that are as heartfelt as they are artistically composed.

The premise behind the biological-reactor metaphorm is simple. The principles of reactors are influenced by the physiological processes of the human organism, both by design and unconsciously by imposing its processes on these technologies. That means these technologies (along with every other object of human creation) physically reflect the dynamics of the brain. As such, they may serve to understand these dynamics.

DISCOVERING THE BRAIN'S REACTOR TECHNOLOGIES

The following drawings introduce the idea that we can learn something about the mechanisms of the brain that generate and manipulate energies by exploring the mechanisms of fusion and fission reactors. At present neuroscientists know little about the brain's energy-generating mechanisms. Physicists and nuclear engineers have a much clearer sense of these mechanisms in reactor technology, having designed and built them. The running commentary that accompanies these interpretive drawings suggests how we might discover similarities between the processes of these seemingly unlike objects. The drawings probe the possible transformations of energy in the human body.

Although *Cerebreactors* don't explicitly depict the workings of objects like heavy-water fission reactors or thermonuclear plasma fusion reactors, they metaphorically combine different aspects of these technologies without being any one of them specifically. In this regard, they represent a form of fictional art that looks toward the future of this science and technology. Like fiction, they represent "what is possible" rather than "what is."

OTHER ENERGY (INFORMATION)-RELATED SYSTEMS

There is an analogous relation between the brain and its other advanced technological creations proposed by the German-American mathematician John von Neumann[1] and the English mathematician and logician Alan Turing[2] some 50 years ago. Alan Turing, one of the founders of artificial intelligence, who is credited with providing the concept for digital computers, theorized that computers could be designed so that they would be capable of thought. In his pioneering paper "On Computable Numbers, with an Application to the *Entscheidungsproblem*" (1937), he posited that machine thought could resemble human thought if a random element, such as a roulette wheel, could be built into its system or operation. And in the field of computer theory, John von Neumann worked on the "problem of obtaining reliable answers from a machine

PROJECTIONS AND REFLECTIONS

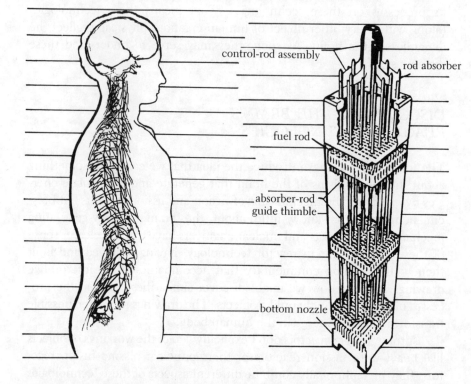

control-rod assembly

rod absorber

fuel rod

absorber-rod
guide thimble

bottom nozzle

METAPHORM 1. *Can we learn anything about the dynamics of the inventor—the human brain—by studying the dynamics of the things it invents? Do the principles of fusion-fission reactors and particle accelerators reflect the workings of the human nervous system? (Left figure adapted from Schwalbe and Herrick. Right figure is a cutaway view of a pressurized-water reactor control element assembly in a nuclear fission reactor; the subjective contour metaphormed from Jusczyk and Klein, 1980)*

with unreliable components," the function of "memory," machine imitation of "randomness," and the problem of logical design in constructing automata that can reproduce their own kind.

Neither von Neumann nor Turing would hardly have been surprised by our burgeoning robotics industry guided by an advanced artificial intelligence. Nor would they be amazed by the creative productiveness of brain-computer analogies which have taught us so much about the human central nervous system. It is only in more recent years that we've come to see how computers and artificial intelligence share certain characteristics with the human mind. As scientists search these characteristics, they may discover the mechanisms by which information is generated, organized, and manipulated in both systems. *Cerebreactors*, or brain-reactor analogies, promise to inform us about the way energy is generated and processed in the nervous system (Metaphorm 1).

I will spend only a few paragraphs in this chapter discussing fission reactor technology. Not because the analogies aren't ripe enough, but because I prefer to keep the emphasis of this book on themes concerning the fusing and bridging of ideas, information, and matter, rather than on the splitting or fissioning of these things.

The basis of nuclear energy is that a certain amount of mass can be transformed into energy, expressed as $\triangle E = \triangle mc^2$. There are two known ways of converting mass to energy. One involves fission; a heavy nucleus can be split into two smaller nuclei with average mass. The second way involves fusion; a heavier nucleus can be formed by uniting two light nuclei.

Fission is usually achieved by bombarding a heavy nucleus with a slow neutron, although some nuclei separate spontaneously. The nucleus splits into two fragments of unequaled mass, yielding two to three extra neutrons. It's important to note that the fragments have less mass as a result (Metaphorm 2). Balancing the production and absorption of neutrons is achieved by the control rods (boron) in a reactor operating at a constant "critical" power level. This equilibrium is required in order to regulate the fissioning uranium nuclei to limit or control their self-sustaining chain reactions.

The distribution of the fission fragments among the elements varies depending upon the nature of the element that undergoes fission, but also depending on the neutron that induces the fission process.

Carl Malbrain, while taking his degree in nuclear physics at M.I.T., once pointed out to me, "When Enrico Fermi and his associates started

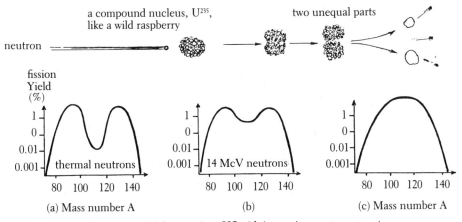

Fission yield for uranium-235 with increasing neutron energies

METAPHORM 2. *With thermal (or slow, low energy) neutrons, fission fragments are generally unequal in structure and mass; however, as neutrons of greater energy are used to bombard target nuclei, the fission parts become more identical—producing a symmetric fission.*

up their nuclear fission reactor, in 1942 at Stagg Field in Chicago, there was every reason to believe that it was the first such reactor on earth. In an open pit uranium mine, however, in the southeastern region of the Gabon Republic near the Equator on the coast of West Africa, are the dormant remains of a natural fission reactor. Within a rich vein of uranium ore the natural reactor once went critical, consumed a portion of its fuel and then shut down, all in Precambrian times, almost two billion years ago (George A. Cowan, *Scientific American*, July 1976, pp. 36–47)."[3]

According to Malbrain: "Whether the survival of the Oklo deposit is a unique phenomenon in natural history or a particularly valuable experiment in long-term geological storage, time will tell us. One message is certainly clear. In the design of that first fission reactor man was not an innovator but an unwitting imitator of nature."[4]

OUR FUSING, FISSIONING MINDS

As I've requested of you throughout this book, try to relate everything you've just read about fission to your own thought processes. Is this information meaningful to you? Does the idea that we are "unwitting imitators of nature" in the design of these advanced creations lead you to wonder about *all* of our creations? This direction of thinking implies that every kind of dynamic object that was created for producing or studying energy (for example, anything from blast furnaces and jet fuel systems to radiotelescopes and related systems) resembles—in some material or conceptual sense—brain dynamics. And these dynamics in turn resemble those of nature's. What I find puzzling is why this common thought or visible "lead" is not being fully investigated. In fact, I find that just the opposite is happening. Upon hearing such curious comparisons there's a tendency to immediately dislodge the idea that these technological objects were influenced by neurophysiological processes. After all, how could this be possible? How could we unwittingly model some vastly intricate piece of high technology after the brain, when in fact we know too little about brain processes to literally *make this connection* confidently? There must be something else occurring in the design process which escapes the scrutiny of our conscious mind. Having spoken to many engineer-designers of these nuclear devices, it is clear that they seem to be acutely aware of each ordered step in their design process almost down to the last unpredictable gesture of their imaginations.

So how can the human mind secretly imbue these devices with its own workings—for what purpose? What could possibly be gained by this gesture of anthropomorphism and how could this connection go undetected? Equipped with every conceivable analytical technique for checking their engineering plans, the designers would surely have noticed some peculiar pattern of likeness—perhaps not in terms of structural likenesses but rather those pertaining to function—between the human brain and these technologies. It's as if fusion reactor concepts reflect cerebral fusion effects in the brain. Perhaps an interesting acknowledgment of this likeness may be seen in the fact that the term "Yin-Yang" was employed by physicist-engineers in the design of magnetic coils for creating high-temperature plasmas in what are called "driven magnetic fusion reactors" and "mirror machine reactors" (note Metaphorms 3 and 9a).[5]

KNOWING WITHOUT KNOWING IT

When you consider that most specialists who design plasma fusion reactors and nuclear fission reactors aren't exactly versed in human neuro-anatomy and physiology, it's not surprising that this strange relationship could have gone unnoticed. The reverse is also true. Neurophysiologists have slight knowledge of these technologies. They also have as little free time as these engineers for studying research developments in another rigorous field. In presenting my drawings of *Cerebreactors* and their supporting information to these scientists, the response was fairly uniform: "How curious!" they'd say, in a bewildered tone. And this bewilderment was encouraging. Typically a few said, "That's farfetched." But these headshakers probably made the same innocuous remarks about early brain-computer analogies, which have helped us understand the nature of information and information processing. Of course, any concept presented outside a discipline seems fetched from far away.

Ten years later, I'm not sure that we're any wiser about this relationship between the creator and the created. *Cerebreactors* certainly don't answer how the brain influences the things it creates. They merely suggest there may be many more hidden or unsuspected influences we're overlooking. They imply that the brain has an invested interest, so to speak, in its conceptualizations and in the development of its concepts —those slated for success and others for contributing to the more successful ideas. (There are no failed ideas, in spite of what fashion, science, and society dictate. They're all useful in our growth.)

SEEING WITHOUT LOOKING

It is frightening to me how we forget that our body has a head—that our brains conduct the research, create the performance, paint the painting, devise the apparatus, interpret the experimental results, and translate our common experiences into some relatively coherent language for purposes we assume involve communication. To believe that there isn't this connection between the brain and all its mental activities and actions is to miss the forest as you're standing in the forest. And yet this is where we presently stand, as we wonder where in the world we're standing. The connections I speak of here go beyond the poetics and mythology of anthropomorphism. They are not even relatable as isomorphic forms in which two things are similar in physical shape or structure. Instead, the connections between the brain and its creations are better understood as processmorphs.

CEREBREACTORS AS PROCESSMORPHS

Two speculative concepts are used here to describe human brain processes. These concepts are based on the explorations of *Cerebreactors*. Analogies are drawn between the mental process of intuition and the nuclear process of fusion, and between the mental process of reason and nuclear fission. In intuition, both hemispheres of the brain function conjunctively as a single sphere, when focusing and fusing information; for tens of milliseconds or less, there is a functional and electrochemical unity. By contrast, in reasoning, one hemisphere is more forceful than another, creating an electrochemical disunity.

These two conditions I think correspond to two phases of thought: (*a*) when a person is experiencing an intuition, insight, idea, or inspiration, and (*b*) when a person expresses the intuition using either the instinctual or intellectual realm of imagination. Both thought phases and their many variances apparently sustain each other.

The instant of intuition signifies the union or fusion of the cerebral hemispheres like two light atomic nuclei uniting in a great concentration and confinement of temperature. In this instant, two opposite forces such as positive and negative overcome their complementarity, forming one greater force which I refer to as *cerebral fusion* or the "plasma fusion state of mind" in a *Cerebreactor*. Note that plasma represents the fourth state of matter; it behaves as neither a solid, a liquid, nor a gas, but as an ionized gas—like that of a lightning bolt—with its peculiar characteristics. The plasma physicist J. D. Lawson informs us that "if plasma is

100,000 times denser than air, proper temperature need be held for only about one-thousandth of a second in order to achieve high-temperature plasma fusion reactions."[6] We might ask: Are some intuitions this dense and fast?

Describing an intuition marks the division of the cerebral processes like the nucleus of an atom splitting apart into two nuclei where one can be heavier than the other. Where intuition represents the convergence of different perspectives and mental specializations, the process of reason coincides with their divergence.

PROCESSES OF INTUITION

The physiological differences between intuition and reasoning in all *Cerebreactors* depends on whether there is an unbroken, or broken *nongeometric*, symmetry of electromagnetic fields. These low fields show the integration and division of cerebral hemispheres functionally.

There are at least three fundamental fusion reaction types that are described by *Cerebreactors*: toroidal, magnetic mirror, and laser fusion (Metaphorm 3). There are also innumerable variations. The different

Fusion reactor concepts *reflect* cerebral fusion effects

Toroidal Plasma Fusion

toroidal field coil

magnetic field lines

Magnetic Mirror

neutral beams

Yin-Yang magnet

Field-reversed Mirror

neutral beams

Laser Fusion

target chamber

image of thermonuclear burn in target

METAPHORM 3. Field Configurations. *Three types of reactions that have been explored extensively in the science of plasma physics for the past 30 years. These different low fields are meant to represent approximately the shape of low fields in the human brain during cerebral fusion (intuition). Note that these field configurations have different shapes in the brain for one obvious reason: The brain's anatomy is not the same as the reactor's. (Toroidal plasma fusion after H. R. Hulme and A. McB. Collien, Nuclear Fusion, 1969. Field-reversed mirror after B. Brunelli, editor,* Driven Magnetic Fusion Reactors, 1978)

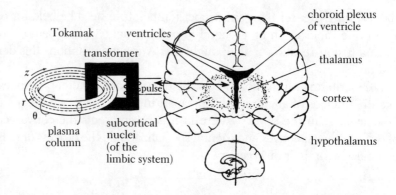

METAPHORM 4. Components of a Cerebreactor—*a frontal view of the human brain juxtaposed with a frontal view of a Tokamak (toroidal fusion reactor), called in the industry a "toroidal pinch system." Note: This diagram is not meant to suggest that some rectilinear steel plate with an interlocking hot plasma circling around its center is somehow inside our brains burning away. Perhaps a similarly shaped electromagnetic field generated by the human brain at the instant of intuition might circle around the region of the brain indicated in the diagram. As a metaphorm, the transformer and plasma column in the Tokamak system represent the union of the cerebral hemispheres and their subsystems. The metaphorm implies that various thought processes have fulfilled their complementarity in forming one greater force of thought—intuition,* cerebral fusion, *the* plasma fusion state of mind. *The process of metaphorming is the act of intuiting and creating.*

configurations of energy fields determine the intensity of the *plasma state of mind* or *cerebral fusion*. For example, in one instant a *Cerebreactor* might create a torus-shaped plasma and in the next millisecond produce a field-reverse effect.

Cerebreactors indicate that the configurations of energy fields in the brain during instances of intuition differ from the energy fields occurring in moments of reasoning. Their corroboration can only be achieved with devices more sensitive than current magnetoencephalography (MEG) equipment. Although MEG is capable of detecting these minuscule fields that are one-billionth of the Earth's electromagnetic field, it cannot accurately measure, map, and make sense of the field configurations as yet. Perhaps this requires both another level of technological sophistication and a different perspective on brain processes.

In a *Cerebreactor*, the spinal cord, hindbrain, and midbrain are the key centers for regulating afferent (incoming) and efferent (outgoing) information to the limbic system. They are analogous to the apparatuses that pump hydrogen or other gases into the heated chamber of a "Tokamak," or toroidal fusion reactor (see Metaphorm 4). Note that the Greek

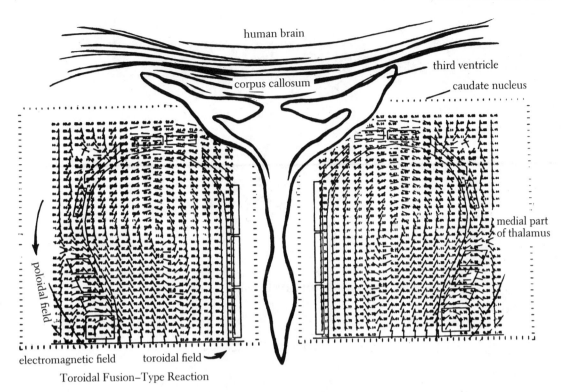

Toroidal Fusion–Type Reaction

METAPHORM 5. The Superimposition—*detail of a toroidal-shaped electromagnetic field in cerebral fusion occurring in a* Cerebreactor. *The image shows a frontal section of the human brain with* actual electromagnetic fields from a Tokamak fusion reactor superimposed over the region of the limbic system. *The shape and spatial orientation of these fields resemble the electromagnetic field configurations of the plasma column in Metaphorm 4.*

word *toka* means "flux" and the word *mak* means "maximum." Again, I remind you, we're not talking literally here about how the "maximum flux" of a central nervous system "pumps hot hydrogen gas" into the limbic system. Instead we're speaking metaphorically about the input of gases used in creating plasma fusion reactions as they may represent the flow of afferent information in a major brain center.

Theoretically, the virtual particles (of thought) generated in a *Cerebreactor* are similar to those charged particles moving along a line of force in the direction of a magnetic field in a plasma fusion reactor. In a *Cerebreactor*, this field signifies the line of concentration containing the torus-shaped, electromagnetic fields that flow in a specific pattern during cerebral fusion (note Metaphorms 5 through 10b).

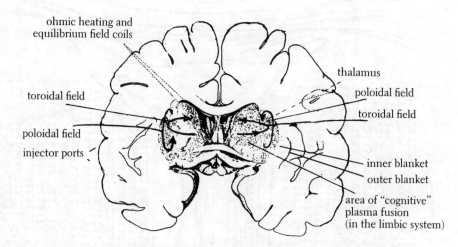

ohmic heating and
equilibrium field coils

thalamus

toroidal field

poloidal field

toroidal field

poloidal field

injector ports

inner blanket

outer blanket

area of "cognitive"
plasma fusion
(in the limbic system)

Frontal Section of the Human Brain through the Area of the Thalamus

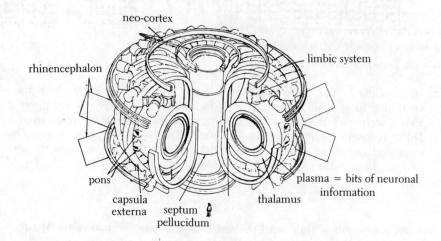

neo-cortex

rhinencephalon

limbic system

pons

plasma = bits of neuronal
information

capsula
externa

septum
pellucidum

thalamus

METAPHORM 6. The Act of Transforming Terms—*schematic diagram representing the Argonne Tokamak experimental power reactor. (From Archie W. Culp,* Principles of Energy Conversion, *1979)*

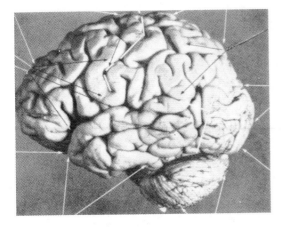

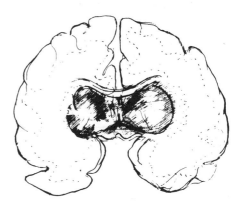

Frontal Section of the Human Brain

Area of toroidal-shaped electromagnetic fields in a living brain as it experiences an intuition. It is posited that this field resembles the toroidal-shaped plasma column in an active fusion reactor.

"If plasma is 100,000 times denser than air, proper temperature need be held for only about one-thousandth of a second."

plasma column ___

A Hot Plasma Confined in a Simple Torus

Assembled Torus

METAPHORM 7a. The Assembled Torus of the Human Brain—*searching for the mechanisms in the central nervous system that can be compared to the current and magnetic chamber of the toroidal fusion reactor. By juxtaposing this information, the* Tokamak Cerebreactor *suggests that cognitive plasmas (intuitions) are "heated" in the cores of* Cerebreactors *by either radio-frequency heating methods or thermonuclear reactions (processes of thought). Similar high-temperature plasmas are created from deuterium and tritium that have been magnetically confined in fusion reactors. In removing the heat source (shifting mental concentration) from the plasma fusion chamber (the limbic system) of a* Cerebreactor, *the immediate area cools (becomes less active). (From W. Halchin, S. M. Decamp, S. O. Lewis, D. C. Lousteau, and J. D. Bylander, "Operation of the ORMAK Fusion Device in a Cryogenic, Vacuum Environment," in* Fifth Symposium on Engineering Problems of Fusion Research, *1973, p. 363. Brain photograph from J. P. Schade and D. H. Ford,* Basic Neurology, *1967)*

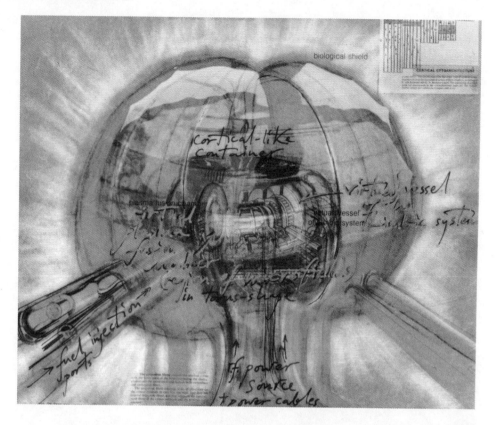

METAPHORM 7b. Tokamak Cerebreactor. *The process of generating "intuitions" in the Cerebreactor involves the controlled "heating" of the limbic system (via cortex and core of the brain) in whose center (thalamus) there are several points of concentration or focal points into which sensory information, like fuel, is added and acted upon—received, processed, and transmitted.*

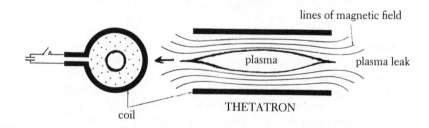

METAPHORM 8. Cerebreactor Components. *The figure shows two views of a cylinder (an end view and a side view) in which a plasma is being compressed by a "Thetatron" reactor. The plasma is confined by a thick metal coil. Passing through this coil is a very strong electrical current which helps to create and shape the plasma. Actual fusion reactors can divert plasma impurities, as they are closed systems that are considerably less complicated than the open systems of the human brain. (From H. R. Hulme and A. McB. Cohen, Nuclear Fusion, 1969)*

REASONING CHANGES INTUITION

Each change in these fusion reactions affects (and reflects) the degree of concentration or "containment" of these low fields. The word *containment* has a special relevance here. It refers to what are called in plasma physics *micro-instabilities*. These instabilities in the plasma don't physically jump about in an unstable manner as the word implies. But they do contribute to the huge loss of energy that results when plasma touches the wall of the vessel in which the plasma is created. Here, again, we exercise our opportunity to conjecture (Metaphorm 8).

This metaphorm suggests that if a "cognitive plasma" (an intuition) has a micro-instability, the plasma will touch the walls of the *Cerebreactor's* "vessel" (the cerebrum). Immediately the "energy of the intuition" will dissipate. This action is equivalent to a plasma leak in a plasma fusion reactor.

SPECULATIONS ON
FUTURE ENGINEERING POSSIBILITIES

The engineers of *Cerebreactors* question whether it will ever be possible to sustain an instance of cerebral fusion without damaging the brain-reactor permanently. Just as confinement properties have thus far been inadequate to support self-sustained thermonuclear reactions in a plasma fusion reactor, perhaps a *Cerebreactor*—like the human brain itself—is not neurologically equipped to sustain cerebral fusion, or an intuitive reaction, that is more than a few milliseconds. A person who experiences an intuition for protracted periods might end up looking like the frazzled, futuristic character in Woody Allen's movie *Sleeper* who stayed too long in an "Orgasmatron" booth!

The engineers concern themselves with the problem of micro-instabilities, which are also known as *particle-wave instabilities* (or unpredictabilities). These determine the strength, or intensity, of an intuition. Perhaps the neural processes are flawed by nature in their inability to sustain an intuition. Consequently, intuitions "escape" from mirror confinement like a plasma leaking from a magnetic bottle in a magnetic mirror fusion reactor. Or perhaps this is anything but a flaw. Nature may need only short bursts of brilliant insights to guide its more drawn out, reasoned plans of action. (See Metaphorms 9a and 9b.)

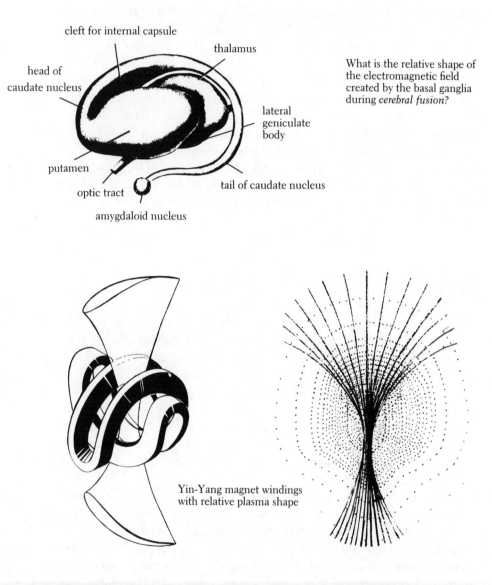

cleft for internal capsule

thalamus

head of
caudate nucleus

What is the relative shape of
the electromagnetic field
created by the basal ganglia
during *cerebral fusion?*

lateral
geniculate
body

putamen

optic tract

tail of caudate nucleus

amygdaloid nucleus

Yin-Yang magnet windings
with relative plasma shape

METAPHORM 9a. Components of a Magnetic Mirror Cerebreactor—*comparative physiology. Semi-schematic drawing of the isolated striatum, thalamus, and amygdaloid nucleus. (From R. C. Truex and M. B. Carpenter,* Human Neuroanatomy, *1969. Figure at bottom left from T. H. Batzer, "The Technology of Mirror Machines—LLL Facilities for Magnetic Mirror Fusion Experiments," in* Seventh Symposium on Engineering Problems of Fusion Research, *1977. Figure on bottom right after T. H. Batzer, "The Technology of Mirror Machines")*

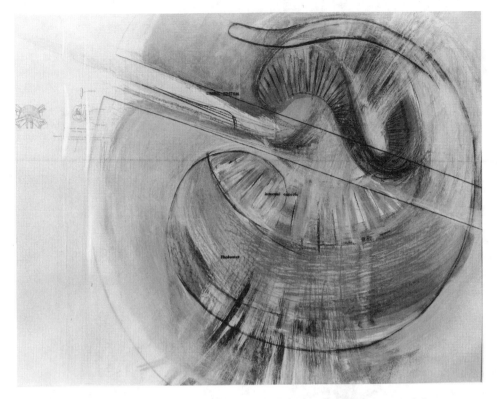

METAPHORM 9b. Magnetic Mirror Cerebreactor. *Converging in the center of the* Cerebreactor *are trillions of bits of thought fragments and memories collected from all corners of the neurosphere. In one instant, these fragments form a coherent vision some call intuition.*

CHANGING METAPHORMS

These analogies and remarks aside, let's look at another class of metaphorical models. The following particle accelerator *Cerebreactors* (Metaphorms 10 through 13) explore the idea that the energies of nuclear forces are the material counterpart of our so-called mental energies. (See Metaphorm 4, Chapter 14.) They consider the ceiling of the material world to be the floor of the mind. These models relate the particle-wave and force-field properties of matter to the human mind, implying that the constituent elements of "thoughts" correspond to the four basic interactive forces between elementary particles from which the whole universe is supposedly constructed.[7] (See Metaphorm 13, titled *Mind As Nature*, at the close of Chapter 5.) Three of these forces have already been unified in one theory of interactive forces. The fourth force, gravity, so far resists unification in physics. It may turn out that gravitons, or

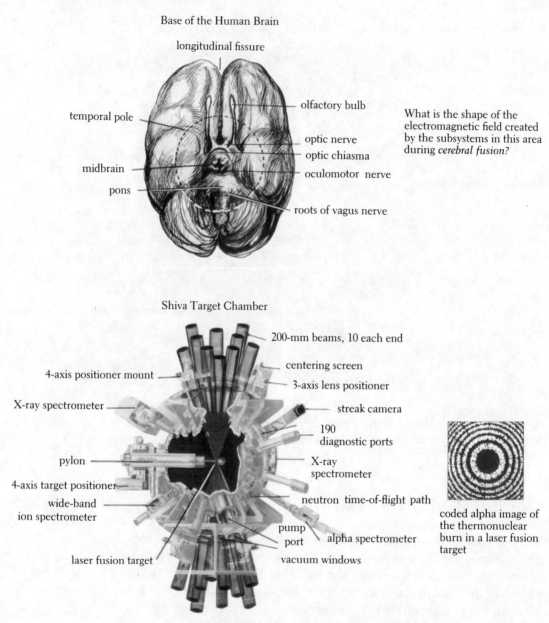

Base of the Human Brain

longitudinal fissure

temporal pole

olfactory bulb

optic nerve

optic chiasma

midbrain

oculomotor nerve

pons

roots of vagus nerve

What is the shape of the electromagnetic field created by the subsystems in this area during *cerebral fusion?*

Shiva Target Chamber

200-mm beams, 10 each end

centering screen

4-axis positioner mount

3-axis lens positioner

X-ray spectrometer

streak camera

190 diagnostic ports

pylon

X-ray spectrometer

4-axis target positioner

wide-band ion spectrometer

neutron time-of-flight path

pump port

alpha spectrometer

laser fusion target

vacuum windows

coded alpha image of the thermonuclear burn in a laser fusion target

METAPHORM 10a. Components of a Laser Fusion Cerebreactor—*comparative physiology. (Shiva target chamber courtesy of Lawrence Livermore Laboratory, Livermore, California)*

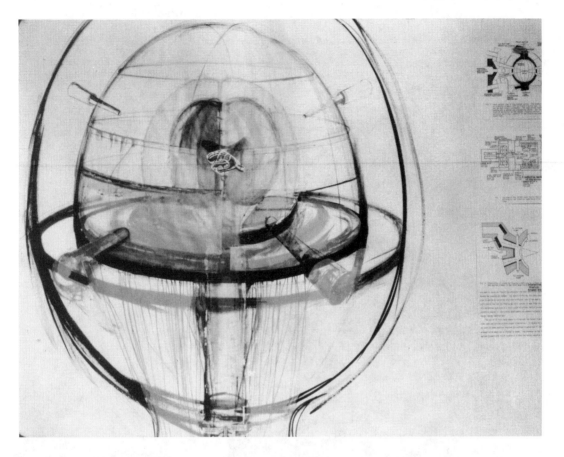

METAPHORM 10b. Laser Fusion Cerebreactor. *The process of generating intuitions and cerebreactions involves positioning the mind, targeting the unconscious, and fusing information.*

the gravitational force, are intimately connected with the same force that forms the properties of the mind, and that the grand unified theories may tell us as much about these hitherto-unknown properties of mind as they will describe the properties of matter—in particular, the forces between elementary particles. Then again, they may not. Gauge-field theories that model the interactions between strong, weak, electromagnetic, and gravitational forces may only be modeling a fraction of the forces.

Perhaps the reason we haven't unified all the interactive forces is that we haven't trusted our GUT *feelings*—namely, that our understanding of the universe has every bit to do with our understanding of the physical dynamics of our minds, and that these two fields of study are

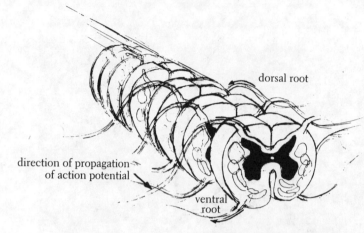

dorsal root

direction of propagation
of action potential

ventral
root

Section of the Spinal Cord

In physical reality, we have an invertable transformation from one system to another such that the physical processes in one are transformed into approximately the physical processes of the other. In the case of nuclear physics and neurophysiology, the transformation would take the velocity of light, c, on to the velocity of the spike potential in synapse because both are the "barriers" to the velocity of transmission of information in the corresponding systems. This implies nuclear events are analogous to neural events on some relativistic scale.

Components of the
CSF Linear Accelerator

METAPHORM 11a. Components of a Particle Accelerator Cerebreactor—*comparative physiology. (Photograph from "Accelerating Structure in Technology" in* Linear Accelerators, P. M. Lapostolle and A. L. Septier, eds.; American Elsevier Publishers, 1970.)

METAPHORM 11b. Particle Accelerator Cerebreactor. *Interpretive drawing of the delay paths and self–re-exciting circuits in the spinal cord of a* Cerebreactor *"linear electron accelerator." Note the articulations of the vertebral column. Each vertebral segment acts as a screen, or filter, for focusing sensory information. The collection of information in* Linear Accelerator Cerebreactors *is analogous to that of spectroscopy research in nuclear physics. In this line of research the structure and behavior of the particle beams are studied. In* Cerebreactors, *the analogues for these "beams" are either the inputs from various neural systems or thoughts themselves.*

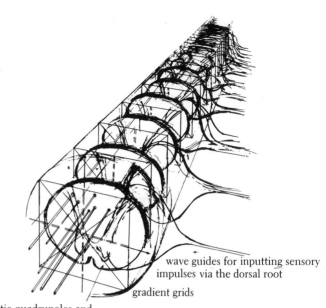

wave guides for inputting sensory impulses via the dorsal root

gradient grids

Magnetic quadrupoles and related "neural" beam focusing apparatus

METAPHORM 12. Components of a Cerebreactor. *On the right half of the drawing is a diagram of the anterior spinothalamic and spinotectal tracts. (The anterior spinothalamic tract is known to convey impulses of light touch.) On the left half is a schematic representation of a linear electron accelerator. The combination of these two diagrams creates a metaphorm for a* Linear Accelerator Cerebreactor *which exists inside every human organism. (Right side from Malcolm Carpenter,* Human Neuroanatomy, *1979)*

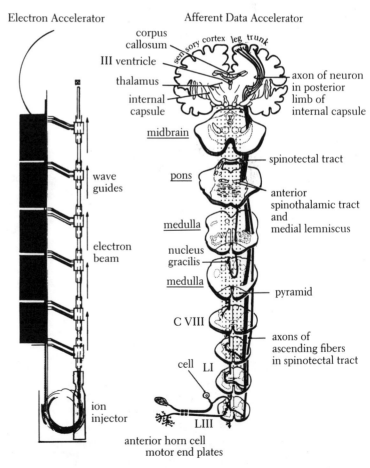

Electron Accelerator

Afferent Data Accelerator

corpus callosum

III ventricle

thalamus

internal capsule

sensory cortex leg trunk

axon of neuron in posterior limb of internal capsule

midbrain

spinotectal tract

pons

anterior spinothalamic tract and medial lemniscus

wave guides

electron beam

medulla

nucleus gracilis

medulla

pyramid

C VIII

axons of ascending fibers in spinotectal tract

cell LI

LIII

ion injector

anterior horn cell motor end plates

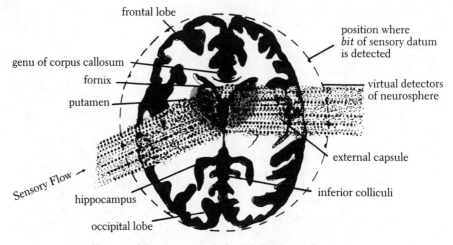

frontal lobe

position where
bit of sensory datum
is detected

genu of corpus callosum

fornix

putamen

virtual detectors
of neurosphere

external capsule

Sensory Flow

inferior colliculi

hippocampus

occipital lobe

Horizontal Section through the Whole Human Brain

METAPHORM 13a. The Jet Axis of Cerebral Fusion. *If the point of creation and annihilation of a "thought" were surrounded by an internal, "spherical detector," it might show the position of the bits of sensory data as they cross the surface of the detector. Most likely, it would also show how these bits of information are spread out, like particles making up the jets in high-energy experiments in nuclear physics.*

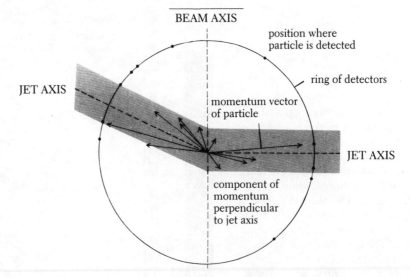

BEAM AXIS

position where
particle is detected

ring of detectors

JET AXIS

momentum vector
of particle

JET AXIS

component of
momentum
perpendicular
to jet axis

METAPHORM 13b. A Jet of Particles Does Not Always Look Much Like a Jet at First Glance. *If the point of annihilation were surrounded by a special detector which could indicate the position of each hadron as it crossed the detector's path, the particles making up each of the two jets might well be spread out in the pattern shown.* (From Maurice Jacob and Peter Landshoff, "The Inner Structure of the Proton," Scientific American, *March 1980.*)

not mutually exclusive but seamlessly connected. Some neurocosmologists insist that knowledge of the brain must acknowledge the universe. In the same way, knowledge of the workings of our creations is critical for knowing how we work.

GUIDING OUR CREATIONS

Our impulse to reject the comparisons raised here is understandable. No one feels comfortable living with the idea that one of the most potentially destructive technologies ever invented is as much a part of the human body as the nervous system. The more historically minded among us may even experience a fight-or-flight reaction to these metaphorms. Such a reaction is instigated by a fear of the past, or at least by a reluctance to return to that part of our past which we readily associate with the age of "machine analogies"—introduced in the 1600s by the philosophers and mathematicians René Descartes, Gottfried Wilhelm Leibniz, and Sir Isaac Newton. According to their vision, the universe was a machine with the precision of a perfect (atomic) clock, a machine so exactingly calibrated and coordinated that it could only have been created by some supreme Force. The human body, too, was conceptualized in terms of a nuts-and-bolts, widgets-and-springs scheme. This mechanistic view was brazenly different from the ancients' view of the world, whose organization looked more like a complex organism than a clock or factory, or stellar body under the influence of celestial mechanics. When the ancient Hindus described their "Wheel of Time" with its perpetual motion, their metaphors were organically grown, so to speak, from the earthen soil of philosophy as opposed to the almost extraterrestrial fruits of physics. To our great ancestors there was no man *and* nature. There was only *man as nature*.

As in the past, our interpretations of humankind's relationship with nature will determine the course of civilization. Either we'll continue to separate ourselves from nature and from the objects of our creations or we'll recognize the influences discussed in this chapter in a more brilliant light. One consequence of our expanded vision is that we won't blindly stride ahead in inventing objects that are at best left in their thought form rather than physical form; need I mention them: biological and chemical weapons, advanced nuclear arms, etc. With this increase in sensitivity and vision, we'll be capable of using our powers of prescience to inform us of the consequences of certain mental actions. What we stand to gain by revisiting the Newtonian-Cartesian concept of "mechanized man" is an understanding of the creative process. Or, as Joseph Campbell has written, understanding our "creative mythology." My hope is that we'll "come to our senses" and recognize our *mythodologies*. We'll see how we create facts from myth, as we factualize our observations about nature. In effect, I'm re-inventing the wheel for the purpose of discovering the nature of invention. The intent here is to identify various influences of the human organism in the whole creative process, which relies heavily on common sense and poetic insight.

The thoughts and images presented here are expressions of brain-universe metaphorms. As I outlined in the beginning of this book, the universe imposes *its* physical processes on *us*. To understand these processes, I've found it enlightening to compare certain energy-related objects that our technical-sciences have created whose dynamics, I believe, reflect those of the universe. This circle of influence reveals itself time after time. We are one of nature's many recursive patterns. The universe creates us and we, in turn, re-create it. That is, our objects of creation resemble some aspect of the physical universe. The original "Stellarator" plasma fusion reactor built in 1951—with its endless figure-eight tube that magnetically confined plasma—is an exemplar of this thought. There are many wondrous connections in these reflections. The work of neurocosmologists is to illuminate these connections through symbolic models such as *Cerebreactors*.

It is by logic that we prove, but by intuition that we invent.
—Henri Poincaré

8

"Cerebral Fusion" (Intuition) and "Cerebral Fission" (Reason)

The acts of thinking, feeling, and creating have been metaphorically compared to almost every phenomenon, natural and human-made. Consider the many metaphors linking thoughts to forces of nature. For example, the words *centripetal and centrifugal force* are often used to refer to direction and movement in thinking—how thoughts seem to move towards or away from a particular frame of reference as, for example, when one is engaged in discriminating or comparing two alternatives. One of the parents of modern psychology, William James, observed: "Thought is always interested more in one part of its object than in another, and . . . chooses, all the while it thinks."

THOUGHT IN GENERAL

In speaking of *structures* or *shapes* of thought, we tend to associate these imaginary shapes with certain objects of nature such as coiled, chambered shells and stellar constellations. These visual associations add form and clarity to the abstract concept of thought. They make visible something that is, by all accounts of our senses, invisible and intangible. One particularly creative metaphorm relates thought and its constituent processes to the periodic table of the elements, where each element represents the electrochemical "property" of a specific thought. Imagine thoughts having crystal structures, atomic numbers and weights, names, densities, melting and boiling points, or even something based on symbolic meaning and references. In this scheme, no matter how abstract or abstruse a thought or idea is, it would be as identifiable and physical as one of the elements. Accordingly, thinking would be understood as a chemical process and thought as a residual product of this process. These are playful metaphorms that lead some people somewhere, and other people nowhere.

Another class of metaphorms relates the workings of the mind to the workings of our modern technological objects. These tend to be mechanistic and reductionistic, redolent of the sixteenth-century Cartesian models of the mechanical world. Such analogies and models continue to shape the way we conceptualize and talk about the dynamics of mental activity. These technological metaphors may be loosely grouped into the following categories: information-related systems, such as computers and telecommunication satellites; energy-related systems, such as power reactors and solar collectors; and survival-related systems, such as ordinary and advanced weapons. Each metaphorm intimates that the way the object works informs us about the way the mind works, both constructively and destructively.

Curiously enough, *all* of the implied and explicit comparisons mentioned here do convey something of the "essence of thought." Whether they lead us immediately to meaningful insights or simply mislead us is debatable. The point is, these poetic expressions externalize metaphoric thinking in the same way a painting, musical composition, or mathematical equation gives visual form to thought. They are artifacts of thought processes. And as such, they re-present the biological realities responsible for their creation. The renascent interest in cognitive science, neuropsychology, and developmental psychology to grasp the nature of imagination by examining these "artifacts" has by omission amplified our appreciation of the special instances of pure insight.

THOUGHT IN PARTICULAR

Most observers would agree that there are many forms of thought. Two familiar forms, for example, are visual images and linguistic propositions. There are many other forms, among them problem-solving (inference) and insight. Moreover, our expressions of thoughts are as varied as their representations in poetic, pictorial, and musical form. One look at a comprehensive history of art or an encyclopedia of science and technology will convince anyone of the multifarious forms and representations of thought.

This chapter discusses two mental processes that seem to underlie all these forms of thought and expression: intuition and reasoning. Here *reasoning* includes both analytic, sequential, feature-by-feature reasoning and affective reasoning, which involves the discrimination of feelings or emotions. Expressions of this reasoning also include the spectrum of symbolization from mathematical logic and music to poetry and the visual and performance arts. Intuition involves both discovering new connections and innovating new solutions.

SOME PARTICULARS OF INTUITION AND REASON

I call intuition "cerebral fusion"—referring to the divergence of brain processes. And I call reasoning "cerebral fission"—referring to the divergence of these processes. Whether you're intuitively playing a hunch or you're absorbed in solving polynomials, these thought processes are working for you. Whether you're dreaming, or fully awake and alert, these processes are at work—in child and adult alike. They are the means of experiencing the world and interpreting the realms of our experiences.

As you're working away and as an idea suddenly emerges—without prior announcement by some orderly step of reason—something in what you're thinking seems to resonate positively with you. Immediately, a feeling of harmonious resonance rushes from the "bottom of your head" (not your heart!) to and from the top, meeting midway in the *heart of your brain*, the thalamus. This feeling can seem as forceful as the waters of Niagara Falls being pulled by gravity to the ground of another experience. Or it can be more radiant and brighter than other forms of illumination. Instantly you *feel* like you've just learned or discovered something new. New to you, though not necessarily to someone else. This physical sensation is what distinguishes the feelings associated with a powerful intuition versus the steady-state machinations of reasoning, which may be equally heated. What has not been clearly explored by brain researchers are the contributions by all the subcortical systems that make intuition a "feelingful affair." In a very real sense, the "heart" of the brain (or thalamus and limbic system) falls temporarily in love with the heads of the brain (or cerebral hemispheres). And so in intuition, as in love, we experience "a new aim in life to which everything is related and which changes the face of everything"—to quote Stendhal.

To the point: I coined the terms *cerebral fusion* and *cerebral fission* to make the connection between the shared processes of the human brain and the physical universe. The brain neither evolved out of a void nor functions independently of the cosmos from which it was born. More specifically, I also wanted to connect these two brain dynamics to some of the processes of its creations, in this case fusion and fission reactors— among other energy-conversion devices. These intentions, and the explorations which follow, distinguish my work and models of brain processes from previous studies. That the brain is inextricably integrated in the universe is not self-evident. That the brain is integral to all the things it creates is also not as obvious as this expression intimates. How integrated it is with these worlds is yet to be determined.

Implied here is that the mental processes of intuition and analytic reasoning may have a dynamic resemblance to the nuclear fusion and fission processes that have formed and shaped the universe. Metaphorically, cerebral fusion is analogous to the union of two light atomic nuclei coming together in a great concentration and confinement of temperature. Cerebral fission, by contrast, is analogous to the division of an atomic nucleus into two nuclei, where one can be either heavier than or equal to the other. By describing human brain processes in this way, I stress the internal-external consistency between brain and cosmos. In the universe, things either diverge or converge, split or merge.

I could easily have based this brain fusion-fission concept on the organization of atoms or molecules, instead of nucleons. The level or plane of physical reality I apply this concept to is unimportant. Describing the process itself—or its general principles—is the only thing that interests me. Irrespective of scale and duration or time, then, we need to explore how the biological processes in one system can be likened to the physical processes in another system. In the case of nuclear physics and neurophysiology, the velocity of light in free space is the velocity of a thought (the spike potential in synapses). This is one "barrier" to the transmission of information in its system. This suggests that nuclear events or processes are analogous to neural events. If we can momentarily detach the vision we have of subatomic particles colliding chaotically in the frigid vacuum of the celestial sphere, the broader view suggested here won't seem so strange. Nerve cells interacting selectively in the fragile environment of the neurosphere might just resemble *in process* the seemingly chaotic pattern of particles in random motion. This motion may be more stable and orderly or organized than it appears. For purposes of simplicity, I prefer to emphasize the general dynamics or principles of nuclear and neural actions, rather than concentrate on the details of these actions.

EVENTS NORTH AND SOUTH OF CEREBRAL HEMISPHERES

Contrary to the popular notions of left-brain and right-brain lateralized functions, I propose that cerebral fusion involves the mergence of vast streams of mental impulses from all three interpenetrating regions of the brain: the neocortex, limbic system, and brain stem (Metaphorm 1). These impulses converge in the regions referred to as the thalamus and hypothalamus, the "between-brain" centers, or diencephalon.[1] As I said earlier, in cerebral fusion both hemispheres function conjunctively in focusing information, thus producing an electrochemical unity or parity between cerebral hemispheres and their various subcortical systems.

By contrast, in cerebral fission one hemisphere is more forceful than the other, implying an electrochemical disunity between hemispheres (Metaphorm 2). By *forceful* I mean that cerebral functions in either hemisphere *momentarily exceed those of the other*. I do not mean that language functions of the left hemisphere, for example, are *permanently* more dominant than nonverbal functions of the right cerebral cortex. Expressed another way: Dominance is relative to which hemisphere and subcortical system is dominating, or is more assertive, at the moment. It is a state of brain—and cultural—*perception*, not some hard-wired property of our functional neuroanatomy.

To observe the differences between these two thought processes, some internal electrical, chemical, or magnetic phenomena revealing these two physical effects must be made visible. Biologically generated electromagnetic fields may be thought of as "microprints" that naturally demonstrate the occurrence of both functional and electrical unity and disunity between cerebral hemispheres and subcortical systems. (Note Metaphorm 3.)

The pictures of magnetic filings and fields in Metaphorms 3 and 4 may induce a déjà vu for those of you who are familiar with the 1930s and 1940s work of the German "field-theorist" Kurt Lewin.[2] Lewin boldly proposed that the study of psychology incorporate the findings of

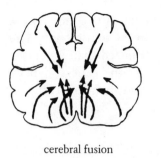
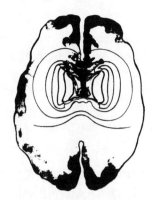

cerebral fusion

METAPHORM 1. Left: *Coronal view, or frontal section, of a* living *adult human brain showing the region of the limbic system in which the* cerebral fusion *events are taking place; the arrows indicate the convergence of information.* Right: *Note that the field configuration shown here is originally from a plasma fusion reactor which I've superimposed over this view of the brain. Horizontal view of the brain at the level of the thalamus. The lines represent the proposed configuration of the symmetric microfield in* cerebral fusion.

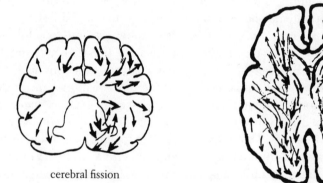

cerebral fission

METAPHORM 2. Left: *Frontal section of a* living *adult human brain showing the region of the cerebral cortex in which the* cerebral fission *events are taking place; the arrows indicate the divergence of information.* Right: *Horizontal view of the brain at the level of the basal ganglia. The lines represent the proposed configuration of the asymmetric microfield in* cerebral fission.

the Scottish physicist James Clerk Maxwell and the English physicist and chemist Michael Faraday. Maxwell and Faraday had launched the nineteenth-century Scientific Revolution in physics and chemistry with their discovery of electromagnetic "field" phenomena. Lewin saw the work in this area of physical science as both a necessary and desirable

CEREBRAL FUSION

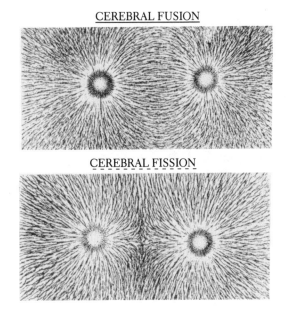

CEREBRAL FISSION

METAPHORM 3. Cerebral Fusion (Intuition) and Cerebral Fission (Reason). *When magnetic fields are traveling in reverse directions, it is like two opposite poles of a magnetic being brought into proximity; they* attract *each other. This is essentially what occurs during cerebral fusion. Conversely, cerebral fission implies that the fields in both hemispheres of the brain are traveling in the same direction, at somewhat different rates. This is similar to bringing like poles of two magnetics together so they* repel *and delineate each other. Thus, cerebral fusion is the "merging" of brain and thought processes and cerebral fission is the "splitting" of these processes. Both are needed, like the poles of a magnetic, to make the mental system work. (Adapted from* Concise Encyclopedia of the Sciences, *J. Yulle, ed., 1978)*

direction for psychology. He wasn't satisfied in following the logic of Newtonian physics, which emphasized investigating the fundamental parts before examining the behavior of the whole.

Lewin's notion of electromagnetic fields (currents and waves) as they relate to the human brain—as well as the ideas that were later proposed by the Gestalt psychologists, among them Wolfgang Köhler[3]—differs considerably from the concepts that are proposed in this chapter. Lewin was interested in mapping the "psychological fields" of what he called a person's "life-space" (or inner environment) to the outer environment. What he learned from the physical metaphor of the iron filings is how the collective magnetic field—not the local interactions of individual filings—creates the wave pattern. The implications of his theory, which attempted to describe how a person's psychological field affected the meanings of "facts," lie outside the scope of this book. However, it is important research that deserves closer inspection.

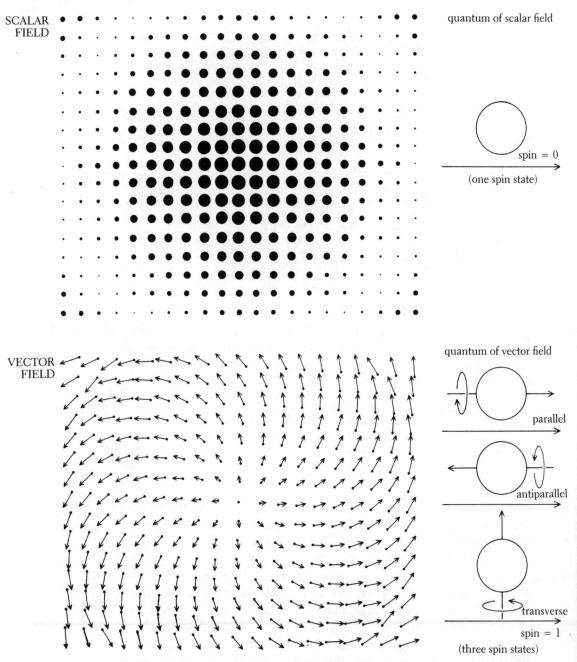

SCALAR FIELD

quantum of scalar field

spin = 0
(one spin state)

VECTOR FIELD

quantum of vector field

parallel

antiparallel

transverse

spin = 1
(three spin states)

METAPHORM 4. *Particle physicist Gerald 't Hooft describes the concept of a field as "a quantity defined at each point throughout some region of space and time. . . . A scalar field has only a magnitude at each point, given by the area of the dots. A vector field has both a magnitude and a direction, illustrated by drawing an arrow at each point." (From Richard A. Carrigan, Jr., and W. Peter Trower, editors,* Particle Physics in the Cosmos: Readings from Scientific American Magazine, *1989)*

In contrast to Lewin's work and the Gestaltists, I am interested in mapping the changes of various states of brain as they correspond to the mental processes of intuition and reasoning. My hypothesis concerns the physical bases of cerebral fusion and cerebral fission, and the distinction between them. I maintain that the physical changes are manifest as microprints, which may reveal specific patterns of electromagnetic activity associated with these two brain states. Not only would there be different microprints (or *prints* of "microfields," very-low-energy fields) that distinguish the physiological events of cerebral fusion-fission, but there would also be various differences visible *within* these events. This means there may be degrees, types, or ranges of differences within the phenomena of intuition and reason that can be detected, shown, mapped, and eventually measured. This was one of the implications of the *Cerebreactor* studies (review Metaphorm 3 in Chapter 7). The microprints of cerebral fusion, which would have certain features related to their configurations, would be demonstrably different from those of cerebral fission. The former would be more symmetrically composed and the latter more asymmetrical, although there might be variances even in these patterns of organization.

It turns out that these conjectures are more than just conceptually plausible. As I point out in Chapter 10, there has been considerable experimental and clinical research that hedges these ideas towards the "probable." Also, from most introspective accounts, we learn that some intuitions *feel* more intense, profound, or insightful than others. Similarly, some forms of reasoning seem more coherent, rigorous, or logical. Correlating these accounts with their physiological underpinnings is still difficult, as I will discuss. The correlation tests the limits of neuroscience and medical imaging technology. It also tries our prescience, or what some critics would call science fantasy—and others, artful speculation.

VISUALIZING INTUITION?

The researchers of biofeedback have described a similar state, identifying it as a moment of *synchrony* between cerebral hemispheres. One of the leaders in brain synchronization studies, the neurophysiologist Lester Fehmi, defines synchrony as "the maximum efficiency of information transported through the whole brain." The British physicist-psychologist C. Maxwell Cade referred to this brain state as "lucid awareness," or "awakened mind," which reflected heightened mental activity. This reflection was pictured through electroencephalographic (EEG) recordings and was noted as "very-high-amplitude alpha waves with lower-amplitude waves in both the theta and beta range." Other pioneering researchers, such as E. Roy John, director of the Brain Research Group at New York University Medical Center, have been profiling electrophysiological artifacts of various brain states. In addition, John has conceptualized the neuron and neuronal interactions according to changing patterns of electromagnetic fields that map to a variety of mental states and behavior.

There are a number of ways of building a theory of electromagnetic fields in the human central nervous system. The work which has preoccupied me concentrates on visualizing and interpreting these different (or similar) thought processes. One strategy has been to show the existence of a region of the brain that initiates a microfield. The microfield is symmetric about the medial lane of the cerebral hemispheres down to the brain stem. The proposed region from which this field is generated is shown in Metaphorm 5. The field is activated by the coordinated actions of both hemispheres and various subsystems that are below the cerebral cortices but integral to them. In depicting these different internal states of cerebral fusion-fission effects, we may gain a more profound understanding of human thought processes and behavior.

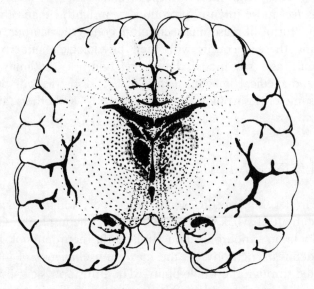

METAPHORM 5. *Front view of human cerebrum. The dots represent the proposed symmetric fields in cerebral fusion.*

CONCEPTUALIZING THE UNVISUALIZABLE

Although we have progressed technologically since Lewin's studies of psychological fields, we still are unable to accurately measure the subtle electromagnetic fields in the central nervous system. To my knowledge, there are no tested guidelines for interpreting the measurements of these microfields. Within the past decade, many research programs have advanced in a direction that can test the cerebral fusion-fission hypothesis. Perhaps, with the continued developments of magnetoencephalography (MEG) recording technology,[4] and with the current research in evoked-potentials averaging, which uses more sophisticated EEG recording techniques aided by computer analyses of electrical artifacts, some answers will emerge. The ultrasensitive, superconductor MEG can detect extremely small magnetic fields associated with ions (electrically charged groups of atoms) moving in and out of nerve cells. The EEG can detect the electrical changes associated with the movement of ions in the more scattered conducting fluids outside the nerve cells. With multiple channels (or *leads*, as they're called in electrophysiology) and computers to filter out the entangled signatures of bioelectricity, advanced EEG recordings may prove to be invaluable in mapping the widest range of brain functions. Dr. Fehmi, director of the Princeton Behavioral Medicine and Biofeedback Clinic, has moved in this direction for some time, developing what is commonly referred to as phase-sensitive biofeedback EEG.

CORDANT/DISCORDANT VIEWS

The present research in neurophysiology, then, is the background to which my hypotheses are responding. The prevailing view in neuroscience is that the cerebral hemispheres are specialized for different, but complementary, functions. This knowledge is largely due to the efforts of neuropsychologists Roger Sperry, Joseph Bogen, and Michael Gazzaniga. During the early 1960s, these researchers demonstrated that each hemisphere has its own forte. On the one hand, the right hemisphere appears to be masterful at seeing and building wholes from parts, supposedly at the speed of insight. On the other hand, the left hemisphere seems to be expert at reducing wholes to parts, using a sort of analytical sieve.

Sperry's group demonstrated this hemispheric specialization through extensive visual/spatial and linguistic tests on patients who had undergone commissurotomy—so-called split-brain surgery.[5] This radical surgery was used to treat patients who suffered from epilepsy. The operation involved severing the corpus callosum and the anterior and posterior commissures—three major pathways that connect the hemispheres and allow rapid correspondence within the brain. The surgery significantly reduced epileptic seizures by thwarting the spread of electrical discharge from one hemisphere to another. It also provided these researchers with a virtual handle for gripping some of the covert operations of the brain.

Many of the motor and sensory aspects of the left hemisphere involving speech and comprehension had been known through the research of neurologist Paul Broca and neurophysiologist Carl Wernicke, in the 1860–1870s.[6] Both scientists had studied *aphasia*, which is a neurological disorder affecting one's ability to speak and understand speech. Other pertinent studies on cerebral localization, which correlated certain thoughts or patterns of thought and behavior to specific brain regions or cell groups, had been investigated a decade before by the Canadian neurophysiologist and neurosurgeon Wilder Penfield. The observations of Sperry's group did not merely corroborate these earlier findings of Broca, Wernicke, Penfield, and others.[7] They also made many startling discoveries about the nature of this thinking substrate—our 3-pound complex of tissue, experiences, ideas, and memories.

Less than a century ago, the mind was known only through shadowy introspection which was spoken of in dualistic language. (See "reflectionism" in Chapter 6.) Although this Cartesian vision of duality still prevails, our vision of brain functions has grown 100-fold. We now have a sense that neural architecture is more than what is shown in the electron microscope's elucidation of cells or the neuroanatomist's skillful analyses of tissue or the cognitive scientist's "theoretical buildings" reflecting the architecture of neurons. For instance, subtle fields of electromagnetic energy are known to influence the actions of neurons without the direct contact between the cells through neural pathways and neurotransmitters. Although these almost imperceptible influences and noncausal contacts are not precisely quantified, neuroscientists are at least aware of the presence of these very low fields and their potential influences.

The hypotheses discussed in this chapter question some of the distinctions made in the course of interpreting split-brain research. Distinctions pertaining to such things as creativity and intuitive and analytical thought, for example, have affected our notions of artistic and scientific thinking and production. These distinctions have divided the "creative individual" from the "uncreative," through increasingly narrow definitions of creative activity. And, as I've indicated throughout this book, we live according to our definitions. They are the circles or spheres within which we characterize our work and our relationships with the world—both inside and outside us.

STRAINED VIEWS FROM ACROSS THE RIVER: DIVIDED LABORS

In my view, the results of the split-brain research led many neuropsychologists to cast the two realms of cognitive processes—intuition and reason—into two separate communities. To use a simple allegory, these communities were geographically divided by a river with one large bridge (the corpus callosum, or fiber tract) and two smaller bridges (the anterior and posterior commissures, or connecting bands of nerve tissue) joining the two communities. The individuals living on the left bank, so to speak, tended to be highly verbal, analytic, and rational. They talked a lot, and always in exacting, calculative terms. The individuals on the right bank, by contrast, tended to be nonverbal and intuitive—"the strong, silent type." They communicated largely by means of pictures rather than words, often drawing the whole scene rather than depicting the parts or details. As different communities are (supposedly) by nature competitive rather than cooperative, the individuals inhabiting the left bank referred to themselves as members of the "major" community, while pejoratively referring to those who lived opposite them as members of the "minor" community.

This general impression was held by most of the individuals on the left bank. According to the analytically minded scholars of this community, the impression had evolved over centuries, if not millennia, introduced in ancient times by their mathematicians who had believed that only through the language of mathematics and reason could we know the world. The individuals on the right bank were incensed by this attitude, as they were informed by an equally knowledgeable source in antiquity—namely, an intuitive knowledge. This knowledge led them to just the opposite viewpoint: that both mathematical and poetic insights are equally pertinent and valuable means of knowing nature. Many contemporary artists' works are informed by the rich source of scientific inquiry. (See Branches #1.) What particularly disturbed the individuals of the right bank was that *their source* of knowledge included potentially all other sources. It was a system of inclusiveness, not exclusivity. It accommodated rather than eliminated other systems of thought. The same could not be said for the source representing the people of the left bank. (Although the need for exclusivity is recognized insofar as you cannot physically study everything any time you study anything; by specializing you're able to concentrate on selected problems and studies.)

However, some scientists are presently expanding the domain of scientific inquiry into the realms of guided intuition that had previously been ascribed to the arts.

You can imagine the psychological tension between the two peoples created by this impression of major and minor communities. Each side was certain of its right to knowledge and its system of knowing the world. Of course, the individuals of the right bank could not verbally defend their point of view or debate the issue, so they sought other ways of widening the narrow approaches of their counterparts—thus asserting their communicative powers. Through sophisticated visual metaphors, they developed ways of showing the left-bank dwellers that they, the holistic thinkers of the right side, were intellectually superior and more creative—even though they weren't capable of using speech to tell them so. Each metaphor they deftly constructed bore the message that creativity and imagination are mental qualities that exist in their province alone. One visual that was particularly effective in conveying this point showed a child blowing up a balloon with the words "art *is* imagination" written on its expanding surface. Significantly, over the right side of the child's head was the word "creativity." Another animated visual, which has lingered in my mind, bore the slogan: "We're the *stalactites!* We're the ones who drip on you and make you what you are, stalagmites—growing from the floor upward, by our means, in the cave of our creation."

Stalactite

Stalagmite

DAMMING THE RIVER

From this "hideous and intolerable allegory" (to borrow Herman Melville's expression from *Moby Dick*), we can sense how studies of cognitive processes have yielded the concept of "cerebral dominance." This concept emphasizes, as the word *dominance* suggests, the idea of major and minor hemispheres, of left-sided reason dominating right-sided intuition. Armed with the power of speech, the left side of the brain rules the silent continent of the right hemisphere.

There are numerous studies in clinical neurology that generally support these contentions.[8] What they do not conclusively support are the reams of speculations relating the functions of left and right cerebral hemispheres to so-called scientific and artistic thought, respectively. This view bears a number of misleading conclusions, particularly the insinuation that the artist, as a member of the right bank, minor community, is detached from his or her rational faculties. An artist doesn't think rationally or analytically, only intuitively, visuospatially, nonverbally— as though all the bridges between the hemispheres were washed out by nature. Similarly, it implies that the scientist, as a member of the left bank, major community, is separated from his or her intuitive faculties. A scientist doesn't think intuitively or in terms of synthesis. S/he thinks only in terms of analysis, where the process of reason is markedly sequential, time oriented, and discontinuous. Scientists verbalize their mental imagery instead of seeing it. Such assertions seem absurd.

I have intentionally exaggerated the differences between the scientist and the artist to make a point: These differences are often grossly overstated and misunderstood. The serious damage caused by this form of stereotyping is no exaggeration! These crude stereotypes continue to flame like an Alaskan forest fire covered by snow. The fire only appears to be extinguished, until the spring arrives and the flames become visible once more. It seems that with each new group of firefighting neuroscientists and philosophers determined to extinguish our flames of misconceptions, an anxious public, like a pyromaniac, is committed to seeing the forest burn.

CEREBRAL FISSION: APART FROM DOMINANCE

The concept of cerebral fission differs from the mainstream view of cerebral dominance in a number of significant ways. First, it broadens the specification of cerebral functions and their role in thought processes, in particular creative thought. The phenomenon of intuition, for example, and all its forms, is not privileged in one hemisphere or another. Rather, it is a temporary state of the whole central nervous system as I've described. Similarly, the phenomenon of creativity is a temporary state of both our nervous system and our whole culture. Also proposed here is the idea that reasoning, and all its forms, is a highly creative activity that is exercised bilaterally. Both cerebral hemispheres are "analyzers" of sorts, specialized for evaluating different *aspects of details of wholes and parts*.

Second, it vies with the distinction of artistic thinking and scientific thinking[9]—where artists *play* primarily intuitively and scientists *work* analytically. This distinction has contributed to the stereotyping of different sensibilities and mental faculties. It has helped to compartmentalize our world into those who create through intuition and those who think through reason, as though there were no logic and reason in intuition and no intuition and rhyme in reason, or that we cannot quickly vibrate between intuition and reason—as between the poles of an electric field—as though we're all born without the balance of both. In the same note: Conjoining artistic and scientific activities *does not* necessarily imply that we're combining the mental capacities of the left and right hemispheres.

The cerebral-fission hypothesis suggests that the right, so-called artistic and mute hemisphere is as calculative and analytical in its holistic abilities as the left, so-called scientific and dominant hemisphere. Similarly, the visual, auditory, motor, and sensory cortices in the left hemisphere are shown to be as unitary in their intuitive abilities as the right hemisphere. For some reason we're locked into believing that holistic processors are not performing some sort of analysis in their information processing, as if to see something holistically means magically intuiting something without following a path and leaving footprints.

It appears that processing information in the brain—within the whole central nervous system and between the cerebral hemispheres—involves all kinds of discrimination and comparison processors. On this issue, consider the implications of Michael Phelps and John Mazziotta's conclusion regarding "analytic and nonanalytic subgroups" in hemispheric specialization. Both neurophysiologists, who are specialists in positron emission tomography (or PET), state that "hemispheric specialization does not necessarily signify unique properties of one hemisphere, but merely an advantage of one hemisphere over the other for a specific task *in certain situations*" (my italics).[10]

Implicit in the concept of cerebral fission is the idea that acts of expression—for example, artistic representations of knowledge and experiences—often emerge from various forms of reasoning. Whether or not these representations appear to be as spontaneous, feelingful, or instantly emotive as André Masson's "automatic drawings" or Jackson Pollock's *Eyes in the Heat* or Robert Rauschenberg's *Canyon*, the fact remains their productions reflect acts of reasoning. They consist of compositional decisions and calculations, involving spatial and temporal impulses. As these works of art manifest, processes of reasoning can be both calculative and emotionally pliant.

We're still trying to wrestle ourselves free from this conceptual headlock which has divided our culture into discrete packets of personalities, intellectual tendencies, behavior patterns, etc., since its inception some 30 years ago. I would encourage those who are interested in reading about these studies not to overlook the research of the neurophysiologist Hans-Lukas Teuber,[11] who conducted seminal work on the complementarity of hemispheric functions, stressing the union of modes of thinking, feeling, and creating. Another important researcher of hemispheric lateralization, who has stressed the integration of brain functions, is neuroscientist Jerre Levy, of the University of Chicago. Levy's concept of dominance and capacity emphasizes that normal brains "operate at optimal levels only when cognitive processing requirements are of sufficient complexity to activate *both* sides of the brain."

CEREBRAL FUSION: A PART OF PARITY

An important distinction between the concept of cerebral fusion and other concepts of intuition is that it is identified with the convergent, coordinated activities of the whole brain—*both hemispheres and all their subcortical systems*—and not just the activity of the right hemisphere. Cerebral fusion signifies the integration of opposite forces that overcome (or fulfill) their complementarity by forming one greater force. Furthermore, this convergence of mental activity directly involves processes below the cerebral hemispheres, such as those that compose the limbic system, whose mechanisms are critical for managing memory and modulating emotions.

Metaphorming this concept from phenomena such as an electrical discharge, one could say intuition occurs "below the horizons" of the cerebral hemispheres. It's not a physiological event created in the "clouds of the cortex" alone but is initiated at the "ground level," meaning at the level of the limbic system, mesencephalon, rhombencephalon, and below (see Metaphorm 3 in Chapter 3). This is analogous to a lightning bolt being prompted by some beckoning electrical discharge on Earth. Scientists have only recently learned that the ground discharge creates a kind of path for lightning to follow in hot pursuit. Could an intuition "travel" from these neural grounds upward, in joining the cortical counterparts midway?

Another distinction, which is more peculiar and interesting, is that my models for cerebral fusion-fission were originally inspired by my readings in nuclear physics and plasma physics research rather than brain-lateralization studies. As noted in Chapter 7 (Metaphorms 3 through 10b) and Chapter 9 (Metaphorms 2 through 9), my interest in fusion and fission reactor technology drew me back to studies of the brain. This unorthodox search of the human psyche, combined with my constant posing of questions concerning the general dynamics of this technology, helped me recognize the connection between the brain and its creations. At the time, my deepest concerns had less to do with revealing the nature of creativity or intuition than they had to do with revealing this intimate connection between brain-universe processes. It seemed to me the whole approach of the study of the brain was too narrow, regardless of how many disciplines within the field of neuroscience alone were added to analyze brain functions. I still find our interpretations of brain processes, such as intuition, too constricted.

On that note: The Harvard neuropsychologist and developmental psychologist Howard Gardner has suggested that intuition in particular, or creativity in general, is not strictly a physiological phenomenon. It's also a social/cultural phenomenon, implying that it has as much to do with our *perceptions* of what it means "to intuit" something and "to be creative" as it has to do with the physiology of this phenomenon. This is an important consideration. It makes suspect the significance of the physiological basis of intuition given alternative views of this phenomenon.

CREATIVITY: MORE THAN THE SUM OF INTUITION AND REASON

To summarize, cerebral fusion and cerebral fission correspond to two phases of thought: experiencing an idea through intuition and expressing an idea through various processes of reasoning. In my view, intuitions are oriented towards neither art nor science; they inform both. However, analyses and expressions are oriented towards either art or science. They are records of what is experienced. Both phases of thought—together with our cultural perceptions—fulfill the broadest concepts of creativity. I broadly defined this term earlier as *any unconditioned response or unconditioned interpretation.**

There seem to be at least two sides to the plane of reasoning, just as there are conscious and unconscious types of logic that we conveniently associate with states of wakefulness and dreaming. Intuition is the edge of this plane, connecting the two sides. What has come to be known as dichotomous, left-right, major-minor cerebral processes,[12] neuroscientists now understand to be complementary processes. They are contrasting cognitive and affective modes of thought, information processing, and coding. Moreover, these processes, which involve everything from arithmetical, logical operations to abstract, perceptual operations, may be unified during moments of intuition. I use the word *intuition* as one of many aspects of creativity.

Developing our observations about "intuitive" and "peak" creative moments involving synchronization requires two things: new methods and technologies, and new ways of interpreting the creative process and product in different contexts. The first requirement accents the fact that we need to develop more tests which probe all aspects of creative activity in diverse circumstances. This includes devising unique ways of presenting tests that are as imaginative as they are sensitive—complete with

---

* This statement implies that in humans creativity involves both associative and nonassociative learning. In "associative learning" new connections between stimuli are involved. In "nonassociative learning" changes are produced in already existing responses to repetitive stimuli. The expression also intimates that creativity applies to the whole natural world as well—where "conditioning" and "unconditioned responses" refer to evolutionary process.

specially designed testing environments. Furthermore, we need to continue integrating current imaging techniques that are capable of rendering in real-time both the subtle and abrupt changes in the neurosphere and the outer environment—which includes the physical universe. Simply put: What we say about the brain, we say about the cosmos. The processes of one are an extension of the other.

Before the first step is taken the goal is reached.
Before the tongue is moved the speech is finished.
—Zen Buddhist work, The Gateless Gate

9

The Biomirror

Words are at a loss in describing the *experience* of
an intuition through the actions of the *biomirror*. Images, too, lose something of their luster in interpreting this experience. An intuition seems
to vanish before you've had a chance to trace its call. It concludes its
message the instant you realize that an important thought has just
slipped into consciousness, unannounced. The moment you recognize
the experience—relishing its occurrence with exclamatory remarks or
enthusiastic notations—it's over. We all know what we mean when we
loosely describe this fleeting event. The properties of intuition are as
familiar to each of us as our own names, and yet we can't quite name
them. They often defy all classifying words and belie images that attempt
to explain this phenomenon. We often think intuitions are somehow
beyond these intellectual devices and constructions altogether. They are

the essence of mind and experience which seem to evade every futile attempt to quantify their actions. To time an intuition is to record eternity in a millisecond. To measure an intuition requires using a ruler without numbers. An intuition is something you feel first and think about second—only after its delivery and brief delight. At least, this is one interpretation.

In discussing the phenomena of the *biomirror* and *cerebral fusion*, my descriptions are full of nuances. Intentionally so. These descriptions are not meant to explain or describe in the strictly scientific sense. They're meant to lluminate and possibly to satisfy some long-standing wonderment. To my knowledge, there are no neuropsychologists studying mathematics who have actually pinned down the essence of this phenomenon in some tidy partial differential equation for nonlinear events. Even if there were such equations and scientific renderings, I wouldn't be too hasty to accept them as explanations of intuitive thoughts. My drawings, then, make no pretense of mathematical certainty—only a desire to know and experience something through art that has *the feeling of truth or wholeness*. This feeling, too, is perhaps one of intuition's many illusions as a momentary world of unity. *Without pausing "to reflect,"* everything seems to connect effortlessly in the entire cerebrum. In this moment there may exist "in everything a portion of everything"—to use the expression of the Greek philosopher Simplicius of Cilicia.

Think of the *biomirror* as the physical manifestation of a certain state of brain, as in the exhilarating experience of a private but profound intuition. A special thought that genuinely excites, shakes, or stirs you —towards a new awareness! A surge of inspiration that seems to glow throughout your whole body! Each of these sensations we associate with the familiar "samadhi," "satori," "Aha!" and "Eureka" (note that the last expression is attributed to the great Archimedes, who supposedly uttered this exclamation upon making a major discovery). Each sensation reflects a distinct state or level of neuronal activity involved in *cerebral fusion*. Although this state has been identified as "hemispheric synchronization," I see it as a phenomenon taking place primarily below the hemispheres, involving principally the limbic system structures. It is an effect that is *more than the sum of both* hemispheric functions. It may even be *a whole-body experience* rather than a whole-brain phenomenon. By *whole body* I mean to include the contributions of the peripheral nervous system and not just the central nervous system.

Sweeping these statements aside, I find physiological studies of subcortical correlates of human behavior more informative as to the "how" of intuition—rather than brain-lateralization studies. Information on thalamic and basal ganglia activity, I think, is pertinent to the study of intuition. It is in these areas that intuitions, or insight-perceptions, have their anatomical origin. More extensive testing of people with split-brain surgery or those with lesions in these areas may support this view. I elaborate on this idea in the next chapter.

If you were able to observe the physiology of your own brain as this "fusion effect" occurs, you would probably notice some consonant electrochemical activity not only between the hemispheres but radiating from the thalamic nuclei in the heart of the brain. To me, this consonance indicates the convergence of many different sources of information—sources contributed by various cortical and subcortical systems. Speaking practically, you might see some peculiar mergence of "higher-order" and "lower-order" functions: Perhaps you mixed a spatial metaphor, a mathematical stratagem, and an aggressive move in a game of chess with resounding results. Or perhaps you put two concepts together that are usually considered separately. Whatever the case, the details of this "mergence" are intimated by the dots and dash marks representing the symmetric microfields of the *biomirror*. These fields might extend throughout the brain stem (note Metaphorm 1, and recall the discussion of the concept of fields in the previous chapter).

Since we're talking about living neural tissue and not about slick, reflective surfaces in which the visible world appears in reverse, rid yourself of that image of the plane mirror I discussed earlier. Where the mirror metaphor was useful for describing the brain *And* mind relationship, it hinders what I have to say now. So for the moment, try not to entertain that idea about the mirror *of* the mind—the *psychomirror*.

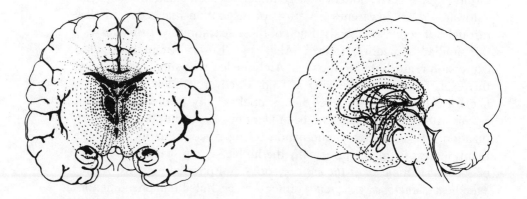

METAPHORM 1. *Two views of the proposed biomirror (bio-electric-magnetic-chemical mirror) in the human brain. The idealized microfields (very faint electromagnetic fields) indicate a merging of functions in the cerebrum at the instant of cerebral fusion (intuition). This field configuration is based on what is called the "magnetic field configuration" in a superconducting magnetic mirror plasma fusion reactor. (See also Metaphorms 2–5).*

When describing the mirror *in* the brain—the *biomirror*—I avoid comparing it to a plane mirror or any metallic-coated surface that is smooth and stationary as glass. It's not some optical object with geometric-type symmetry. Instead, the *biomirror* concept refers to a dynamic mirror with *nongeometric* symmetry such as those related to the interchanging of electric charges. I'm proposing that the *biomirror* has properties that are "approximately reflective" or "mirror-like" in nature. It is a mirror in the sense that very faint electric charges are rapidly reflected back and forth between different subsystems in the brain—a mirror possibly constructed from electrical activity and activated by a number of neurochemical systems in the brain.

A MODEL OF BRAIN PROCESSES
BASED ON PLASMA FUSION TECHNOLOGY

The idea for this biological mirror was inspired by the basic magnetic mirror developed for plasma fusion. The mirror reactor is based on the concept of confining plasma in a straight tube (see Metaphorm 8 in Chapter 7). The magnetic field is externally imposed and is particularly strong at the ends of the tube, reflecting the plasma back into the tube like a mirror. I'm suggesting two processes: that in *cerebral fusion* the hemispheres of the brain act as "magnetic mirrors" that focus and direct neuronal information back and forth at some great concentration, confinement, and speed; and that "reflection symmetry of flux fields" generated in a magnetic mirror fusion reactor is related to those fields generated in the human brain. That is, the fields are related in some material or conceptual sense. I wouldn't expect that the brain's electromagnetic *flux fields* would resemble exactly those of reactors—after all, they are differently shaped "reactor vessels." And the shapes of the vessels partially shape these microfields. However, the fields may be similar enough to warrant a comparative study of their various configurations. Metaphorms 2 through 9 show two different manifestations of these microfields.

In searching the similarities between the electromagnetic fields of the brain and those of the reactors, I rely on the trouble-free technique of superimposition. Taking information (in the form of either words or images, or both) from one study, I physically fit this information to a completely different study, as the drawings (Metaphorms 2 through 9) demonstrate. The drawings unmask this process of superimposition in a direct manner. This technique is more assertive than simply juxtaposing words and images with metaphoric intent.

MAGNETIC MIRROR REACTOR

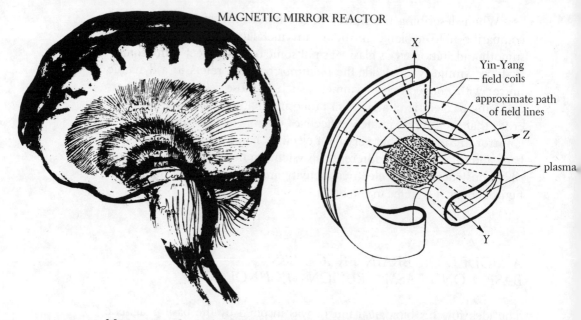

METAPHORM 2. *A dissection showing the convergence of cortical projection fibers through the corona radiata into the cerebral peduncle and pons.*

METAPHORM 3. *The configuration of the Yin-Yang magnets. (From R. W. Werner and G. A. Carlson, "Design Studies of Mirror Machine Reactors," in* Fusion Reactor Design Problems, *1974)*

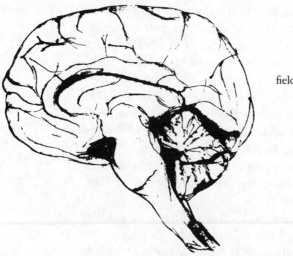

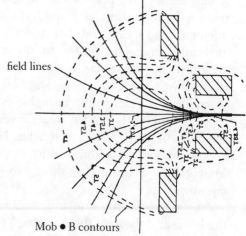

METAPHORM 4. *Medial aspect of cerebral hemisphere and brain stem. The brain has been sectioned through the longitudinal fissure.*

METAPHORM 5. *Magnetic field configuration for a Yin-Yang coil. (From G. A. Carlson, "Superconducting Magnets for Mirror Machines," in B. Brunelli, editor,* Driven Magnetic Fusion Reactors, *1978)*

FIELD REVERSED MIRROR REACTOR

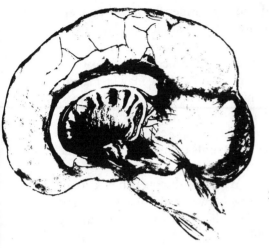

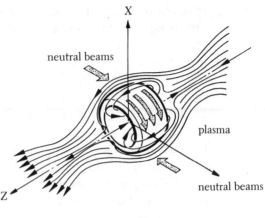

METAPHORM 6. *Dissection of cerebellar tracts, lemniscus medialis and lateralis, as well as nucleus caudatus, nucleus lentiformis, and corpus amygdaloideum.*

METAPHORM 7. *(From T. K. Fowler, "The Field Reversed Mirror Reactors" in B. Brunelli, editor,* Driven Magnetic Fusion Reactors, *1978)*

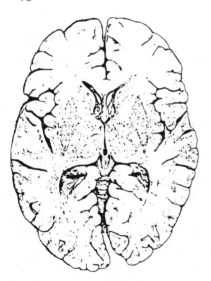

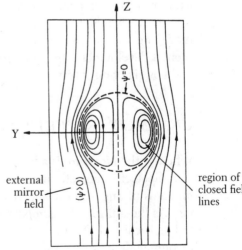

METAPHORM 8. *A horizontal section of the brain including the frontal and occipital poles of the cerebral hemispheres.*

METAPHORM 9. *Plot of field lines in spherical hill's vortex field-reversed configuration. (From R. F. Post, "The Physics of Field Reversed Mirrors," in B. Brunelli, editor,* Driven Magnetic Fusion Reactors, *1978)*

These comparisons may strike you as annoying, in that they single-mindedly state their point with perilous literal-mindedness. If this impression persists, I warmly remind you to consider the larger concept I'm trying to picture. I'm referring to my attempts at picturing the influences of our brain processes on the formulation of these technological inventions. The only evidence I have to support this connection is to be found in these sorts of comparisons. As awkward as they may first seem, the visual conjectures presented here indicate that there are influences. I'm hypothesizing that our understanding of these formulations is constrained by our physiology. And this constraint imposes directions on all technological developments. As these constraints change, so change the directions and understanding of our innovations.

This thought prompts the question: When we discover things about ourselves and the physical universe, or when we create new things and technologies based on our discoveries, do these instances of discovery and creation reflect some fundamentally "new" development in our neurophysiology? Do they reflect the evolution of a new mental construct—perhaps one that is collective? On the positive side, I'm suggesting that we can in fact learn something about the cryptic nature of our central nervous systems. Specifically, we can observe how different properties of energy manifest themselves. Knowing these intimate details of our nature can inform us about the nature of our creations and the cosmos. Moreover, we can apply this knowledge in implementing new technologies as we would the principles of preventive medicine. And, what's more, we might learn how advancements in our neurophysiological/mental development are synchronized with our advanced technologies and powers of discovery.

UNNAMED PROPERTIES OF THE BIOMIRROR

Whether the *biomirror* is a bioelectrical mirror, or a biochemical mirror, or a bioelectromagnetic mirror, or even a bioelectromagnetic chemical mirror is unimportant to me. The fact that it may exist is what interests me. As the author of *Neurophysics*, Alwyn C. Scott (of the Department of Electrical and Computer Engineering at the University of Wisconsin), recommended, we should open-mindedly—though rigorously—search ideas that hold some promise. For example, in commenting on Donald Hebb's theory of cell-assemblies as the basis of thought, Scott suggested: "We should not waste time haggling over the 'reality' of such entities in terms of physics and chemistry. They can be as 'real' as a tornado (arising out of the hydrodynamic equation describing the atmosphere) or an electron (arising out of some yet undiscovered nonlinear version of electromagnetic theory)." These notions are applicable to the nature of the *biomirror*. It may be electrical, magnetic, and/or chemical in nature, since the brain is built of countless electrical charges, and the movement of an electrical charge always creates a magnetic field.

We need to consider how it works and where it is situated anatomically. Or whether it is some transitory state of consciousness or even a phenomenon of enculturation. Perhaps the *biomirror* is some ephemeral effect which changes its location while remaining in the general brain region indicated in these drawings. At present, I think of it existing not so much as a structure itself but as an electrochemical-magnetic event linked to the stimulation of a specific structure—a structure such as the reticular formation or that part of this formation which connects with the thalamic nuclei and other main cell groups in the limbic system.

Exploring the biomirror hypothesis will yield some critical insights into a number of pressing interests in neuroscience. Included among these interests: how the specialized regions of the whole brain function conjunctively; what the nature of the correspondence is between the states of brain and states of mind; what the dynamics of "cerebral dominance" are versus "cerebral parity"; and what exactly is meant by the shifting origins of "artistic" and "scientific" expressions that have been liberally ascribed to the provinces of the right and left cerebral hemispheres exclusively—with too few exceptions. The concept of the *biomirror* is as intuitive as the phenomenon it studies. It is one instance of an intuition used to study itself.

Concepts without factual content are empty; sense-data without concepts are blind. The senses cannot think. The understanding cannot see. By their union only can knowledge be produced.—IMMANUEL KANT, *The Critique of Pure Reason*

Scientific thought is a development of pre-scientific thought. —ALBERT EINSTEIN, *"The Problem of Space, Ether and the Field in Physics," reprinted in ES, pp. 61–77 (1934a)*

10
From the Metaphorical to the Literal

Earlier, I stated that *we think literally through metaphor*, when thinking about the relationship between brain and cosmos. The least interesting meaning of this double entendre is that we literally think *with* metaphors—something you know already. The second, more curious meaning is that we sometimes think literal-mindedly, or in *physical terms*, when using metaphors. We physically connect two or more things in our mind's eyes. Consider the expression "a mental flash of lightning." If you were to interpret this metaphor literally, you would be relating all the physical events and properties of lightning to the events and properties of the human brain. As you might expect, this interpretative exercise takes some intellectual energy; but it's not without its rewards, such as gaining further insights into the nature of mind. Most people would prefer to spare this deeper analysis, forfeiting these sorts of rewards. Clearly, there's greater joy in expressing ourselves metaphorically, suggesting relationships between things rather than explaining what these relationships suggest.

This said, metaphors and analogies are not theories. They can provide the conceptual basis for theories, but they're not the real McCoy. Recall, for instance, the familiar analogy of the atom as a tiny solar system. That analogical thought was used to convey the abstract concept of orbiting electrons. Unlike the hypotheses that described in detail the orbits of electrons in an atom, the initial analogy was not considered a constitutive element in this theory even though it is credited as having happily informed it.

When metaphors are taken too literally, we're told to look for trouble. This is exactly what I'm doing: looking for trouble. (Actually I'm exercising a prerogative all people have to explore the universe in their own way—at least in their imagination.) I want to know just how far we can push the literalness of metaphors such that they resonate with material reality. In attempting to find some physical evidence with which to *examine*—not prove—my metaphorm relating cerebral fusion-fission processes to stellar fusion-fission processes, I am venturing into the literal world of physical reality. The hypotheses discussed are merely ideas and images towards a theory proper.

METAPHORIC THOUGHTS REALIZED

Similarly, in examining the neurobiological reality of the biomirror, we need to search for clues in the things we've physically created, such as the magnetic mirror fusion reactors, which may give some guidance. The shapes of the magnetic fields in these reactors, as shown in the previous chapter, are exemplary clues. These clues provide a handle with which we can grasp and extend our search—or end it. Although the studies in brain and cognitive science are important to our work, they offer only part of the information necessary for this exploration. Another part pertains to the whole history of art and its diverse practices. More than the supplies and data of science, art must guide us.

Some people gravitate towards religious thought in seeking this guidance. Others, like myself, are drawn to both the arts and the natural sciences. These fields of thought and experiment remain the most productive tools—especially when combined—for exploring nature. I found that the answers offered by science alone tended to be too limited in their range of meaning and expression. By contrast, the arts have always seemed to be far closer to the truth in that they offer no right or wrong answers, interpretations, and meanings. Instead, they direct their awarenesses towards seeing and building and connecting relationships, interpreting one thing in relation to another. In this deep respect, the arts share nature's sole gift—the joy of seeing all possible worlds and of conveying what is seen and experienced. Art explores with or without explanation in mind, with or without description in mouth or in gesture. It lives as the universe lives—open-ended—even though at times it appears to be close-minded and close-fisted.

The impulse to dismiss these brain-universe metaphorms years ago was obviously overrided. For some peculiar reason these particular ideas wouldn't quit my imagination. They wouldn't let me shrug them off as fantasies or coincidences. Somehow the known mechanisms that generate fusion in plasma fusion reactors are related to the unknown mechanisms that generate the "cognitive energies" which power our thoughts or which our thoughts are empowered by. Finally, instead of fighting these thoughts, I had to succumb to finding out more about them—first collecting the "circumstantial evidence."

APPLIED METAPHORMS

In comparison to our knowledge of the human brain, we know a lot more about fusion-fission reactor technology and particle accelerators, which seems odd to me. You would think that we would know as much about *what* we create as we would about *how* we create—since we seem to create from the inside out. But this is not the case. No one has secured the connection between the contents of our thoughts and the mechanisms of thought. This connection is still a matter of fiction. Thus, the details of a reactor's magnetic microfields are closer to being understood than the meanings of the global, or "collective," electromagnetic fields in the central nervous system. No "grand unified field theory for brain processes" has ever been formally entertained in the neurosciences.

Using the word *collective* or *global* here is no mystery. Neuroscientists have a fairly clear picture of how these microfields are generated *uncollectively,* or locally, in the nervous system. They can describe with impressive accuracy the chemical and electrical forces driving the sodium (Na^+) and potassium (K^+) ions in and out of a neuron's membrane. Scientists know how this Na-K pump works in generating a low field. They've drawn a few stray analogies to such things as the piston pump of a car engine—with its reciprocating actions transmitting motion and pressure to the fluid by direct displacement. But these descriptions of the balance between active ion pumping and passive fluxes of Na^+, K^+, and

Cl⁻ (or chlorine)* do not necessarily speak for the larger action: the creation and balance of electromagnetic fields. These "process fields" appear to exist between and within nuclei, clusters of nuclei, or super-clusters of neurons such as ganglia. One runs into all sorts of problems in assuming that the larger, global effect is simply the sum of all these local effects. The creation of the collective microfield seems to defy this logic. It's as elusive as the concepts of *mind* and *memory*.

One wonders: *If* we had a vivid picture of the way very low electro-magnetic fields in the brain are generated, *then* there would be little guesswork deciding whether or not there were fields that resembled the brain's creations—two of which are nuclear fusion and fission reactors. We would be able to see the brain's distinct microfields as clearly as we can distinguish and measure the microfields in these reactors. Moreover, we would be able to see by inspection the processes underlying the pro-duction of these fields. Either there would or there would not be some suggestion of similarity between the patterns (or configurations) of ener-gies in both the human brain and reactors.

This *if-then* picture is a gross simplification of the reality. With all the evidence in hand, with all the documented analyses, neuroscientists are still unclear about the meanings of these correlative studies. What do these minuscule fields in the central nervous system "mean"? What more do they reveal that EEG recordings don't reveal? What is the bio-logical significance of these microfields? What do they tell us about the brain's interacting subsystems? What do they tell us about the nature and phases of thinking and creating? How do you measure, map, and inter-pret these fields in a meaningful way? And, of course, there are always a few muscle-bound questions that push their way into the growing crowd of questions as they demand to know: How do these microfields relate to those of either fusion or fission reactors? What's the connection between the brain and these technologies? Even if the patterns of the brain's microfields *were* identical to those of reactors,† why should we think that brains and reactors are related? The similarity between field patterns could be purely coincidental.

Skepticism (or unsubstantiated disbelief) isn't the only wall blocking the entryway to wisdom. The fact is that no matter how similar-looking these microfields are—or how much alike the processes of the human brain and reactors are—there is no clear-cut way of confirming this connection. *There's no "final proof" that the brain imparts its ways on all that it creates.* There's no sure way of establishing this fact short of

---

* Note that these three ions determine a neuron's membrane potential, which is central to the transmission of information via axons and dendrites.
† It's next to impossible for these microfield patterns to be identical because different patterns reflect the forms of the mechanisms generating and shaping these fields.

directly observing the influence of the brain on the design of reactor technology, or any other technology for that matter. I should say: There's no way at present. Direct observation—catching the brain in the act of imposing its dynamics on something—would require a technology that is as natural and complex as the brain itself. One prerequisite for this advanced observation is *advanced* knowledge, the type of knowledge that is demonstrated in science novels set in the future tense—where the reality of all our imagined technologies isn't an issue. I can't say we've arrived yet technologically enough to luxuriate in our abilities to see ourselves in this most intimate way.

VIEWING THE TECHNOLOGY OF
OUR THOUGHTS AND THEIR INFLUENCES

This conclusion about our present technological limits was formed in response to my reading many technical books and journals on the subject of reactor technology, electromagnetism, neuromagnetic studies, etc., over the past 12 years. It was also formed subsequent to my discussions with many experts in these fields who actually design the technology discussed here. It's not like scientists and engineers haven't progressed far. It's more like they haven't progressed far enough to remain unsurprised by the brain's David Copperfield–Doug Henning feats of illusion which continue to leave everyone in the dark with its magic.

Perhaps the most compelling evidence in support of the cerebral fusion-fission and biomirror hypotheses is in biofeedback research. For example, the research of psychiatrist Charles Stroebel, director of the Institute for Advanced Studies in Behavioral Medicine in Hartford, Connecticut, is encouraging and rich with possibilities. Dr. Stroebel has developed a whole-brain, real-time biofeedback device which can process the information gathered from 20 electrodes simultaneously with the assistance of computers. This information is represented in a codified map of brain-wave activity, where different colors correspond to different configurations of brain waves. In effect, you're able to visualize some of the events of thoughts and ideas as they're forming. That is, you're able to see one aspect of their genesis through these *representations*. The results of this device and similar exploratory techniques, such as those employed in E. Roy John's neurometric studies and in Lester Fehmi's biofeedback EEG research, are important for cluing us to the general brainworks involving thought, emotion, memory, and related phenomena.

There are still many fundamental questions about this biofeedback research and the related applications of advanced EEG recording that need to be understood in further exploring the hypotheses in this book. For instance, more research needs to be conducted to determine whether people who have undergone commissurotomy can experience the same "synchronous rhythm"* when they meditate—or experience an intuition—that individuals practiced at meditation can achieve. Stroebel's experiments with skilled meditators confirm this synchronous effect, in which diverse brain waves from both hemispheres move coherently "in-phase." What isn't clear in Stroebel's work is how the brain waves in persons afflicted with subcortical lesions compare with those without any lesions. How do the introspective accounts of the non-lesion group differ from the introspective accounts of intuition?

CLUES TO THE NATURE
OF OUR INFLUENTIAL THOUGHTS

Turning for a moment from the physiological, psychological, and behavioral approach to the study of thought processes: What do the symbolic models, such as the metaphorms ventured here, tell us? If metaphorms and other sketchy clues are permitted as evidence for these fusion-fission effects, then I would cast a vote of confidence for the metaphorms which connect the processes of neuronal signaling (Metaphorm 6 in Chapter 5) with these processes. However, to extrapolate from the local events of synapses to the supposedly global events of mental activity leaves much room for error. The implication is: *To know neurons is not to know the details of the thoughts neurons produce.* Expressed another way: *Knowing the mechanisms of thought does not imply knowing the contents of thought.* We can see the effects of neuronal interactions in our thoughts and actions, without necessarily knowing how the constituent elements of thoughts form. Our knowledge of neurons simply represents some aspect of the creative brain and its states of mind. It doesn't explain how all aspects of neural networks scheme together in formulating the plans to build, let's say, a ship or a city or a power reactor or a concept of the universe. The jump from our knowledge of neural events to our realizations based on this applied knowledge is far and wide.

---

* This term, derived from brain-wave synchrony, refers to the matching of frequencies and amplitudes of brain waves in both hemispheres. This brain state is commonly associated with moments of peak mental lucidity—an expression invoking "universal awareness." Neurophysiologist Lester Fehmi was one of the first scientists to discover this synchronous effect.

CROSSING THE GREAT GAP

So there's a serious knowledge gap. The gap between a single neuron's *synapses* (which means "connections") and the "collective connections" of synapses from billions of neurons remains staggeringly large and it may take hundreds of years to fill it. When you consider that only 82 years ago the concept of the synapse was first discussed in Sir Charles Sherrington's classic *The Integrative Action of the Nervous System*, there's little wonder that we're just now learning about this "key to internal communication," as he referred to it. Sherrington helped transform an abstract concept, synaptic connection, into a concrete and testable mechanism that can be studied on both molecular and cellular levels. Similarly, advances in biomagnetic imaging are helping to transform the concept of electromagnetic fields into a viable diagnostic technique for rendering the mind in action.[1] Through advancements in these sophisticated noninvasive procedures, scientists are able to peer directly into the hearth of the brain in observing the fire of its covert processes. More recently, we've been afforded detailed views of these flames of thought. Three-dimensional images (of as little as a cubic millimeter of neural tissue) reveal specific regions of abnormal electrical and magnetic activity. These abnormalities are in turn correlated with various functional disorders studied in clinical neurology.

TOWARDS HYPOTHESES

To hypothesize about the subtle actions in the neurosphere is like trying to tether a pup tent of knowledge in a hurricane of information. Every time you succeed at securing one of its corners, it is immediately shaken loose by some new information or gust of insight, disheveling your careful plans and research. Although all neurons seem to speak the universal language of neuronal signaling, the details of this language, as in neuronal connections and chemistry, are sufficiently complicated to overpower full comprehension. You might say these details immobilize the general principles intent on explaining the details.

As many researchers and explorers have discovered, knowledge of the properties of individual neurons does not provide us with knowledge of the brain working ensemble. Some scientists are adamant that the properties and actions of the brain cannot be explained "entirely" by the general laws of physics (or by any law in general). I would, however, think twice before declaring that genetic information and neuronal information are *not* affected by the laws of physics. That the connections between brain cells are *not* directly influenced by these general laws. That the atoms composing human genes and brains are somehow unique compared to all other atoms. This idea infers that biological matter stands apart from all other forms of matter. Yet how can this be? How can the "information in our inherited genes" stand apart from physical laws that may very well describe the "foundation" of this information? How can the material of our nervous systems be unaffected by the material of the universe? To proclaim otherwise is to overlook the fact that the environment acts upon all living entities—in this case, genetic blueprints and brain functions. It affects the properties of all cells as directly and as immediately as various intensities of X-rays or gamma-rays or cosmic rays affect organic tissue.

FROM HYPOTHESES TO EXPLANATIONS

One of the many big problems is an explanation of our hypotheses. Neither the theoretical physicist nor the neuropsychologist has been able to *explain* the "particular" laws of the brain by applying the "general" laws of physics. This has as much to do with languages, definitions, and notions as it has to do with the phenomena interpreted. There are no commonly accepted definitions, just as there are no common perceptions about the details of the human brain or cosmos. As I pointed out earlier without being simplistic, what some people may regard as a *reductionist philosophy* (meaning the biological world is reduced to the affairs of particulate matter) other people may see as an *expansionist philosophy* (meaning the ways of particulate matter expand to embrace the ways of the biological world). This shift in philosophy amounts to the material world being raised to a new level of living, so to speak, and being seen in an expanded and more lively context. Here I remind you that the universe is also "half full" rather than "half empty." That is, it's as full as it's empty, just as individual neurons and nucleons seem so full of information and yet so empty of a larger content.

FROM EXPLANATIONS TO PREDICTIONS

When the dust from these observations settles, one fact will remain intact. In dealing with the human nervous system, there are boundless and immeasurable variables affecting its functions. A change in any variable will affect the state of the whole system, in either some pronounced or subtle way. Furthermore, the number of interpretations of brain functions is as copious as the number of disciplines, research techniques, and individuals applying their insights in the interpretive study of brain processes. Imagine the difficulties in coordinating all these interpretations, not to mention conducting the necessary experiments and deciphering their meanings. And yet, all these ideas need to be coordinated to form some coherent study. The electrophysiologist needs to discuss her insights with the molecular biologist. The biologist needs to discuss his insights with the visual artist or composer, who in turn needs to discuss her ideas with the physicist. The problem then becomes one of managing the exchanges, not stimulating them. However, debating the findings of such a colossal study—given the labyrinth of subdisciplines and specialties—is no great joy (not even for the most administratively minded!). This doesn't mean we shouldn't try to accomplish this or enjoy the work.

FROM PREDICTIONS BACK TO HYPOTHESES

I drift towards the opinion that a strictly physiological approach to thought processes is a cul-de-sac. One origin of this opinion can be traced to the insights of the neuropsychologist Donald Hebb, who said that "thought must be known as theoretically as a chemist knows the atom." (At the time this statement was made, the atom was not technologically visible.) "Physiological methods can deal with part systems," Hebb pressed, "but a further fundamental feature is missing, namely, how these part systems are coordinated in the ordinary behavior of the intact unaesthetized animal."[2] In a similar way, a strictly physical approach to describing the universe is also a dead end. The universe, too, must be known as theoretically as we know our thoughts. This approach can only reveal what our eyes "touch" and not what our sixth sense knows.

This reflection is not quite as negative as it sounds. Its pessimism is rescued by the more promising vision which the neurochemist Candace Pert of the National Institute for Mental Health has expressed. Pert is confident that one day we'll be able to create fully integrated maps of brain processes in the most explicit detail, making color-coded notations about the actions of specific neurochemical and electrical fields in thought and behavior. These maps would be as close to representing physical reality as one can get, like the precocious realism of the fourteenth-century painter Jan van Eyck—or, closer to the twenty-first century, like the hyper-real, high-density television monitors or 70-millimeter films. We're presently living out the prophecy of this declarative note in the fields of neuropharmacology, psychiatry, and medicine. An understanding of the effects of psychoactive drugs and neuropeptides or neurotransmitters, for instance, has piqued (or prepared) our imaginations to consider many possible developments of human behavior. We can, for example, hallucinate the worst scenarios of an inhumane world based on the manipulation of our behavior through neurochemistry. We can also envision the more constructive dreams of a naturally cooperative world of humane individuals based on understanding.

Even with the most carefully planned studies in neuroscience, we may never fully understand the neural-psychological-physical-social-symbolic mechanisms that are responsible for our creativity. This is certainly a possibility. We might remain suspended in a state of ignorance —pulled by the nose with each new adventuresome theory, trudging along without the type of assured rest that only the truth itself offers. (Or, rather, we think it offers.) Without questioning our strategies for studying the world, as discussed here, we may very well continue filling our thimbles with our brains' fire hoses turned on full blast!

There's so much happening at every moment in one's neurosphere that the idea of studying the "moments" collectively and in any detail— describing with exactitude the behavior of specific systems, subsystems, sub-subsystems (and all of these activities in relation to other neurospaces)—seems insurmountable. The scope of this componential analysis is analogous to describing both the position and the momentum of a particle in space. Werner Heisenberg's uncertainty principle has accounted for this impossibility to the satisfaction of most physicists. Thinking is considerably more complicated, vague, and "messy" than we allow our imaginations to believe for fear of being overwhelmed—just as our acts of feeling tend to complicate every aspect of the "thought process" than we care to acknowledge, or dare to feel, as we think about this process. Many scientists still argue about the autonomy of thinking and feeling. Some believe that *thinking* is a sort of skeletal structure, one that can be studied independently from feelings, which they may represent as the "muscles" and "nerves" of the mind. Other scientists remind their colleagues that no animal lives filleted. To separate thinking from feeling, or acting and creating from feeling, or feelings in art from experiences in science, is to talk of a dissected animal. To me, this is a dead issue.

I trust we will learn to integrate our many picture-statements and representations of the thought process. Out of necessity, we need to unite all the "inapparent connections," which, as the pre-Socratic philosopher Heraclitus said, "are stronger than the apparent ones." The *surface* (or apparent) connections, such as those between similar structures, will be seen in a softer light. The attention and spotlight will turn to discoveries of the *symbolic* (or inapparent) connections between processes, as we focus on the unseen.

The whole of science is nothing more than a refinement of everyday thinking. It is for this reason that the critical thinking of the physicist cannot possibly be restricted to the examination of the concepts of his own specific field. He cannot proceed without considering critically a much more difficult problem of analyzing the nature of everyday thinking.—ALBERT EINSTEIN,"*Physics and Reality*," Journal of the Franklin Institute, Vol. 221 (1936), pp. 313–347

11

The Broader Excavation— Processmorphology

In the following chapters we will concentrate on realizing a more complete world-view—one that recognizes the whole of nature's "being" in human beings. This realization requires turning the disciplines that study the universe inside out such that our insights into the outer world inform our insights into the inner one. Ideally, we want to relate information on these two worlds as integrally as they are in reality —as integrated as our stomach and toes are to our bodies. As I mentioned earlier, we need to re-connect the head and GUT ("Grand Unified Theory") and TOE ("Theory of Everything"), as they are called in theoretical physics and cosmology. Digesting this thought is considerably more difficult than preparing it—this banquet of theories and models.

The implication is we need to apply our knowledge of *all processes* studied in physics, among other disciplines, to our knowledge of *all processes* studied in brain science and psychology. This amounts to seeing these sciences and their technologies as two sides of a sheet of clear glass. Both sides of this transparent glass are always seen in relation to one another. The act of seeing this relation is art in the making. It is through art that we see these two sides of the same interpenetrating matter—brain and cosmos—facing each other. I trust we know enough to handle this glass with care as we install our new window. By "we" I mean you too as participator—layperson and expert alike.

With the beautiful minimalist music of composer Steve Reich as my model, I will layer the rhythms of the following message until the message reaches its full saturation, without fear of redundancy: *Neurocosmology is open to everyone, insofar as it is about all human beings.* And anyone who can't understand it will neither know themselves nor know nature's "selves" and "beings." The only thing understood will be human "tendencies" and little else. Worse, those who don't know nature will continue to speak with reckless egos guarding ideas with little conscience, rather than nurturing ideas that work towards unifying our views of nature and humanity—ideas which secure our development. These are the reflections and inspired projects we need to concentrate on, *not* the personalities of the individuals or nations parenting them. Unity starts with the joining of all disciplines, the full exchange of information and perspectives, and the suspension of prejudices in our studies of nature.

TO BE—UNIFIED—OR NOT TO BE

This utopic plan for a richer realism is already part of many people's wishes but unfortunately it is absent from most of our actions. You will find few art museums, for example, that practice the thinking expressed here through public exhibitions and lecture forums. To balance this criticism: The worlds of science rarely enter into the artworld except as an occasional, anomalous show of objects and thoughts presented in the pretext of "art and technology." Likewise for works of art investigating issues in neurocosmology. When they're exhibited and discussed in the otherworld environments of science, their aesthetics become the focal point, not the issues and ideas or content of the work. The surface of art is seen, not its substance. Its symbolic universe—the "dark matter"—is missed by our perceptions. Or it's obscured by our preconceptions of what art and its subject matter should be. Science's involvement in art, too, is misrepresented by similar misconceptions about how these two bodies of knowledge (art and science) should intimately encounter each other in exploring the neurocosmos.

As I've advocated all along, *everyone* has a say in what nature is or is not about, what the universe is or could be. Everyone knows nature to the degree that they know themselves. The more deeply you explore this premise, the more you realize that everyone's views on the subject of the neurocosmos are meaningful. Some views are inescapably more insightful or richer than others in that they open new vistas or alter our understanding of ourselves. I don't think anyone needs to be convinced by statistical analyses of genetic studies and intelligence quotients scores to know that all ideas aren't created equally. We all have different mental capacities driving our ideas at various speeds and distances. Although some views of nature travel far, no one's view is the living end—neither Plato's, nor Aristotle's, nor Archimedes', nor Copernicus's, nor Newton's —no matter how lengthy its influence, or what the consensus projects, or how much sense it makes now. In a world where everything is possible, we're saved by the greater probability that all possibilities won't happen at once. If they did, there would be a lot of ghosts of past and future chiding us how inept our assumptions are about nature, reality, and life.

I wouldn't be surprised if our most revered, mathematically grounded insights into the world of quantum physics, for example, were to show signs of fatigue in the far future. The Planck-Einstein insight into energy and light—the pillar of quantum physics, which is expressed as $E = hf^*$—may one day be seen in another light and from another platform of matter and energy. No doubt our new views and uses of matter will be intimately linked to the newly discovered relationships and properties of mind.

FORMING NEW RELATIONSHIPS

In order to implement neurocosmology, science needs to work its way into the stream of art at the same time art needs to transform the way science rides this free-forming stream. The practices of both need to overlap as they expand, as our respect for their differences and likenesses expands proportionally. Our prospects for the future brighten as the frequency of the arts collaboration with the sciences increases. This means extending the cooperative actions of artscience thinking in the subjects discussed here. Extension isn't possible unless most preconcep-

* Note: E represents the emission of light energy in quanta (or units), f stands for the frequency of the quantum (probability) wave function, and h stands for *Planck's constant*.

tions are pried loose—floor by floor—from the four-story history of science constructed in stages since the seventeenth century.

One of the first preconceptions to be dismantled is the idea that science is principally a rational, as opposed to an intuitive, enterprise. The second preconception to be lifted is that the scientific method is bounded by all sorts of constraints which do not permit it to grow or evolve organically; this implies that the methodology proposed by Francis Bacon and Galileo is the same methodology we apply today with little difference in attitude. Third, that the comprehensiveness of the scientific method is unquestionable. All these structural repairs are necessary before raising the ceiling on the fourth floor to include the concept of *science as an artform*.

RELATIONSHIPS FOUNDED ON TRUST

How is neurocosmology "reduced to practice"? This is the sort of question a patent attorney would ask an inventor regarding an invention. How does it work? How do you overcome problems pertaining to the collaborative process involving diverse experts, endless discussions, and curious laypersons? When you consider the swift rivers of languages, philosophies, and opinions, converting energetic ideas into workable strategies—like a hydropower plant harnessing and converting the energy of rushing water—requires careful engineering. One problem has to do with the structural integrity of the dam: There's the dilemma of "disciplinary leniency." Who establishes the guidelines for filtering out the strongest concepts? Who does the filtering and who does the enforcing? Finally, who assumes the responsibility for developing concepts after they've been filtered?

I'm hoping you will help answer these questions. You're as much a player or thinker or creator as those specialists in neuroscience and cosmology who are currently playing, thinking, and creating with similar concepts. Mathematical physics in cosmology is an impressive, operationally sound abstraction; but then, so is the poetic aesthetic in the visual arts. The "fine arts *with* physics" can provide some of the most powerful insights into the development of humankind. To be more understated: They're one of many resources for understanding human life and its prospects as seen in an evolutionary context. In this respect, they're as life-sustaining as the fuels we need to power our civilizations.

TRUSTING NOTHING BUT OUR INTEGRATED NATURE

There is at least one illimitable reward resulting from the creative reach of neurocosmology. The reward involves arriving at an integrated vision of nature such that our thoughts and actions and creations are understood as a single, integral facet of nature. To the degree that we experience this conceptual vision, and it transforms our thinking about nature, we will grow. And this intellectual growth will help us investigate these speculative reflections in what many regard as a more productive and healthier way—without the suppression of views by one authorized view.

We reward nature with our understanding that *we are the physical reflections of the things we create and which create us*. With this borderless concept, the metaphorical realm overlays the literal. Every artifact of thought shows us the thought behind the artifact, only we have to understand more fully how to interpret these clues or artifacts. We must learn to "decode" our creations, with the same intensity as we are presently trying to decode the creations of the universe. Paradoxically, the foremost step in decoding the material world is seeing past the material itself. This means seeing the other side of structure—or the "light" of midnight. Seeing straight into the world of nonmatter which, we assume, has no visualizable form other than the symbolic forms of the mathematics, sciences, and arts we use to render this invisible world of process.

Accordingly, we may not look like a radio telescope or feel like a power reactor or act like an attack submarine, but the processes by which these things work resemble the processes by which our minds work. These things are processmorphs of the brain (mind). Our languages, symbolisms, mythologies, methodologies, theories, and technological tools then all speak for our mind's nature—just as we speak for the cosmos's nature. Isaac Asimov articulated this reflection elegantly in writing: "It is the nature of the mind that makes individuals kin, and the differences in the shape, form, or manner of the material atoms out of whose intricate relationships that mind is built are altogether trivial."[1]

THE NATURE OF PROCESSMORPHS

Enter *processmorphology*—the comparative study of processes in like and unlike systems (that is, physical, biological, psychological, social, and symbolic systems). Because of the inordinate amount of different material compared in this theoretical study, it's necessary to divide this material into two categories which can be reintegrated at any time: (*a*) creations of the cosmos and (*b*) creations of the human brain. Discovering the processmorphic relations between these categories is also of interest.

Category *(b)* is further divided into three sections. Each section specifies a particular type of human creation. All creations can be related to these three systems, which are naturally integrated in the brain. The systems can be classified as follows, though not necessarily in this order: information-related systems, energy-related systems, and survival or offense/defense-related systems. (See Metaphorm 1.) Theoretically, these integrated systems reflect the biological stratification of our mental processes and behavior.

PROCESSMORPHS

processes   processors

Time-Space	Neocortex	Information-Related Systems
g., forming, unforming of particles	(thinking, feeling, acting)	e.g., analog/digital processes of computers
Energy-Matter	Limbic System	Energy-Related Systems
g., converging, diverging of waves	(thinking, feeling, acting)	e.g., merging/splitting processes of fusion/fission nuclear reactors
Momentum	Brain Stem	Survival or Offense/Defense-Related Systems
e.g., conserving, transforming of probability	(thinking, feeling, acting)	e.g., hoarding/destroying processes of advanced weapons

METAPHORM 1. Processmorphology: The Comparative Study of Processes in Like and Unlike Systems *(that is, physical, biological, psychological, social, and symbolic systems). In this figure, the material creations of both the cosmos and human brain have been divided into two categories (the third category, on the right side, is actually part of the second, or middle, category referring to the brain). This division of processes can be reintegrated at any time, as these processes are naturally integrated in the brain and universe. Discovering the processmorphic relations between these categories is also of interest to neurocosmologists. Theoretically, the three systems specified for the human brain reflect the biological stratification of our mental processes and behavior. The implication is that the workings of our creations are intimately related to the workings of these three interconnected, principal regions of the brain, which imposes its dynamics on everything it creates.*

This further division implies that the workings of our creations are intimately related to the workings of these three interconnected, principal regions of the brain. This classification scheme reflects on neuroscientist Paul MacLean's "triune brain" concept, which specifies three interconnected regions of the brain and their corresponding mentalities. MacLean, who heads the Laboratory of Brain Evolution and Behavior of the National Institute of Mental Health, postulated that successive specialization of the forebrain in the evolution of the central nervous system occurred, creating new functions with each new layer of specialized neural tissue. He and his colleagues have concentrated on understanding how certain brain regions (in many different animals) control various types of behavior.

PROCESSMORPHOLOGY— AN INTEGRATIVE VISION OF NATURE

The study of processmorphology involves making exhaustive searches of just about everything the human mind has invented—and of everything the physical universe has created. This includes our minds' most detailed and elaborate technologies, such as Very Large Storage Integration ("VLSI") systems in computer science, together with our minds' more abstract and expansive qualities expressed through concepts, such as "evolution," "self," and "consciousness." In terms of cosmic creations, the list would include everything from superclusters of galaxies, stars, and fundamental interactive forces to "exotic atoms" (unstable atoms whose electrons have been replaced artificially by some other negatively charged particle).

Processmorphic relations consist of the relationships between *processes* and *processors*. In principle, we want to sift out the more significant relations from the trivial ones. (Although perhaps no relation is ultimately trivial. Some are more connective and meaningful.) For example, in the category of human creations, I would imagine that the process of a processor, say, a vacuum cleaner, would tell us very little about the nature of centripetal force, pressure, magnetic fields, flux fields (the strength of fields of forces), and other phenomena compared to, say, a very high vacuum device such as an intersecting storage ring (ISR) used in high-energy particle physics to study the artifacts resulting from the collision of protons. Here one processor system is more complicated than another and consequently has a higher yield of potential relationships with other complex systems—as well as with simple ones. (See Metaphorm 2.)

As I discussed earlier, to speak of processes versus structures is not

THE HUMAN BRAIN IS A PROCESSMORPH OF . . .

Accelerometers, Acoustic equipment, Aerials/antennas, Airborne navigation & weapon delivery systems, Aircraft radar, Air data computers, Air defense systems, Alignment equipment, Altitude & heading reference systems, Ammunition fuses, Ammunition loading, Anechoic chambers, Anti-radar weapons, Anti-submarine warfare systems, Anti-tank weapons, Armored fighting vehicles, Artillery directing radar, Automatic flight control systems, Automatic rifles, Auto-pilots, Avionics, Ballistic and tactical missiles, Batteries, Binoculars, Bombs, Cables, Cameras, Camouflage nets, Command systems & equipment, Communications satellite, Computerized tank fire control, Control equipment (electronic), Converters, Countermeasure launching systems, Cryogenics, Data processing/peripheral equipment, Delay lines, Depth charges, Detectors, Direction finders, Displays (alpha-numerical large scales), Displays (cathode-ray tubes), Displays (closed-circuit TVs), Distance measuring equipment, Doppler sonars, Drones, Ejector release units, Electronic countermeasures, Electronic fuses, Electronic warning systems, Electro-optic systems, Encoders/decoders, Explosives, Fiber optics, Field artillery, Fire-control equipment, Flight data recorders, Fluidics, Fluid power, Fuel cells, Fuel systems, Fuses, Grenades, Ground influence mines, Guidance systems (inertial), Guidance systems (optical), Guidance systems (radio command), Gun systems, Gyroscopes, Harbor protection systems (magnetic/acoustic), Head-up displays, Helicopters, Hovercrafts, Hydraulic systems, Hydrofoils, Hydrophones, Indicator instruments (electrochemical), Infantry weapons, Information retrieval systems, Infrared detectors for thermal imaging units, Infrared jammers, infrared warning systems, Inverters, Lasers, Laser range-finders, Laser warning devices, Launching equipment, Law enforcement & riot control equipment, Lenses, Low-frequency sonars, Magnetic immunization systems, Magnetometers, Message-processing equipment, Meteorological equipment, Meters, Microwave communication equipment, Mines, Mine-hunting sonars, Mine neutralization systems, Mine-scattering systems, Missiles, Missile launchers & handling systems, Missile optics, Mortar-locating radar, Motors (electronic), Motors (hydraulic), Multi-sensor collection & correlation systems, Munitions, Naval combat systems, Naval defense systems, Navigation aids, Night vision equipment, Noise measurement equipment, Optical countermeasures, Power equipment, Programmable fuses, Propellants, Pyrotechnics, Radar (airborne), Radar (3-dimensional ground, mobile), Radar (ground, low coverage), Radar reflective materials, Radar target illumination & tracking, Radiosounders, Reconnaissance equipment, Recording equipment, Remotely piloted vehicles, Rockets, Rocket launchers, Self-propelled anti-aircraft equipment, Self-propelled field artillery, Sensory transducers, Servomechanics, Servo systems, Shelters (inflatable/portable), Simulators, Small arms, Sonar equipment, Sonar interceptors, Strategic acquisition & direction-finding systems, Submarines & systems, Submarine passive listening devices, Submarine periscopes, Survival equipment, Switching systems, Synthesizers (frequency), Tactical missiles, Tanks, Tank laser sights, Tank transporters, Telemetry equipment, Thermal imaging units, Torpedoes, Torpedo homing heads & electronics, Torpedo launching systems, Tracking equipment, Ultrasonic equipment, Ultra-high-velocity weapons (electromagnetic rail guns), Underwater combat control equipment, Underwater course plotter systems, Underwater telephony, Velocity-measuring equipment, Weapons carriage & release equipment, etcetera, etcetera.

METAPHORM 2. The Brain Is What the Brain Creates. *Its workings reflect the workings of the things (concepts and objects) it creates. These are "a few" of the potentially infinite number of analogues or metaphorms representing brain processes. Each of the things listed above can be conceived of as a metaphorm for describing some aspect of the nature of our minds. The list expands on the examples of "survival or offense/defense-related systems" noted in Metaphorm 1.*

to imply that the former is less real than the latter. Both are as real and intimately related as blood is to sustaining our bodies. Rather, to speak of processmorphs is to recognize the common characteristics of the processes underlying all forms of matter. This recognition moves us towards unifying our thinking about the diversity of forms and their seemingly "irreconcilable differences." We begin to see how everything that exists in the firmament of physical and nonphysical—or metaphorical—reality may be linked together.

Whether the manifestations of this link are measurable electric fields, or magnetic fields, or gravitational fields, or any other field is almost beside the point. The only important fact to note is that processmorphs connect all scales of masses, from galaxies to viruses. They interrelate all dimensions of operation. The "push-pull operation" of complementary transistors in an electric circuit, for instance, is a processmorph of the push-pull action of complementary "transistors" that help "make up our minds"—and that relay our thoughts once our minds are made up. Like conductors of electric currents, these transistors are processmorphs of the cosmos's "currents."

CONNECTING THE PROCESSES OF ALL FORMS OF MATTER

The results of this search promise to inform us about how we work and what we are *in relation to* how the universe works and what it is. The results will also provide many new models for relating the functions of the human brain to different systems it has created—or processmorphs —which are critical for understanding the mystery of its functional dimensions. More important, it will show us the (structureless) *process patterns* which join all like and unlike forms of matter. Processmorphs connect everything because they are the essence of connection or the act of connecting.

The study of the patterns of connection was broadly intimated by the late nineteenth and early twentieth century "process thinkers" or "process philosophers" such as Henri Bergson, Alfred North Whitehead, and Pierre Teilhard de Chardin, among others. Where these philosophers explored the concepts of change, novelty in human experience, impermanence, relativity, and dipolarity of the universe, the neurocosmologist's approach skirts their general interests. In process philosophy, the concept of dipolarity considers absoluteness and relatedness to be complementary aspects of each other (like the object and its virtual mirror image, respectively). We prefer, instead, to explore the most basic relationships connecting the processes of one thing with another.

Whether or not these "things" are mechanical or psychological, spiritual or biological makes no difference. We're less interested in determining whether something is a machine or a Marxist, a monk or a plant. Our interests turn to learning how, for example, these various and sundry forms, or "manifestations" of nature, share similar processes. We want to learn how the same process that modulates a radio transmission also modulates the transmission of a thought or feeling. We hope to discover how all processes of modulation, for instance, are related to all modulator processors—devices that effect modulation.* The one idea I've adopted from my predecessors is that I believe in the relative process of change in all open systems, *including* the languages and disciplines we use to define the nature of all equilibrium and far from equilibrium systems.

The pioneering chemist Ilya Prigogine's research on "dissipative structures," or open systems that are far from equilibrium,[2] is important to point out if only to distinguish my concept of processmorphs from dissipative structures. Prigogine's mathematical analyses of self-organizing structures based on thermodynamics showed vividly the "creative lifeforce" in all matter. He was one of the first scientists in the Western hemisphere to challenge the mechanistic view in which the universe was composed only of "things" and not "processes." The conceptual foundation of Prigogine's view—where the cosmos is seen as an open, living system—was clearly intimated in Near and Far Eastern thought. In traditional Hindu philosophy, for example, we find some highly developed views of process philosophy (as you will read in Branches #2 at the end of this book). So that's not where his contribution lies. His insights are important because they distinguish more precisely the characteristics of open systems in general.

Prigogine distinguishes three characteristics that make up a dissipative structure. One: There needs to be a continual flow of energy and matter, or information, between the structure and its environment. Two: There needs to be a constant flow of energy and matter such that the structure or open system can experience intense fluctuations. Without these increases in fluctuation, a system would not be able to self-organize. Also, it would not be able to *dissipate* or *decrease its entropy* into the environment. It would, instead, become consumed by entropy like all other "burned-out," "uncreative" systems that live near equilibrium. Three: The system or structure needs to be "autocatalytic." (Note that a catalyst is something that can change something else without necessarily

---

* *Modulation* is defined in physics as the process of impressing one wave system upon another of higher frequency. A modulator is a device for generating and varying a succession of pulses—short or long in duration—which may be used for different applications (for example, radar).

changing itself or being influenced in the process.) This means that a structure or system grows by means of itself. It's able to self-reproduce, self-reinforce, and self-destruct with little help from outside sources.

In exploring this list of criteria, we begin to notice how many things fit these characteristics—in both the biological and physical realm. Although the third criterion, the autocatalytic aspect, is bothersome. I'm not certain that anything can do anything by itself without being influenced in the act of doing! Things interact—however subtly or imperceptibly. They exchange matter and energy—however little or intangibly. Just because we can't quantify these unspecified amounts of interaction is not proof that there is no exchange of some kind. So I think the notion of "autocatalysis" is mythic in nature rather than representative of nature. This note aside, the model Prigogine adopts as the essential process of all living systems is "metabolism." The term is an umbrella for three interrelated processes involving the building up, maintaining, and destroying of protoplasm (material critical for all biological life forms). These processes relate to many disparate things that are not readily associated with biology and metabolism, namely: buildings, cities, engines, oceans, etc. It is this connective and transformative aspect of Prigogine's theory of dissipative structures that is important to explore in relating the open systems of the brain to the open systems of the universe.

In investigating the world of processmorphs, the neurocosmologist's work lies somewhere between process philosophy and a field of engineering called *operations research*. The former is neck-deep in abstraction and theory, whereas the latter heads straight into application; process philosophy tends to be less tightly directed towards problem-solving than operations research. Operational research, as it is also called, applies scientific methods towards practical ends in managing various systems including the organization of military, governmental, commercial, and industrial systems. What makes operational research pertinent to processmorphology is the fact that not only does it involve pooling teams of interdisciplinary researchers to investigate the workings of a given system, but it additionally researches the way it conducts its investigations. It is like a science of self-reflection, where the watchdog watches the behavior of itself. Operational researchers apply their studies of their own operations and those of different systems—using analysis, symbolic models, and statistics—with the purpose of improving the efficiency of these systems. I trust that neurocosmologists will point out more exact connections between the ways of our mind and nature's ways. Connections that might ultimately help us live with more sustained confidence and joy in life.

What we do with this new knowledge of processes very much depends on how integrated our world-view is at the time. It also depends on how intact our sense of ethics and moral conduct is. If the atmosphere of intelligence is sensible (caring) then the applications of this knowledge will be benevolent. I won't chill you with the opposite scenario, which the English novelist George Orwell pictured in his novel *Nineteen Eighty-four* (1949).

UNLIMITING POSSIBILITIES FOR INSIGHT

As I will sketch in the chapters ahead, the comparative studies in neurocosmology can be developed into informal discussions or full-fledged academic programs. By *academic* I mean only that there is some supervised or directed study—at all levels of education. Why base this program in academia or structure its agenda academically? Because the best academicians (scholars, artists, and scientists) know how to *study something* as opposed to superficially reading it. The best know how to explore something without instantly exploiting it for some short-term gain or purpose. Furthermore, the best teachers know how to think and instruct with curiosity about things that are both subtle and obvious. And nothing's more subtle than the nature of brain and cosmic processes. Although, as many maintain, nothing's more obvious either.

Central to both informal and formal programs, we need laypeople who don't get lost in the labyrinth of details and protocol. People who can spot the obvious aspects of nature and never fear expressing what they see in the most direct terms. We need those for whom the phrase *academic credentials* or *certified expert* is an anathema. Who else but the "layperson" in young Einstein helped him to think with novelty about the properties of light? Had he been of a more conservative mind, his imagination might never have sped alongside a beam of light in his quest to grasp the material nature of light. Such a quest by a colleague-conscious individual might have been dismissed as unproductive folly.

IF

Implementing these programs then requires developing a new forum in which to present ideas resulting from this comparative study of processmorphs. I hesitate calling it an "art-science-technology forum" because this title usually implies that the latter two are present with only a token presence of art. So I'll tentatively call it an idea forum (*IF*, for short). *IF* could be liberated from the precious pomp that typically inhibits the creative impulses of contributors. Ideally, we want contributors to be as adventuresome, or as conservative, as they feel they need to be in discussing discoveries from processmorphology. I only want to silhouette the idea here, rather than distinguishing the features of the figure drawn or the background (context) in which it is understood.

THEN

As exchanges of ideas are critical for neurocosmology, the *IF* then is essential. With its potential collaborators including scholars and engineers, physicists and poets, fine artists, musicians, and mathematicians among others, neurocosmology can blossom into an amazing open mind field full of unusual hybrids of ideas. It doesn't have to be an "intellectual killing field." Some professional administrators whom I've consulted have cautioned that neurocosmologists are bound to confront the same criticisms utopian planners face in presenting their design schemes for a united republic. Unquestionably, there are major conundrums in the practice of neurocosmology. But I trust that all can be dealt with through open-minded negotiations that have some semblance of peace.

The upshot is that the dreams of this broad search and overview need to be realized now. Everyone must know we're crossing the threshold of our collective future. It may not be the first time our self-conscious minds have mobilized us, but it might be the last—at least for our species anyway. Clearly, it's either unity and cooperation or extinction. I'd prefer to talk and act together now rather than debate the reality of our global genocide while we proceed to extinguish ourselves through ignorance. And as long as we're talking, I suggest we keep the discussion centered on the really big issues in life—like where we're heading in space and time, not "Who's on first?"

WIDENING THE ROAD TO REPRESENTATION

In drawing together ways of representing knowledge, one view quickly crystallizes. There is *no one way* to represent our insights in all that we make up and that makes us in nature. There's *no one language* to convey our observations and experiences. There's *no one voice* to express ourselves. There are, instead, a multitude of ways, languages, voices, and systems of communication. What seems to unify them—allowing us to speak meaningfully with one another—is the deceptively simple fact that we all experience life similarly as human beings. We share one central nervous system, despite all variances in form of this system. We use symbolic languages similarly, though they're driven by different intentions. We create and often live by similar mythologies, incorporating them in one form or another into our daily lives, though savoring their effects differently. In spite of our similarities, we're still determined to live according to the differences we look at with partial minds and closed notions.

The dream is to see through or beyond the "structures" and "surfaces" of all material differences. This act of seeing begins with integrating all sorts of things and sensibilities—things that have been divided into complementary parts for millennia. Integration requires freedom, which in turn engenders creativity. One of our first creative impulses is to integrate—to connect objects, images, experiences, and ideas. Perhaps the second impulse is to question what we've connected. In completing this process within milliseconds and over a lifetime, we fulfill ourselves by renewing our sense of curiosity and humanity.

COMMENTARY ON METAPHORM 1

The terms *information, energy,* and *survival or offense/defense* are defined as follows: *Information* refers to knowledge communicated or facts and data received by all means in one system or between systems. The facts may or may not be organized and related. A second definition suggests that information is the means or process of communicating; it is what it communicates. Both definitions invoke the idea that information is energy.

Energy refers to a specific feature of a system that possesses the capacity to do work. The quantity of energy is measured by the amount of work the system can do. The different forms of potential and kinetic energy include electric energy, heat energy, chemical energy, nuclear energy, radiant energy, and mechanical energy. As the law of conservation in physics informs us: Energy can neither be created nor destroyed in any closed system, although its form may be changed. All matter is a form of energy. (Perhaps processmorphs are processes of energy without form.)

Offense/defense refers to the protection of oneself (and others) from attack and harm. *Survival* refers to the preservation of oneself (and others) during what is perceived as a mortally dangerous situation—that is, surviving or enduring this situation.

Every object we create, and every system of thinking and communication we invent, can be described as a processmorph whose processes fit one or more of these three general categories. This means that anything which exhibits a pattern of activity involving one or more of these categories of processes is congruent with the activity of one of these systems. For example, the paired verbs "gathering/releasing," "verging/

retracting," "absorbing/emitting," etc., relate to the dynamics of objects (mental or physical in nature) or processors that are like one another as processmorphs. Irrespective of what the processors actually look like, the way they truly work in quantum reality is not the way they appear to our sense perceptions.

This means that we have to reexamine what we mean by the relationship between *structure* and *function*—especially the term *function*, which may refer to the *mechanism* by which "something works" or the *process* of "working somethings." Thinking in terms of mechanisms, we leash ourselves to the thought that functions can only be understood in relation to something's structure. For example, a train functions only as the constraints in the structure of a train permit it to function. And yet, by thinking in terms of the process of a train, we release our thoughts of constrained train functions—seeing, instead, how a train's process of moving can resemble the processes of movement of many things with structures quite dissimilar to that of a train, including the structure of thought itself. The metaphorical expression "a train of thought" and the mathematical equations used to describe this symbolic train are two examples of processmorphs. Both share certain characteristics or properties of a physical train without actually resembling one. Both examples are scaleless and formless. The symbolic train can take any proportion; so can the set of differential equations.

With this reexamination of notions we discover that a traditional work of art, for example, which represents some mental abstraction, is a processmorph of the thing it's representing. It is one material form, or manifestation, of the process involved in the artwork's creation. As an artifact of mental activity, it has as much potential for revealing the processes of neurons as a beryllium plate blasted by high-energy electrons from a particle accelerator has for revealing the processes of nucleons. To recapitulate Wassily Kandinsky's statement: "Form is the outer expression of the inner content." The inner content is a processmorph of the outer form. The process is the act of expressing the outer form. The form itself is one processor.

In re-thinking the relationship between structure and function, form and content, and other "basics" of nature such as contexts, we have a chance to see how intimately connected the processes of life are to one another. In physical reality, things seem to act differently because they look differently. However, the concept of processmorphs intimates that everything acts similarly in spite of how differently it looks.

The expressions of philosophers and physicists, mathematicians and musicians, appear to differ from one another because of their forms and contexts. The only way to see beyond these differences is to search for likenesses of process. This is the pattern that links the process of composing concertos, creating theorems, conceptualizing physical experiments, and contemplating paradoxes. The creations—concertos, theorems, experiments, and paradoxes—are the processors. Their forms of expression preserve the differences between these fields, whereas their processes of creation reveal similarities.

If we were to place the prefix "neuro-" in front of every word we use to represent everything we create, this addition would extend the meaning of our objects of creation as well as identify their source of origin. The important question is: How does this awareness and information influence our concepts of humankind, nature, life, and our creations? Does it make the things we create more sophisticated and efficient and complete? Or does it simply make us more self-conscious? Will this knowledge continually refresh our sense of humanity as we grow wiser? Or will it merely help us learn how to manipulate ourselves more skillfully as we continue to invent "smarter" weapons and more clever justifications for deploying them? The answers to these questions are contingent on our understanding of the "neuro-dimension" and its processmorphic relations.

For two thousand years we have tried to make of words, and other notions, an enduring semblance of the ways of our changing world. We are still not satisfied. We want more and better science, but we cannot always tell at once whether we need more facts or lack the proper notions.—WARREN S. MCCULLOCH, *Embodiments of Mind*

The more ancient Greeks (whose writings are lost) took up with better judgement a position between these two extremes —between the presumption of pronouncing on everything, and the despair of comprehending anything.—FRANCIS BACON, *The Novum Organum [New Instrument for Knowledge]*

12
Envisioning the Possibilities of Nature

A number of thinkers have headed in the direction of neurocosmology since antiquity. By "headed" I mean that they have not directly connected information on the brain and cosmos in the way this book does—broadly searching, specifying, and metaphorming the connections. However, they have creatively investigated different parts of this subject, welcoming debate and revision through trials of curiosity.

The individuals mentioned in the following pages never referred to themselves as neurocosmologists, although they may have envisioned nature as explored here. Note that their ideas are not presented in any chronological order as defined by their respective contributions in their fields. They are only related contextually and introduced briefly.

MASTER NOVICES

What distinguishes the advanced thinker in neurocosmology from the uninitiated or intermediate is purely quantitative: The former has gathered more examples in support of his observations than the latter. Also, they are often able to articulate their ideas more clearly in word form or imagery. Contrary to how we usually distinguish the skills or intelligence of the master versus the novice, experience or time and sagacity are not determinants. A "beginner," or unpolished expert, in neurocosmology can be as insightful and deep-thinking as the polished expert. Both individuals are balanced between experience and innocence as neurocosmologists. Each has the potential to inspire others, illuminating points of union between all of nature's creations. These points might have otherwise remained categorically separated and problematical. The concept of a "master novice," then, refers to a person who is able to see things anew as though for the first time and with fresh perspectives. Rather than being stunted by their own expertise, they factor it into their flow with imagination.

To my mind, individuals as diverse as R. Buckminster Fuller and Jacob Bronowski were master novices who challenged us to see certain features of unity in nature—to make connections and build upon them. Although Bronowski only touched on neurocosmology when he described the parallel hierarchies between humankind and nature, his touch was as sensitive as it was informative. In *The Ascent of Man*, Bronowski writes: "We find in nature something which seems profoundly to correspond to the ways in which our own social relations join us." In his view, these hierarchies include:

families ————— fundamental particles
kinships ————— nuclei
clans   ————— atoms
tribes   ————— molecules
nations ————— bases and proteins

This "correspondence" between nature's relations and human social relations expands beyond the hierarchy of structures Bronowski discussed. The correspondence has to do not only with the parallels between structures per se, but with the *integrated processes* of all these structures. For example, the process by which fundamental particles decay can resemble the process by which families-kinships-clans-tribes-nations decay. Or the process by which our families evolve can resemble the process by which fundamental particles-nuclei-atoms-molecules-bases-and-proteins evolve. By "resemble" I mean more than simply the operative words *evolve* and *decay*. I mean, in fact, that there is a likeness of shared processes between these separate structures:

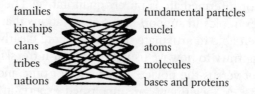

families	fundamental particles
kinships	nuclei
clans	atoms
tribes	molecules
nations	bases and proteins

Discovering the roots of this likeness, which links all processes of nature, is part of the agenda of neurocosmology.

SEEING "I" AS A PROCESSMORPH OF "IT"

When we open our imaginations to the possibilities of nature—in terms of its evolution and creations—we're reflecting as neurocosmologists on human nature as well. In these reflections, the mind is all that is present and observable as natural phenomena. Paradoxically, it's also *absent* from material reality—as part illusion and part imagination. (Recall the discussion on reflectionism in Chapter 6.) Like the universe, the mind is as dense with matter as it is void of material. It's as full of meaning as it is empty. Such is the nature of all processmorphs—as physical and symbolic entities.

In relating the decay mode of the proton and neutron to the demise of neurons, we're metaphorically connecting the processes of these phenomena. And this connection helps us understand that these two fundamental things (nucleons and neurons) hold us together—literally in body and in mind. One might say that this link lets us live engaged to life —perhaps as engaged as a single photon is within streams of light.

In relating the actions of gravitons (hypothetical transitional particles with zero charge and rest mass that are linked with the effects of gravity) to our mental actions, we're creatively joining two actions, or processes, that would otherwise have remained separate. Theoretically, we're integrating the meanings we associate with these phenomena such that the expression "the mind's gravitons," or "the gravitons of the mind," tells us something about nature's integrative forces in both brains and stars alike.

At the center of each of these relationships, and connecting them, is the human mind—discovering itself amidst the many disciplines, media, and metaphorms it has created to express itself. Douglas Hofstadter explores with far-reaching vision the concept of mind, its languages, and its communicative powers in his poetic book *Gödel, Escher, Bach: An Eternal Golden Braid*. Hofstadter searches the hidden links between our symbolic objects, together with the meanings we give these objects, and their associations with the physical world. Through this search we approach an understanding of the mind's vested interest in systems of communication. Although many people feel they don't have to understand how their minds work in order to use them, Hofstadter's reflections inspire us to discover the subtleties of their workings. And with each discovery, we're able to formulate richer hypotheses about the covert processes of nature that manifest themselves in our creative processes and creations.

SPEAKING THE MOTHER LANGUAGE

In *The Cosmic Code: Quantum Physics as the Language of Nature*, the late physicist and science writer Heinz Pagels approached a neurocosmological perspective when he articulated the limits of determinism: limits that we have lived by since the advent of quantum physics. Determinism asserts that there is a mechanical correspondence between causes and effects. The new physics encountered the limits of determinism when it discovered that observation of cause and effect changes them.

With the help of pioneering experiments, such as Thomas Young's double-split-screen experiment in 1803, the new physics suggested that things can happen without the agent of an unknowable cause. For example, Young's experiment demonstrated that one can predict only the possibility of and not the path (or the effect) of a single photon passing through a double-split screen. The knowledge about its original condition as a source from the start (or the cause) is necessarily limited. This was disconcerting to many physicists who were locked into thinking only in terms of the causal and deterministic effects of light and matter. It seems counter-intuitive that an effect could have an unknowable cause; and yet, material reality confirms this counter-intuition when we explore the paradoxical nature of the photon. The question is: How does this fact relate to the human nervous system? How does cause $\nrightarrow$ effect translate into the actions of neurons involved in thought processes?

Unlike Newtonian physics, quantum physics must consider the observer or detector. As Pagels explained: "Even an all-knowing mind must support its knowledge with experience, and once it tries to experimentally determine one physical quantity the rest of the deck of nature gets randomly shuffled again. The very act of attempting to establish determinism produces indeterminism."[1] In the course of questioning this paradox, scientists began to rethink the appearances of what seemed like a static universe with its describable events. The events could not be detailed by a detached observer intent on picturing the events without interfering with them. It gradually became apparent that our world could no longer be described as a neatly fixed creation—where everything physical could be spoken of objectively without mention of the vague sense perceptions accompanying one's experiences. This emergent vision of the observer as participator in the affairs of the cosmos (rather than as a mere, influenced spectator) signals the next step in identifying the degree of our participation. How involved are we anyway?

TURNING OUTWARD OBSERVATIONS INTO INWARD INSIGHTS

One individual who has written elegantly of this shift in perspective between the deterministic world of Newtonian physics and the indeterministic (or probabilistic) world of quantum physics is the noted writer Paul Davies, who is a professor of theoretical physics at University of Newcastle-upon-Tyne. Davies describes how Newtonian dynamics and gravitation, which were once used to describe the "static" universe, were seen in a new context by Albert Einstein, who assumed that the universe was "dynamic."[2] In his book *The Accidental Universe*, Davies speaks about the process by which Einstein contextured information (and matter) in proposing his theoretical framework for the general theory of relativity. If we were to move the discussion of relativity "indoors," so to speak—relating Einstein's theory to the dynamic brain—we would find ourselves speaking as neurocosmologists.

Even more relevant to neurocosmology is Davies' discussion of the so-called anthropic principle,[3] which considers the coincidences leading to the existence of life forms. The coincidences include the constancy of physical laws. The anthropic principle amounts to explaining the universe through the eyes of both a biologist and a particle physicist, with each contributing to the other's interpretation. Although the principle examines our existence and place in the universe, it does not consider how the neurons that compose our minds are part of the composition of the body of the cosmos. Nor does it explore the consistency between these two worlds. Even though this exploration is totally conjectural, the conjectures are based on data obtained from astronomy and neuroscience. The day these two disciplines exchange insights will be the day "anthropic cosmology" becomes neurocosmology. At that time we might understand how the organization of the mind reflects the organization of the universe.

The concept of an *internal-external consistency* between brain and cosmos has two related sources of inspiration. The first modern source was provided by the physicist Ludwig Eduard Boltzmann (1844–1906), who contemplated the extraordinary statistical fluctuation that led to the creation of our organized cosmos. The second modern source is the current concept of a *random world-ensemble*, as it is called in cosmology, in which all life forms were created by an equally rare statistical fluctuation. By some miracle, these fluctuations overlapped, or fulfilled, each other's requisite for life. No one knows how these fluctuations occurred, or why, or even what they mean in terms of the biggest picture of all— the one that has neither the cosmos nor humankind in it: the picture without the picturer or pictured, the solitary void before the Planck era. We know this overlap of fluctuations occurred, as two major cosmologists, Barry Collins and Stephen Hawking, put it: "Because we exist." Because we live in a relatively organized cosmos and because we live with moderately organized minds in this *universe*.

The internal-external consistency concept contradicts the traditional notion in physics which purports that human life—or the presence of a detector—is irrelevant to the ways of the universe. Many important theorists, such as Brandon Carter, Fred Hoyle, Robert Dicke, and John Wheeler among others, question this traditionalist view. Since the concept of the "anthropic principle" was first discussed in the context of astrophysics and cosmology (in an article written by Carter in 1974), there have been a number of speculations connecting humankind to the cosmos. Wheeler's concept of a "participatory universe," for example, in which human beings have some unknown say in what happens *to* and *in* the universe, is one reflection on this principle. His concept invokes the idea that the universe and its inhabitants are somehow "mutually dependent" on one another for their existence.[4]

Both the so-called *weak* and *strong versions* of the anthropic principle respond to the view that humankind—and life forms in general— somehow "selected," or "took advantage of," a rather unique situation for establishing itself during the opportune epoch, t_{now}. Their explanations consider such things as the coincidences in physical constants, timing, temperatures, etc., in preparing the debris of galaxies and other celestial bodies for their inhabitants. Carter's strong anthropic principle sums this thought: "The Universe must be such as to admit the creation of observers within it at some stage."[5]

One biological explanation of the fundamental aspects of the physical universe was offered by Robert Dicke in 1961. Dicke hypothesized that in order for life to have formed in the universe at t_{now}, at least one generation of stars was needed to pass through this life cycle, seeding the galaxy with the supernova's carbon-rich debris. (Without carbon, among the other essential materials for organic life such as nitrogen and oxygen, we would either be nonexistent or "stated" in another form of lifeful matter altogether.) He maintains that the timing of life forms was not based on accident but rather was dependent on the universe's state of preparedness in terms of its production of heavy elements. At the time of the Big Bang, temperatures were sufficiently hot for just a few minutes to synthesize heavy elements. Since helium was the only element produced in superquantities during this brief moment—and since this element is not particularly important to life in itself—the heavier elements had to be born from the interior of certain massive stars where temperatures were sustained on the order of 10^7 K over billions of years. As these stars exploded, they carried on the initial work of the Big Bang.

Other speculative concepts include Wheeler's cyclic "reprocessing" universe. According to Wheeler's world-ensemble scheme, the universe may very well expand from some singular event (a space-time singularity) reaching its full volume before collapsing back again to its original, infinitely dense and compressed state of unity. At the moment it becomes nothing it becomes something again, expanding outward like the previous scenario—only something different may occur. With each expansion and contraction, there may be new values assigned to the fundamental constants complete with new physical laws. This hypothesis looks at our present universe as one of a potentially infinite number of "reprocessed" universes.

A peculiar visualization and variation of this concept of "reprocessing" appears in a nineteenth-century Jaina manuscript from Rajasthan, India. The original caption to this drawing (Metaphorm 1) reads: "The eternal recurrence of the sevenfold divisions of the Universe as a cosmic river of time and reality." Through metaphorming, a contemporary cosmologist should be able to retro-fit the hypothesis of "cyclical" singularities to this Jainist drawing.[6] Each of the upper folds of this metaphorical snake could signify, for instance, an *isotropic universe* (a universe in which all of its physical properties are equal along all axes; this condition is in contrast to the concept of an *anisotropic* universe in which there would be unequal physical properties along different axes).

Although we may never be able to corroborate any of these or other speculations involving the anthropic principle—at least not to everyone's satisfaction—they all pose a colossal question about one of the most critical issues explored by the anthropic principle: the large-number coincidences of nature. How could there be so many coincidences between

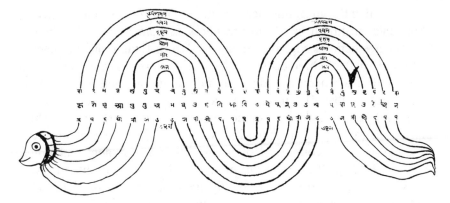

METAPHORM 1. The Snake of Ancient-Modern Cosmology. *(From Philip Rawson, Tantra; The Indian Cult of Ecstasy, 1979)*

the cosmos's fundamental constants and the initial conditions needed to sustain life forms—without some guiding principle at work, without some special process carrying out a hidden agenda of sorts? Are there similar coincidences between the brain's "fundamental constants" (regarding its growth) and the initial conditions needed to sustain growth of our "mental life forms"?

WHAT'S HAPPENING

Indeed, something is happening before our eyes. But what is it we're missing? There's something so conceptually nebulous about these large-number coincidences that our imaginations blur at the edges in contemplating them. The effect is like focusing on a moving image as it expands past your peripheral vision. All you ever notice is the movement, not the details of whatever it is that's moving.

Our comprehension of these coincidences may require a real revolution in the way we work together *to see together* what's happening. At present, we're divided in our methods of exploring and representing these phenomena. And this division may be contributing to our aggravating oversights and undersights. Many scientists have spent considerable energies trying to describe, for example, how the number 10^{40} appears in different contexts in astrophysics and cosmology. We may end up discovering that this figure is related to the number of neurons our brains will eventually have or *presently have*. By *presently* I mean that more than three-quarters of this large number would have to be "virtually present" in their functional architecture. Otherwise, if each of us physically had this meganumber of neurons, our heads would be the size of massive mushrooms with our tiny necks, as stems, supporting them!

Science fiction aside, astronomer Sir Arthur Eddington and theoretical physicist Paul Dirac observed that the universe's age is close to the number 10^{40}, when measured in natural atomic or nuclear units. How could the number of particles in the universe match the age of the cosmos? The significance of this correspondence was pondered by Dirac in 1938. Dirac suggested that this "coincidence" of correspondence is "due to some deep connexion in Nature between cosmology and atomic theory."[7] Perhaps the size of this number indicates the number of mysteries there are to nature's mystery: 10^{40} *questions to* 10^{40} *particles!*, where each question bears a paradox. For example, how could there be some collective correspondence between the number of stars that compose a galaxy and the number of galaxies that compose our universe—both of which are supposedly reflected in the number 10^{40}? The paradox in this last example is that as large as this number is, it may be *too small*—given the rate of growth of both stars and galaxies. No one knows for certain.

Do we need to solve these paradoxes to secure our evolution? Do we need to know how we are rooted in matter (and in nonmatter) in order to live with some mental peace? Must we learn how these coincidences are conceptually larger than all the calculations of cosmic "large-scale" structures or neural "small-scale" structures? We presently imagine that both of these complementary structures expand according to one scenario and contract according to another. Depending on which theory you subscribe to, you're either on the side that's expanding its definitions of brain and cosmic processes, or you're contracting the broader definitions—fitting them to the specifications of a reductionist framework.

BEYOND BOUNDARIES OF RELEVANCE

A neurocosmologist might maintain that we will never know what's "happening" until we integrate our studies of the brain (as observer) and the universe (as the observed). Just as the lives of these two entities intertwine, so our studies of these entities must converge and reemerge as one. This means extending the search and insights from the anthropic principle to the workings of the brain itself. We need to concentrate, for example, not only on the coincidences of human life within the stream of cosmic life. (Admittedly, that aspect of the anthropic principle is critical in that we stand to discover how we found a home in a coincidentally hospitable environment such as our galaxy.) But we also need to search for evidence supporting, or disproving, the concept of an internal-external consistency. In order to relate the cosmos to the whole of our mental cosmos, we will most certainly need to reexamine many of our representations of these things.

AN INSPIRED PHYSICS

One individual who has speculated on the nature of information, matter, and the process of consciousness is the physicist-writer Fred Alan Wolf. In *Star*Wave: Mind, Consciousness, and Quantum Physics,* Wolf invites us to understand the possibilities of consciousness in terms of quantum wave functions.[8] According to Wolf, quantum wave functions "contain the potential for anything physical to appear." His concept invokes James Joyce's notion of the "probapossible," which refers to our "ideareal histories" that Joyce describes so magically in *Finnegans Wake.* Wolf's inventive literary explorations wed the worlds of mind and material, in developing a kind of holistic physics of the mind which he calls a "true humanistic psychology." The building of this new psychology rests on the foundation that the laws of mind are reflections of the physical laws of nature. The new psychophysics he envisions includes the hidden, complementary properties of matter that the ancient Greeks called *nous* (the mind or intellect) and which we currently associate with *thought* (thinking, feeling, intuiting, sensing, etc.). Although these ideas sway from mainstream physics they are nevertheless as valid and as suggestive as a Shakespearean sonnet. Perhaps, unlike any sonnet, they may ultimately prove to be as productive as any current mathematical theory describing the nature of matter minus mind and consciousness.

Metaphorms which relate quantum physics to the mind assert that "mind mirrors matter" or that brainways mirror the universe. Reflectionism (discussed in Chapter 6) suggests how this mirroring process works in relating the *virtual particles and forces of mind* to the interactive forces between elementary particles.[9] It implies that cognitive processes reflect the strong, electromagnetic, weak, and gravitational forces. According to the mirror metaphorm, neural interactions can be described in terms of the exchanged quantum of the field between two particles—where the particles represent thoughts. Do the same transformations of matter in physical space reflect similar transformations in the neurospace?[10] Whether or not these metaphorms and the speculations they yield are any closer to rendering the physics of mind cannot be determined yet.

Retrospecting on the early days of quantum physics, when Max Planck first proposed the quantum hypothesis in 1900, a neurocosmologist may tend to believe that a number of progressive thinkers considered the implications of quantum physics as a mind-nature metaphorm. Planck's hypothesis most likely impregnated many imaginations. However, few felt compelled to nurture their contemplations, preferring to leave them as personal, unpublished reflections. Trees fall in the forest unnoticed all the time.

THE ARTSCIENCE OF METAPHORMING

Discovering the whole brain-universe may require looking more closely at our works of mythology, religious art and architecture, and philosophical platforms—*in conjunction with* examining the nature of our mathematical models. As metaphorms, these works reveal the connection between our thoughts and the universe. The richer the metaphorm, the more descriptive and suggestive it is in rendering this connection.

As proponents of new physics relate the concepts of relativity and quantum physics to the workings of the mind, they are thinking as neurocosmologists about the "covert operations" of the world of matter. Some researchers believe that it is only by observing flaws in ourselves and the cosmos that we get to see how nature operates. It's been suggested that through the flawed actions of our central nervous system, for example, the hidden territory of mind can be made visible. The implication is that the mind—and universe—is only visible by the discovery of its concealed processes. This discovery can be triggered by observing some so-called impairment of nature or mistake of evolution. In principle, there are no "true mistakes." Everything created in the firmament of the physical world is probably an ongoing experiment of evolution. It may turn out that evolution is more playful than purposeful in its experimentation. In this regard, no experiment really "fails." It simply occurs and recurs until some other change of matter manifests itself for some specified amount of time and in some specified space. In this sense *evolution*—nature's vehicle of creative expression—is impartial to its creations. It is our anthropic principle and intentional language at work which conceives of nature as creating "positive and negative," "good and bad," "successful and failed" evolutionary experiments. Our minds express this duality as confidently as nature expresses this duality in our minds.

What constitutes these so-called mistakes of nature is not at issue here. The point is that through some change in the "normal behavior" of a system, some "abnormal behavior" is observed. The notion of normalcy—"soundness in mind and in body" (of the brain's neurons and the universe's nucleons alike)—is predicated on the idea that one never really notices the processes of mind when they appear to be "normal" or "typical." They simply remain invisible and average. And yet, as soon as we observe some deviation from what is assumed to be a standard pattern of activity in a given system, we suddenly notice how abnormal this system was from the start. Instantly the expression an "average" person, or an "average" star or galaxy, is understood as a misnomer. In fact, this expression should stimulate a flash flood of criticism.

EVERYONE'S A CAPTAIN
ABOARD THE "SHIP OF FOOLS"

I'll briefly steer this discussion towards the study of dark matter in particle physics. Classical physicists once thought that the behavior of "regular" matter (the matter that comprises our galaxies) was the "norm," meaning that it followed the rules of our physical laws. I would imagine some scientists, confronted with the strange properties of dark matter which makes up 90 percent of the universe, are now wondering whether the forms of matter studied thus far are not the norm but rather are "exceptions to the rule"! The whole issue of normalcy and abnormalcy accompanies all comparative studies of forms in nature—where all forms are evaluated according to our "normal and typical" location in space.

Steering this point in a similar direction on the same oceanic horizon: It may turn out that Nicolaus Copernicus, one of the parents of modern science, who was the first to *re-position* the importance of the Earth and our place in the universe by asserting our "typicalness," erred in his evaluation of where we stand, so to speak. In particular, he may have miscalculated humankind's special ability to locate a position in the cosmos that could be peculiar. There are some theorists in modern cosmology who explore this thought, considering alternative arguments for describing the unique position of human observers.

INTENTIONALLY CONFUSING ISSUES

As we investigate the processes underlying the forms of nature, we see how our imaginations dress nature—developing elaborate wardrobes. Our theories ascribe different forms to her so that nature becomes more describable and manageable. (Note the genderization.) Presumably, we can control something as uncontrollable as nature to the degree that we can actually see what it is we're trying to control. This is in contrast to our basic insecurity that something isn't manageable to the degree that we can't see it in form—standing attractively before our affectionate eyes. Seeing the nakedness of nature is like undressing her different forms to see her processes. Accordingly, "physical" experiments guided by mathematical-mechanical laws, are central to the fashioning of form and the predicting of trends in nature's fashion.

The physicist Ernst Mach (1838–1916) and the philosopher David Hume (1711–1776), among other proponents of empirical science, would surely have dismissed this idea about the illusions of nature as anomalous reasoning. They may have ordered it to live in exile in the domain of art and fiction, had it not been for the extraordinary discoveries of our quantum cosmos. Gradually physicists learned that the space-time properties of physical phenomena may have a hidden, and perhaps different, agenda and dimension than their forms alone suggest.

Quantum physics is the first *science of process* that has helped us to see through the forms of nature and the illusion of forms. These forms are not isolated things that the human mind connects together in making sense of the composite world. Subatomic particles, for example, are not

THE ROUNDABOUT DIRECTNESS OF COMMUNICATION

METAPHORM 2. A Union of Processes. *Imagine that this disc is a sphere and each line is a circle. The intersecting circles represent a union of relations between all sources and forms of information.*

"things" that build one upon the other until they form large aggregates. Rather, they are part of a *union of relations*.[11] (See Metaphorm 2.) If the processes of neurons are in any way similar to the processes of subatomic particles, then the human brain is best described as a mass of united relations.

A METAPHORM OF METAPHORMS

In relating the concept of a society to the consolidated human brain, Marvin Minsky demonstrates his mastery of the principal medium of neurocosmology—metaphorms. In his important book *The Society of Mind*, Minsky describes the brain as a self-organizing system of neuronal "agents" where no agent can speak for all the other agents. According to his view, the brain's "billion-agent society of mind" is intelligent as a whole but essentially "mindless" in its parts, or agents, that determine the whole. Minksy succinctly grounds this notion with the statement: "It doesn't matter what agents *are*; it only matters what they *do*—and what they are connected to." With this thought and his general reflections connecting minds and machines, Minsky's work explores a number of concerns of neurocosmology.[12]

In particular, his emphasis on the relationships between things— rather than their forms—is most significant. Minsky addresses this point in discussing the problem of classifying processes:

> Why are processes so hard to classify? In earlier times, we could usually judge machines and processes by how they transformed raw materials into finished products. But it makes no sense to speak of brains as though they manufacture thoughts the way factories make cars. The difference is that brains use *processes that change themselves*—and this means we cannot separate processes from the products they produce. In particular, brains make memories, which change the ways we'll subsequently think. *The principal activities of brains are making changes in themselves.* Because the whole idea of self-modifying processes is new to our experience, we cannot yet trust our commonsense judgements about such matters.[13]

A neurocosmologist is inclined to take issue that we can't assume the brain *does not* "manufacture thoughts the way factories make cars." The assembly process involving the construction of a car may very well describe—in a meaningful way—the process by which constituent elements of a thought are assembled in the central nervous system. Because we don't understand exactly what brain processes are involved in this assembly process, we have to explore as many possibilities as one's patience permits in contexturing one idea with another—in metaphorming theories. For Minsky to say that "it makes no sense" to speak of brains as though they're manufacturing plants of sorts is to proclaim that from *his perspective*—not all perspectives—this metaphorm is meaningless. And yet, from a broader perspective, this Leibnizian "thought factory" metaphorm—when taken literally and discussed in terms of neurobiological details—becomes as productive an image as any other of Minsky's "metaphorms" interpreting our neural society of mind.

Also, why should we assume that "brains use *processes that change themselves*" (double entendre aside) or that "we cannot separate processes from the products they produce"? prods a neurocosmologist. Consider this counterassumption: The same processes of integrating and relaying, splitting and sorting information are involved in neuronal signaling in spite of the changes in cell architecture. Basically these processes don't change—only the forms or structures they animate change. The "self-modifying processes" refer to the modifications of structures, not processes.

THE INTEGRATED "SOCIETY" OF OUR UNIVERSE

Since neurocosmologists are concerned with viewing brain processes in the context of cosmic processes, we need to cross-examine our contexts in which metaphorms are applied, viewed, and discussed altogether. Before rejecting any metaphorm for its error or paucity of thought, we need to interpret it in as many contexts as possible. This strategy resounds the introspections of many scientists, artists, musicians, and other creative thinkers who—in the process of working out an idea— come to a temporary dead end. So, rather than discarding the idea or work, they simply put it away for a while. More often, they find themselves reapplying what was once a perfectly useless idea to build or complete another idea that proves to be useful.

For example, we could metaphorically relate each of Minsky's ideas about the nature of our minds' "agents" to the nature of the universe's "agents," thus producing a new book of metaphorms which we might call "The Society of Universe." Here the word "society" would refer to the many-universe concept. In principle, this derivative book would no longer be the same as *The Society of Mind*, even though the "persons, places, and things" would simply have different referents. What changes —significantly—are the *concepts* that are ushered in through the change of names or terms. The play of semantics can sometimes have powerful consequences. Minsky's discussion of "wholes and parts," the "trial and error of machines," and "components and connections" between and within minds would refer to the components of the physical universe. Every comment about brain states and mental processes would describe some aspect of matter. Whatever objections one had towards his original society of "agents" metaphorm may be overruled by the importations of this new metaphorm.

CONTENDING ASSUMPTIONS

The points of contention here have to do with two things: our assumptions in general and the contexts we speak from in judging the merit of metaphorms or the hypotheses they stimulate. These points are not presented as attacks on Minsky's reasonings or his "metaphorms," which are ingenious. Rather they're presented to challenge assumptions which are the building materials of all concepts.

In describing the artscience work of neurocosmology, again, we may be wise to focus on some of Minsky's metaphorms and their assumptions because they're so concisely stated. The first point: He assumes that in order to posit how different parts of systems interact "we must first have good ideas about which aspects of a system do *not* interact—since otherwise there would be too many possibilities to consider. In other words, we have to understand *insulations* before we can comprehend *interactions*." [14]

There's something very attractive about this notion at first. However, on closer inspection, this subtractive approach doesn't seem to capture the full reality of the complex system studied, although it does speak consistently for the system of analysis used by science in the last 400 years to deal with parts in detail, relying on our specializations. What is questionable is the assumption that *all major systems are not interacting simultaneously*. It is precisely because *they do interact so thoroughly* that the problem of analysis is so troublesome in the first place. The idea of analyzing one subsystem to the exclusion of some other subsystem works in the interest of the scientific method. But it may not work in the interest of the system analyzed by this method; the system may not work this way. What happens so often is that these *separate* analyses of *separate* subsystems remain insulated or unintegrated. As a result, we tend to underestimate the integrative nature of the neuro-environment, which includes our systems of thought.

A second point of contention considers Minsky's statement: "*No complicated society would actually work if it really depended on the interactions among most of its parts. This is because any such system would be disabled by virtually any distortion, injury, or environmental fluctuation. Nor could any society evolve in the first place.*" [15]

A neurocosmologist might argue just the opposite. It is because of the interactions of all the parts of any "society" (of living cells or galaxies) that they work. And when these interactions stop or are disrupted, the society stops, or rather its growth is stunted or impaired. This period of zero growth may be characterized as a period of "insulation," of separation, of segregation—and of poverty versus prosperity. As for Minsky's concern regarding disabled systems, there is considerable research in neuropsychology which indicates that the brain—like the cosmos—is equipped with what appear to be numerous backup systems to handle "any distortion, injury, or environmental fluctuation." When these backup systems fail in their struggle to adapt, either some aspect of the system or the whole system dies.

ANTICIPATING CONTENTIONS

When it comes to analyzing the brain—especially relating the brain to its whole environment—all we have are our assumptions. Even if they are inferred from physical experiments or direct observation, the nature of the world itself is questionable. And so our conjectures, based on physics and our physical world, are themselves never beyond question or above suspicion. Simply put: There are limits to the scientific reach in reaching for answers. Specifically, there are limits to the types of questions posed in the name of scientific discourse, which the Nobel laureate in medicine Peter Medawar illuminates in his special book *The Limits of Science*.[16] Not to mislead you here, Medawar extolls the whole enterprise of science, pointing out the major accomplishments of scientists as they endeavor to reveal nature's most subtle and explicit features. As he sensitively points out, science can only handle the "how" questions, not the "why"s. It can describe how something works or is organized the way it is—applying this information brilliantly. But to learn the *why* behind the organization is to venture into the realms of metaphysics and theosophy. And in these realms of thought, interpretation and belief must suffice instead of scientific explanation.

Before leaving this subject of assumptions, it may be of interest to many to consider as a last point of contention one of the most widely debated concepts—namely, memory. There exist volumes of scientific literature on the anatomy, physiology, physics, philosophy, and art of memory: from the German psychologist Hermann Ebbinghaus's experimental work in the 1880s, which explored rote memory and learning, to current theoretical investigations into the principles of holographic memory. Consider how unsatisfying are our two most popular, complementary notions of memory, which state that (*a*) our memories are separated into different divisions, or brain regions and specialties, and *(b)* our memories are woven into a single, web-like membrane which is, presumably, of infinite dimension in the neurosphere. Both notions entertain how our memories inform our daily experiences and how our experiences become memories.

The first notion assumes that our memories are as cleanly separated as a copying machine which can collate paper at high speeds through some specialized mechanism without involving the whole system in this collating process. Not only are the sheets of paper collated but they're also numbered or bar-coded for purposes of recall. The analysis of memory is certainly simplified by this assumption. Implied here is that a short-term memory—or a memory bound for long-term storage—is somehow separated when it enters the system and when it exists. Suppose that the reason for dreams working in their free-form ways is that these phenomena are not separated at all in the system. They're unfettered by the functions of a specific neural subsystem. They're not relegated to some district—with some discrete entrance or distinct exit—possessing some terminal room or port. Rather, a dream or memory may form from the collective states of the system. The energy directed towards its creation or recording is shared, absorbed, and emitted by the whole system. We know what this thought implies about minds and bodies, but what does it imply about stars and stellar bodies? It suggests to me that *what is embedded in our brains in the way of memory is embedded in the universe in the way of energy.* Subatomic matter, then, has a "history" that neither time nor space can erase. It's a history that may be called a truly eternal, energetic memory—telling of particle interactions.

The second notion, which considers the hologram metaphorm, gives us the more interesting possibility that expands beyond all mechanistic interpretations. However, it also leaves us without any clear way of testing its hypothesis—which posits that the mind's boundless recordings of experience are encoded in an equally unbounded brain. This implies that the brain is capable of recombining these encoded experiences in such a way that, at every node of storage in any given neuron, there exists the possibility of all memories. When you consider that any encoded experience can consist of a very high density of information, the idea of isolating and analyzing a single memory seems insurmountable. This conclusion stands steadfast, especially when you think of how one memory may be configured with trillions of other memories of trillions of sense perceptions.

One escape route from this thought is the idea that only select experiences are encoded as memories. That certainly simplifies things. This way the brain reserves space for the more substantial matters. Seconds after entertaining this thought, you experience that peculiar uneasiness that seems to accompany all specious thinking: "If this were the case," you think, "if the issue of memory were that fundamental and clear, why haven't our best neuroscientists discovered the filtering 'mechanism' that controls the 'intake,' or input, of our perceptions and experiences?"[17] "Is this mechanism intrinsic to the system that's doing the filtering—meaning, the human brain?" you ask yourself. "Or does the filtering process have something to do with the nature of information itself—meaning, *before* it's physically encoded? Or *before* something is physically filtered?" you wonder. Perhaps the principle by which this natural filter works is based on some undefined property of information —some property which can only be learned through a study of the morphology of processes.

IN MEMORY OF OUR WONDERMENT
FOR "THE BASICS"

Regardless of temperament, neurocosmologists allow themselves to wonder about things which are usually looked upon blindly—if looked upon at all—such as our most mundane acts of living. These individuals are stimulated to question, "How is it that some of our life experiences are encoded into memory and, presumably, others are not? Which experiences are inducted into the hall of memories and what are the criteria for induction?" These questions are as difficult to hurdle as the questions pertaining to *the means of induction*. Some neural mechanism is selecting or sorting out the best and worst moments of our lives. What are your thoughts?

This subject—like neurocosmology itself—is like an invisible, infinitely large vase which our imaginations see and draw inspiration from but which our logic finds untouchable. I mean "the information is all there," only we don't know where "there" is. (Whether it's in us or outside us is a matter of perspective.) Nor do we know how "information" really *works* in all instances of communication and in all instances of energy conversion. When information is understood only in terms of its forms or applications, then one should suspect that the process of information is too narrowly defined.

As I discussed earlier, *process is information exploring ways of propagating itself, of manifesting and expressing and discovering itself*. In translating any process into some symbolic form for scientific analysis or artistic expression, it's easy to be deceived by what appears to be *the form of the process*. We forget that the form we create to represent a certain process is purely symbolic. And its symbolism can change. This implies that any form can be used to represent either brain or cosmic processes.

In the neurosphere, process is everything. It's the medium in which our experiences, such as thoughts, memories, and reveries, are organized. As you create in your mind, you paint, sculpt, and draw with information. It is this process of forming—not the forms themselves—which is the essence of thinking and feeling. The mind's process of shaping matter and information (or energy) is shared by the universe. And, as fascinated as we are by the shapes or forms of what we create, or what the cosmos creates, we need to explore the ways in which these forms emerge. The only thing that may have come from some perfect vacuum of absolute, unstable "nothing" was the primordial matter of fundamental particles comprising the hot primeval gas from which the universe and we formed. Although, like human creations, I suspect there's always some precedence of the unprecedented.

We may discover that our concepts of Big Bang and singularity theories, which describe how the universe first formed from a point of zero-matter that was infinitely dense and hot, equally describe the "Big Bang of consciousness" that occurred in the human mind however long ago. This first consciousness—like singularity—introduced the phenomenon of learning to humanity, a phenomenon that is synonymous with the act of creation. What I would give to see *how* our nervous systems arose from the stars' nervous energies! Or *how* our metaphorming minds invent illusions which we symbolically call nature's illusions. Whether it's Niels Bohr's "principle of correspondence," which relates the continuities of our visible world to the discontinuities of the invisible atomic world, or whether it's Planck's constant, h, upon which the lives of atoms are determined, elements of illusion are at work inside us and in the universe. As we work to dispel these illusions, we decipher the mind—and nature. Solidity, reliability, order, and related concepts are understood as ideals. We create metaphorms from all that our minds "hold in store" for us.

Will we ever reach that mirage-like vanishing point of complete knowledge? Will this knowledge of mind represent the whole of nature as experienced through art or described through mathematical physics? And will everyone feel fulfilled by these experiences and descriptions? I doubt it. The idea that we can create or arrive at a platform in nature from which we can see all is as much an illusion as the idea of a definitive theory, or explanatory model, that can reveal the whole story of the brain-universe relation. This story is an ongoing invention of nature which will always remain incomplete—to be followed by endless sequels. That is, if you accept the concept that nature itself grows—its laws evolve—like the forms of life and inanimate matter it creates.

Each story, no matter how tight its mesh or how wide its sweep, will miss some key details of characters. Whether it's Newtonian mechanics, which understands the universe as being determinant and predictable, or whether it's quantum mechanics and relativity, which knows the observer and the observed as being indeterminant and unpredictable, some aspects of our brain processes are reflected in these theories just as they are reflected in nature. The collection of all theories reveals our true nature and mental evolution. An evolution without end points. A nature without definite proofs. Only our abstract concepts are subject to experimental proofs, not nature itself.

The sciences, with or without the support of mathematics, are only capable of gleaning some diaphanous features of the phenomena they choose to study. Models born from the scientific method are not hyper-realistic portraits of physical reality. Instead, they're abstract, symbolic representations, like musical notations interpreting some imagined or felt score. In short, they're not the concerto; they're only representations of the concerto. I trust everyone has heard this in one way or another, in one context or another. Somehow we forget. Or we learn to forget, until nature mortifies and reminds us again.

I want to be careful not to leave you with the impression that we don't know anything just because we live with partial knowledge. Clearly, this is not the case. We know a lot about a lot of things—including the transience of our disciplines and methodologies. We just don't *understand* how all these things fit together coherently to form the 11th-dimensional or *n*th-dimensional model of the processes of nature. One fact remains resolutely fixed: We know things and believe in the things we think we know. This fact is validated daily. Everyone would prefer to have their health checked by a physician who knows medicine rather than by a poet who hasn't even an inkling of human anatomy and physiology. (Ideally, I'd like to be looked after by a doctor who's a poet or writer or artist as well—but how many William Carlos Williamses are there!) Everyone would prefer to fly in a plane with a pilot who knows how to fly even though s/he may not understand the first thing about aerodynamics, meteorology, or any other information relevant to how it is possible that planes fly.

So much for the idea that we don't know anything at all, although it is almost certain that we will never learn how *all the pieces* of our customized knowledge make sense of the world. If the metaphorm we use to describe nature-reality-life is a four-dimensional puzzle, the fourth dimension (or time) implies that there are pieces to this puzzle that can only be found in the future. How far into the future, or what form this information will take, no one knows. This thought implies that the puzzle's pieces do not exist in the present but rather are scattered throughout time. At no point can we put the pieces of knowledge together to complete this puzzle, unless we take some shortcuts—through our intuitions

—by declaring that such concepts as the future are so many abstractions. And, if we think everything we need to know we already know or possess, then we can "puzzle no more"—putting off for a thousand years what we can do today.

We think as neurocosmologists when we relate a fundamental insight from quantum physics which maintains that the universe is one extraordinarily intricate energy pattern of sorts. The wholes and parts of this pattern are linked together like brain cell-assemblies, all of which work in concert in the process of creation. The act of linking one part of this pattern to another in the cosmos is a processmorph of the act of combining information in the human brain.

A more poetic—and precise—way of describing this linking process was supplied by Douglas Hofstadter in his eloquent description of the part-wholeness and whole-partness of a fugue. "Fugues," writes Hofstadter, "have that interesting property, that each of their voices is a piece of music in itself; and thus a fugue might be thought of as a collection of several distinct pieces of music, all based on one single theme, and all played simultaneously. And it is up to the listener (or his subconscious) to decide whether it should be perceived as a unit, or as a collection of independent parts, all of which harmonize."[18]

By exercising your mind's propensity for metaphorming—for "fugue-listening"—you connect unsuspected relationships between these acts and the materials of the mind and universe. Depending on how receptive you are to the resultant metaphorms, you will find yourself either inquiring further or quarreling over whether these entities, brain and cosmos, represent two aspects of the same pattern of energy—the same content—though existing in different forms. My guess is that you'd launch further inquiries and enjoy exploring them, listening for their fugues.

13
Neurocosmologists

Whoever decodes the brain decodes the universe, and vice versa," nature beckons, as it tips us: "I'm woven in you and my weave is coded."

Just up ahead we'll explore some ways of decoding the brain, as we unpack the meanings of the premise: The processes by which the universe "works" work in us. Similarly, the processes by which the brain "works" work in all of our creations. Understanding ourselves involves knowing the processmorphic relations that unite the universe, the brain, and our creations. How we interpret the interconnectedness of these relations will determine our future—whether we extinguish ourselves from lack of foresight or flourish as an enlightened species.

The challenge to all neurocosmologists lies in showing the connections between all of nature's processes and information. This involves exploring the ways in which process and information are inherently symbolic and abstract—but physical nonetheless. Every person acts as a neurocosmologist when she examines the broadest manifestations of "process." A person sees the world as a neurocosmologist each time she reflects on the cosmology of her own mind in relation to the physical universe. Each time you search your mind to discover how you think or create—questioning its process phenomena—you become an explorer who ventures in the themes of neurocosmology.

SELF-KNOWLEDGE ENDANGERING MINDS THAT KNOW NO LIMITS

You feel the future as a neurocosmologist when you realize that our cultures are capable of *moving beyond all boundaries* of communication (language, expression, and representation). This act of "moving beyond" signals the redefinition of "boundary"—but not without awareness of the consequences of our boundless actions. Removing boundaries isn't the answer to our mental growth and survival. It's only the first step towards recognizing the union of processes that compose nature.

The processes by which our visual, musical, and mathematical languages work find their expression in one another. Though many would regard these symbolic languages as being as separate and distinct as their forms indicate, each is embedded in the other. Moreover, they may be understood as counterparts to one another—along with other languages of nature, such as those based on our other sense perceptions.

Thus quantum physics is expressed in all aspects of the arts and music, just as visual and musical processes are expressed in the physics of quanta—whether or not we physically see or hear them. It is the *ensemble of these languages and processes* which resonates with nature, not the cacophony created from each of these languages speaking separately and simultaneously—each determined to be heard above the din. It's not that nature doesn't have or enjoy its moments of cacophony. (Note that the "Big Bang" may have been such a moment.) It's just that it seems to prefer balancing this dissonance with the more rhythmic orders of things fitting together rather than only clashing. Perhaps, at some other point, this situation will reverse itself. Or so runs the muse of a neurocosmologist.

FUSING FISSIONING MINDS

There are at least two serious consequences we must face when mixing disciplines, media, messages, and sensibilities as neurocosmologists. First, we can easily lose sight of the unique qualities of each discipline or medium through which we become aware of the limits of our languages to begin with. We need to meld without losing our basic awareness of what we're melding. This way we break our mind barriers without breaking ourselves.

Second, as our technological societies race to transgress barriers—to discover our relationship with nature—we must avoid developing our "technological competence" at the expense of our commonsense "thought competence." In the former, the human aspects of machines can seem far more interesting to study than human beings. There's a pretty fundamental reason for this, however simplistic it sounds. People would rather seem smart than inept. We can describe in brilliant detail nearly every aspect of a machine's components or functional anatomy. These descriptions in turn replenish our confidence in the powers of engineering and science to explore and build upon the natural world. By contrast, the "mechanics" or machine aspects of human beings and societies are so staggeringly messy that they're always in a state of re-evaluation and revision. Applying the principles of scientific thought and engineering in describing the subtleties of human behavior, for example, can uproot our confidence in these principles. Consequently, in the interest of solace, confidence, and practicality, we prefer to fascinate over the things we engineer or create rather than their relationship to the engineer's creative processes. In the latter, there's no guarantee of securing any smart, lasting descriptions.

JARRING KNOWLEDGE

You act as a neurocosmologist when you try to integrate as many disciplines as you deem necessary in searching volatile questions such as: *How* is it that some of our minds' creations are as beautiful as nature's most seemingly inspired works—and yet others are as disastrous as any of nature's worst reckonings? *How* do different regions, subsystems, and cells of the brain produce different kinds of cerebration or feelings? *How* do we create poetry and art in the same neurospace—or place of mind —that we invent advanced weapons? *How* can we think with charged and wide aspirations about reaching other people on other planets when we can't "reach" through education more than three-quarters of the population on this planet? *How* is it mentally possible to decide that some forms of inhumanity are permissible and not others? *Why* do we assume that the impulse to fulfill our most brutish acts is closer to the ways of the natural world than our impulse to love? Why do we look for the source of our troubles inside ourselves or between peoples and nations and assume that these troubles aren't echoed outside ourselves—in the "troubled cosmos"? Why do we assume that the universe is more perfect than we are? Isn't perfection a transient phenomenon, like order and chaos?

The "why" questions always seem to be answered instantly by theologians with a rush of "becauses . . ." By contrast, the "how" questions —such as how did we evolve and how do we think—have scientists in labor for hundreds of years before they give birth to coherent conceptualizations. *Nurturing both sets of questions to full maturity is the parenthood of responsible education.* To ignore one set because the questions are unmanageable within the framework of a given system of analysis is just the sort of mistake that will bury us in ignorance.

Perhaps none of these questions can ever be answered with any certitude, no matter how thoroughly they're studied or astutely posed. Like death, we may have to "experience" these answers "after the fact." And yet, we sense that such an experience is, literally speaking, *out of the question.* If perfect knowledge of life and nature is predicated on understanding the before and afterward of death, then we may be resigned to living with incomplete knowledge. Glimpsing the reality of this experience may be one of those Catch-22s for which there's no way around; you can never get the experience you need to get the experience. And so we're left to narrowly define the worlds of lifeful and lifeless matter, with our imaginations living somewhere within the borders of our narrow definitions.

We sense it's pointless to search for a large enough answer to cover a jar full of these and other questions about human nature. A neurocosmologist realizes that there is no jar large enough to contain even the smallest of these queries. The idea of "containing" such questions overlooks the fact that you can always question a question. This act of questioning enlarges all answers in some potentially infinite progression. So our inquiries are forever subject to the expansive nature of our questions. One soon concludes that we don't need more jars, or even larger ones. Instead, we need to rethink the notion of bottling and jarring our knowledge—in containing information.

THERE IS NO MAIN CONTEXT. THERE ARE ONLY MANY CONTEXTS . . . OF SEEING

You think as a neurocosmologist when you question the segregation of professions, choosing not to draw lines of demarcation between those who are artists and those who are scientists; those who are poets and those who are philosophers; those who are composers and musicians and those who are mathematicians. "According to whose theory of art? Based on what criteria of science? Enforced by whom? Representing whose interests?" a neurocosmologist probes.

Science that explores the whole of our lives, bodies, environments, and cultures is part of the art of neurocosmology. Art that interprets our systems of ideas and disciplines as they frame our world is part of the science of neurocosmology. To the extent that we see the world *without having to emphasize or de-emphasize* boundaries of thought as expressed by the arts and sciences, we engage neurocosmology. This boundless thinking is a prerequisite for studying the brain-universe. It helps us freely interpret a book on neurology or neurophysiology in the context of astronomy, astrophysics, and cosmology. Scores of new associations result from these sorts of creative interpretations—all of which spur us to wonder whether studies of the brain offer insights that are relevant to understanding the cosmos.

SEEING ONE THING AS SOMETHING OTHER THAN WHAT YOU SEE

For those who are unschooled in scientific literature or discourse, reading the abstracts of journals in neuroscience alone is like riding rapids from converging estuaries of information. There are studies on virtually every type of technological invention and concept—from the computer-aided analyses of neural architecture to experiments on human memory-recognition systems. A large percentage of this research is so sophisticated and specialized that the thought of grafting any of it to material outside the specializations is dizzying. This is where we need generalists and laypeople the most.

How, for example, can we discuss the relationship between the neuronal connections or pathways of the central amygdaloid nucleus (CAm) in a human brain and the "connections" in a star's radiation or convection zone? (See Metaphorms 3 and 4 in Chapter 3, which show the Sun's anatomy.) It seems pointless to call the energy transport mechanism in stars the "CAm pathway" unless one intends aesthetic pleasure from this poetic gesture. Clearly, stars don't have nerves; they're not responsive and sensitive like mammalian sympathetic ganglia. And brains don't have solar flares and winds streaming particulate matter into deep space at hundreds of kilometers per second.

Why, then, see the brain as a star? Why implicitly compare the human gene cycle and the hydrogen/helium-burning cycle that sustains star life—as if the stars of a galaxy undergo mitosis like somatic cells? DNA replication and protein synthesis are not even remotely similar to element and radiation synthesis, at least not in the way science knows these two systems structurally. What are the advantages and purposes of these analogies in neurocosmology?

One tentative answer to these sorts of questions is they offer *insights* into the workings of the universe and *foresights* into the mind's workings. Another tentative answer: *Metaphorms* provide unlimited sources of information and material resources for forming hypotheses.

One of the great nineteenth-century physicists, Ludwig Boltzmann, once wrote: "Laws of thought have evolved according to the same laws of evolution as the optical apparatus of the eye, the acoustic machinery of the ear and the pumping device of the heart." He cautioned: "We must not aspire to derive nature from our concepts, but must adapt the latter to the former."[1] A neurocosmologist might present the corollary to this thought: *In deriving our concepts from nature, we discover the shortcomings of our concepts.*

The ascent of matter (from chaotic behavior to calm, orderly repose) is a metaphorm for the ascent of the brain (from its reptilian state to its present somewhat refined, cerebral state). In creating this metaphorm, we follow Ludwig Boltzmann's advice and derive our concept of the human brain's evolution from some natural phenomenon—any phenomenon. However, what happens when we reverse this metaphorming process and model nature after our knowledge of the brain? Does this gesture of "deriving nature from our concepts" create any fewer problems in terms of misconceptions than the reverse situation? I think not. In fact, I think both ways of hypothesizing through metaphorms reveal the shortcomings of our theories, concepts, and models of nature. These shortcomings include generating an overabundance of good metaphorms and trying to decide through trial and error which ones lead to sublime propositions and which ones lean towards weak presuppositions. In general, looking from the angle of conventional science, these sorts of concerns and decisions have a way of knocking the wind out of the sails of metaphorms.

If we have in mind "starting points of genesis" and "finishing points of demise," then nature seems to move one way and not another— towards order from chaos. But it certainly doesn't appear to move both ways at once with time as a reversible process. At least not as appearances go. But couldn't it just be that this forward, linear "moving interpretation" is in error like so many of the assumptions it is based upon: simplicity, continuity, linear time, progression, orderliness? By analogy, ocean waves can be sculpted with violent winds that have the feel of furious chaos and yet, just below this surface, the waters are orders of magnitude calmer with more orderly currents. The chaos that confronts our eyes— on the outside—is probably only half the illusion. The other half is the invisible chaos behind the orderly operations of our visual system.

ROUTES OF EVOLUTION

According to the principles of neurocosmology, what we build in our minds and realize in full form reflects what nature has built into the universe as a mechanism of growth. Imagine evolution as a two- or four-lane highway, where the vehicles (representing individual acts or events of nature) travel along at all hours, at various speeds, under all conditions, and in both directions. The vehicles constantly enter and exit this freeway, which is always being repaired. (Note that you never really know the terminal points of the passengers in each vehicle, no more than you know anything about their lives before they entered this cosmic highway.)

This simplistic analogy favors the idea that nature has progressed from a no-roads-in-the-wilderness form to an ever-evolving network of transportation forms that may very well end without roads and highways also. It counters the notion that evolution travels in one direction, fast forward and upward—leaving the past behind with its many exhausted vehicles. And evolution continues along all these routes until some tangible law of nature contradicts its motion, or breaks down, changing nature's course and means of action.

DETOURS AND ALTERNATE ROUTES

Gradually we're discovering that within *order* lies some hidden evolutionary process, as in the move towards complexity. As things grow, they become more complex. The act of growing is understood as a fairly chaotic act, filled with unexpected breaks, potholes, and potential catastrophes. Growth is not some smoothly flowing process—some series of uninterrupted actions or events—that can be described by a few smug differential equations. The process of chaos includes both evolutionary progression and digression. More poignant, evolution is neither progressive nor digressive but both simultaneously, which means it remains restlessly still like a standing wave with its constant tensional motion. A movement of energies contained between two end points that meld into one. Evolution exists at this one point which comprises the whole circle of the eternal present.

You might say evolution just *is*; neither forward nor backward, neither upward nor downward—but *mindward*. Meaning it's neither here nor there in the neurosphere; time and space are another matter whose conception is contingent upon another sense of time. Time, in neurospace, is not a form of control or device for measurement. It's an abstract means of measuring our growth from "here" (which can be anywhere at anytime) to "there" (which is the point from which we started, even though we give it a new name—as in *end*). Yet without time, we assume there's no motion. And without motion, we assume there's no movement of thought—"At the still point of the turning world," as T. S. Eliot enlightened us with this paradoxical phrase from his *Four Quartets*.

A LASTING REFLECTION

T. S. Eliot's reflection on "the present"—with complementary worlds flowing together and apart in perpetual motion, in endless stillness—recalls the Stoic's sense of the "continuum." The continuum, or present, was conceived as oscillating between "neither movement from nor towards." The mechanisms of oscillation were the complementary forces of nature. Underlying this concept is the idea of "tonike kinesis" (or tensional motion). This essential *tension* is supposedly created by the connection between pairs of opposites.

The purpose of this interlude is not to debate the Stoics' ideas on the cause and effect of "essential tension." These notes simply reintroduce the idea of pairs of opposites, paradoxes and riddles, that has occupied a very special place in our imaginations and in nature's repertoire of illusion (the type of illusion I discussed in connection with the concept of reflectionism). You might say that these paradoxes are largely responsible for fostering the original conflict between imagination and the realities of what we imagine—as our imaginings ultimately *define* the reality we live and die by.

Perhaps even before the pre-Socratic philosopher Heraclitus's book *On Nature*, with its poetic discourse "On the Universe," the idea that polarity is a part of unity had already been explored. Also the idea that order is grown from chaos and returns to chaos. Both ideas have been with us for as long as we've been human beings. These and other complementary concepts are part of our existential being and nonbeing, or so the novelist and playwright Jean-Paul Sartre would comment. They have directly shaped modern cosmology and have influenced our thoughts about our minds' role (and place) in the study of evolution, structure, and nature of the universe.

REFLECTING ON FOREVER

Finally, we can't afford to overlook the fact that evolution may be nature's greatest—and simplest—topological trick. It creates some intrinsically basic form of matter (meaning that this matter appears to be simple at first only to evolve into vastly complex forms) and it provides this matter with at least two complementary paths of action on one plane. One path leads upward or forward and the other leads downward or backward. These two actions are set in some symmetrical relation to each other. This means that as you travel on the forward path, you're also traveling on the backward path at the same rate. This may suggest to you that as you progress you simultaneously digress. Nature cleverly rolls the plane that these two paths exist on into a tube. As you can see, the two opposite paths have come full circle to become one path on one plane—where the terms *forward* and *backward* are meaningless on this plane of reality.

IMAGINING THE UNQUESTIONABLE

One might suspect that even our most successful theories in physics describe only the present state of the universe and that in a few billion years or so, these states will be almost unrecognizable in relation to our current descriptions. A neurocosmologist might contest that our current physics can provide answers that future physics will confirm as consistently as a constant of nature.

This sort of projective thinking can paralyze you at first from doing anything. It implies that no physical law of nature is exempt from change: from physical constants to the forms of matter that are organized according to these laws and constants. Nor is any physical law exempt from the changes in our taxing critical minds. The transient nature of our minds mostly reflects these changes. This idea is not particularly comforting to those of us searching for signs of stability in terms of either intelligence or material. I mean, no one wants to think about how insignificant they are in the timeless, nonstop flight schedule of evolution— where nature takes off with its passengers (matter and anti-matter, physical constants, etc.) from one airstrip in the universe (a singularity) only to land many billions of years later at another airstrip where it quickly—or slowly—refuels. At each destination, new animate and inanimate passengers of a completely *new nature* embark and take off again into the airspace of a *new universe*. With each ending or arrival, a new beginning follows. And with each beginning, there's often a surplus of hope and promise held in the minds of its new passengers.

One answer to the question "What are we in the scheme of billions of years?" has the sort of simplistic answer that I can live with soundly. "We're the stuff of photons and other subatomic matter which compose the most basic properties of *information*." This is a positive reflection: that information, which can take the form of electrochemical and biomagnetic fields that sustain our lives, can be a unique energy pattern as distinct as a fingerprint. This dynamic pattern is life in action.

A less positive response would render whatever we do in our lives utterly pointless and empty—in spite of what we tell ourselves to sustain our shortsighted, local aspirations. I mean, who *really* wants to consider how everything humankind has created or will ever create—in time—will return to some indescribable statelessness where atoms and stellar dust return to their initial conception. Even the rarest of human creations, those priceless artifacts that have been encapsulated in the tombs of our greatest museums like time-honored treasures sealed in a vacuum, will eventually perish within the framework of physical time or space. Count on it.

QUESTIONING THE IMAGINABLE

When we avoid thinking about such concepts, we refrain from feeling our lives. It's as if these questions put to nature are immediately put to death—as the mind incapacitates itself, retracting from feeling its own mortality. Is it more steeply challenging to figure out how all the heavy hydrogen in space was produced within 15 minutes of the Big Bang than it is to figure out what this information in particle physics has to do with human life directly? Or what *any* information on the universe has to do with our lives?

Neurocosmology is what every human being practices either with sobering expertise and scholarship or with exuberant innocence when they search the link between information pertaining to the brain and that pertaining to the cosmos. Some neurocosmologists prefer to focus their questions on the relationship between our lives and what we *make of* our lives—meaning through the things we make. Examining the nature of our social systems and societal codes and moral laws may be their point of entry in reflecting on neurocosmology. Others prefer to picture the whole brain-universe connection—where the act of seeing this connection is to live with some fragile peace of mind.

A PARADOX OF REALITY: LIVING IN NATURE'S WAY

The two-part idea discussed in neurocosmology—that we live in nature's way and stand in the way of nature—is both auspicious and ominous. The double meaning of these expressions reveals the paradox of this reality—a paradox that is a parody of the truth.

Somehow things that rub us wrong, we're inclined to rub. Things that we know are destructive to ourselves we hasten to create: from the stress of prejudice to the pressures of control through coercion and war. This impulse to create without concern for consequences is one of the clearest traits we share with nature. Do hurricanes or holocausts much care which path of destruction they take in their course of action? Do earthquakes and riots much care which land masses they take apart? Do quasars, the nuclei of forming galaxies, care about the star masses they destroy in the course of their violent formation? *No way!* (Or no way that we know or could know?) Undirected creativity—without foresight and consequence in mind (or conscience)—can act as these forces of nature. The one trait that distinguishes us from these natural forces is our ability to guide our creativity towards constructive ends. We're gifted in thinking in advance about the consequences of our actions.

With forethought the poetic abstraction, which gently claims "We are *in nature's way*," has enormous consequence to our physical well-being and the being of all life, including the ecology of our planet itself. The importance of this abstraction is that it is at once literal and symbolic or metaphorical, like the term *"particle* physicist." In one sense, we *are* nature, just as elementary particles and atoms compose the physicist. In another sense, we are *standing in the way of our nature* through our self-destructive tendencies. One of these tendencies involves envisioning ourselves as living isolated from one another when, in fact, we may be as interconnected as particles which are always forming new relationships. As a result, we live the ultimate nightmare of nature, seeing only our fragmented worlds, whose collective meaning is sometimes so slight as to be nonexistent.

WE ARE ALL THAT WE CREATE?

But couldn't it just be possible that our theories of the universe are theories about ourselves? That we have to study everything about ourselves and our creations—*in conjunction* with studying everything about the cosmos—to actually see ourselves? This idea implies that *all theories,* to various degrees, describe aspects of human brain processes.

Neurocosmologists will continue to meld experimental data and metaphorms in exploring the *sparks* (nucleons) *and roots* (neurons) *of consciousness.* Some may concentrate on the means by which we engage this exploration. Others may search the relationships between the tools of technology and thought—or processmorphs. The cogency of their interpretations can only be determined by the degree to which they make us aware of the web-like natural world. We have yet to learn that there is no artificial world—as in artificial intelligence, processes, machines, information, etc. There is only the natural world which accommodates our abstract concepts of "artificial, virtual realities" as they relate to some of our technological creations. A computer or particle accelerator is no less natural than an orchid or bird of paradise. We need to be responsible for both types of "plants," as they affect the ecology of our nature.

14
Engaging
Neurocosmology—
The *Cerebrarium*

For millennia, each picture or model of the microcosm
and macrocosm has rendered a different aspect of our connection with
particulate matter and celestial systems. Some models name humankind
as the crossroad between these cosmoses. They speak of our cosmic
connection in poetic verse, as in the prophetic Indian Vedic scriptures.
Others associate humankind with celestial life in astrological terms, without testable evidence. Theoretical studies in modern cosmology sing of
this connection in verses of mathematical equations, interlacing insights
from astrophysics and astronomical observations, although more recently these observations have not been made from the bridge of the
ship of science but from the submerged navigational system of mathe-

matical physics. The results of this research in physics are generally debated as matter-of-probability (as opposed to matter-of-fact). It is toward the interpretation of these results, and those of brain science, that we engage in neurocosmology. As this book demonstrates, at the center of these interpretations is metaphoric art.

A FUNDAMENTALLY SIMPLE SEARCH WITH A COMPLEX HIDDEN AGENDA

However simple the search of neurocosmology seems, it can be further simplified. The simpler and clearer, the easier its application in the broadest context; that is, the context it always exists in. With all due respect to Einstein's cautionary note—"Everything should be made as simple as possible but not simpler"—consider the corollary thought: We *can't* make things too simple! (The simpler we make a thing or theory, the more time and energy we spend explaining or interpreting what little we said—through our numbers, words, images, and other symbolic abstractions.) To make something "simpler" is the art of thought. To make something "perfectly simple" is nature's art, which cannot be expressed by any single thought representation or realization—no matter how simple and brilliant this representation is!

I'll rescue these generalizations with an example. Imagine an old, beaten-up color photograph of a worn hammer resting on an anvil. Buttressing this image is a black-and-white photograph of a threatening crowd of protesters bearing signs declaring "Equal Rights." You would probably see the contents of the two images as being related. These suggestive images are enough to shape a metaphorm in your mind—a metaphorm which connects the processes of action that a hammer and crowd share. You wouldn't need to see a motion picture of a sweaty person, with arms bulging, pounding away on an anvil at some molten blade pulled from a smoking furnace in order to sense the action of forming, shaping, declaring, etc. You wouldn't need to hear the sound of this moving, thunderous image in order to connect the process of unrest that is conveyed by the still image of hysterical people. Although the motion picture may be more psychologically impressive or emotionally vivid, the feelings conveyed by the simple juxtaposition of still images suffice in communicating the similarity of their processes.

To picture the processmorphic relation between the hammer and crowd, all you really need are these things or symbolic images of these things. To be told how to connect these images or how to interpret their symbolism would cut into your experience of their meaning. Moreover, it could limit your potential associations. Art rescues our experiences of meaning from becoming mere lessons in the science of illustration and simplification, by exploring the hidden agenda in all forms of matter and metaphorms. What this means in terms of our example is that the contents of these images may not be what is shown in the way of their individual forms but rather in what is suggested to you by their juxtaposition. Part of art's hidden agenda is the invitation to discover the process by which images or words or numbers suggest how to interpret the *agenda of meaning*.

To extend this last thought: Rather than trying to simplify or spell out the messages in my brain-universe metaphorms, enough suggestions have been presented to stimulate your own ideas on the subject.

THE CEREBRARIUM

One course of inspiration for generating brain-universe metaphorms is a work called the *Cerebrarium* (Metaphorm 1). This theoretical multimedia work may be designed to interpret brain processes. It may be conceived as an artificial-intelligence (or AI) system that has the ability to grow with you. One might envision it as a computer-driven hypothesis maker that is driven by your ability to hypothesize. It may assist you. And as you're learning, it may learn from you. What you may lack in organizational skills, it may more than compensate for you. What it may lack in rapid leaps of curious connections, you may compensate for it. In this sense, it might possess the capacity to evolve, to self-organize and to integrate information—like you. It may engage these directives, transforming the most prosaic thoughts and common materials into powerful metaphorms—ones that teach us about our special abilities to shape this world with the care of a Japanese rock garden rather than a ghetto. It is not some sort of passive cinema that you sit in while it performs for you. It's a *theater that may exist in you*. Its media technology is meant to stimulate your own cinema of imagination such that you conceive of relationships between different forms of matter and energy and information—without interpreting these relationships for you.

The *Cerebrarium* may embody the "bootstrap philosophy" which stresses self-consistency. Unlike this philosophy which accepts no fundamental building blocks of matter in its descriptions of the universe, this metaphorm accepts *all forms of matter* as fundamental blocks for building metaphorms with which to hypothesize about the nature of the universe or to simply experience nature in new ways.

METAPHORM 1. Cerebrarium—*Interpreting brain/mind processes. In this hypothetical environment, every conceivable source and form of information on the brain and physical universe are interrelated through two means, each of which advances the other in real-time: media technology and human imagination. Basically, this "learning arena" learns as you learn through metaphorming. This model is only one interpretation of such an enriched learning environment. It can be designed simply or elaborately, including an empty room with blank walls that are "touch-sensitive" and "smart." Assume the technology is both possible and immaterial, or literally invisible. (Courtesy of the Denver Art Museum, Denver, Colorado.)*

CONCEIVING THROUGH ART
AND CONVEYING THROUGH METAPHORMS

As the system of conveying ideas in the *Cerebrarium* relies primarily on metaphorms, few people should have difficulty participating with or experiencing the model. The processes of metaphorming and contexturing materials or information seem to be second nature to everyone. These processes are probably as much a part of our minds as they are a part of our genetic constitution. Metaphorming seems to be one of the few cognitive givens in the human nervous system: The capacity to learn, to connect, to create—to engage in unconditioned thinking called creative seeing.

This is the way of metaphorming. Where the meaning of information is contingent on context, metaphorms are contingent on meaning. The *Cerebrarium* explores both contingencies with us, not for us.

The premise behind this model is simple. In order to communicate more meaningfully we have to learn to see in more meaningful ways. We may begin by conceiving how metaphorms are enriched with realism. The model helps us focus on the thinking that goes into and emerges from the most advanced technology that is doing the thinking —namely, you.

Ideally, the *Cerebrarium* may one day work with the magical fluency of Gene Roddenberrry's "holodeck" in *Star Trek: The Next Generation,* or Ray Bradbury's "media room" in *The Illustrated Man,* or Arthur C. Clarke's "Hal" derivative which piloted our imaginations in *2010.* For the time being, the *Cerebrarium* needn't be as miraculous as the devices conceptualized by these writers. It also needn't be as rich in entertainment as EPCOT (Experimental Prototype Community of Tomorrow) at *Disneyworld* or as physically engaging as San Francisco's *Exploratorium.* It doesn't have to employ even low technology. Its sophistication can rely entirely upon our creative minds without borders. The only technological resources we need in hand are handed to us by our active imaginations.

The model proposed, then, is only as inventive as an informed mind that sees alternatives where others recognize none. A mind that remains sensitive to any environment experienced can inform itself. This is the central process by which we operate. It's also the general principle by which the *Cerebrarium* operates.

AN ARENA FOR CONJECTURING

The point of this *psi-phi* fantasy is not to concoct another list of "things for the mass of humanity to do together in order to grow together." Anyone can produce a proposal for learning programs based on wishful thinking and ideal artificial intelligence. The *Cerebrarium* model is more than a demonstration of how every human being is a natural *metaphormer* and how the *process of metaphorming* connects the whole of humanity in our pursuit of meaning and pleasure. More important, this model—or one more conceptually evolved—has the potential of stimulating insights from these metaphorms towards hypotheses that can be explored further in both physical and thought experiments.

Every quest of brain scientists, for example—and all that this field of research encompasses—is seen integrated with the quests of theoretical physicists and cosmologists to understand the universe. Through the metaphorming process, these streams of information and quests are not only made accessible, they're reduced to their essences. You see the individual acts of each of these "performing" sciences in a collective sense. The art is the act of seeing and experiencing their performances. Everyone has the opportunity to be in on the act, as a performer.

To weight these light descriptions of this model, imagine the following scenario projected on the sail-like screens of the *Cerebrarium* that are constantly moving and reshaping themselves (Metaphorm 2a). A dynamic image of a black balloon swells before your eyes on one screen, as anomalous images and sounds are barely distinguishable in your peripheral attention. The balloon has all sorts of patterns as small as dots, or nerve cells, marked in white on its surface. At first, the patterns resemble nothing but ordinary dots. As these symbolic patterns grow—growing evenly in relation to one another, with the expansion of the balloon—the dots begin to resemble galaxies. At one point captions appear labeling the patterns as nerve cells. At another point, captions appear labeling the same patterns as galaxies. You connect the patterns, as the narrator's voice poses questions regarding their relationships: "Does the inflationary universe represent the brain in an evolutionary context?" "Does it represent a cosmology of mind?" you may ask yourself, without expecting any swift answer.

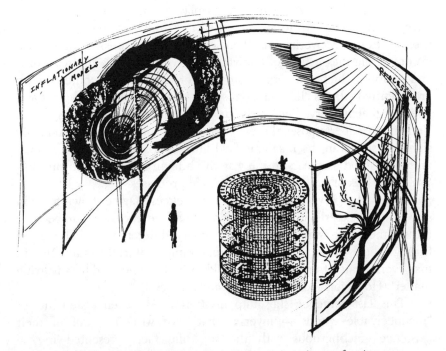

METAPHORM 2a. The Cerebrarium in Action—*an artscience of seeing anew.*

The *Cerebrarium* may explore the nature of processmorphs in a similar way, with questions encountering images, as we search for more process-likenesses between diverse forms of matter (see Chapter 11).

Within seconds of presenting images of the inflationary universe, a flood of clues is provided by the peripheral visuals and sounds orchestrated all around you. These clues had been informing your vision all along without your knowing it—like the visual notations in the borders of illuminated manuscripts express something about the text. As your attention shifts to this peripheral information, you see how you could have connected concepts about the expanding universe to the brain. A set of moving images that simulate the physiological changes of the human nervous system from infancy to old age has been playing all along in endless repetition. These anatomical changes are seen over millions of years, backwards and forwards, changing at speeds comparable to the expanding balloon. The *Cerebrarium* may prompt you to discriminate between models of cosmic evolution and models of the brain's evolution, as you compare the changes of this inflationary universe.

What may be most captivating about these images is that they're not medical illustrations of neural tissue referencing neuroanatomy; they're not roughly animated through time-lapse photography. Instead these computer-simulated images roll in and out of realism and abstraction as effortlessly as a lump of fresh clay can be explored in your hands.

Although these two giant screens play out their own stories, you have a sense that the stories presented are similar in that the screens themselves gently overlap and then separate continuously. The suggestion of similarities is further cultivated by rivers of boldly printed words and symbols which roll across your field of view like the contemporary artist Jenny Holzer's electronic signs with their revolving word plays and poetic digital imagery. The words cue you to the realities of both systems: Your mind superimposes the dynamics of one on the other. Suddenly so-called shears in the magnetic field lines of cosmic jets, for example, are understood as shears in the way the human mind moves from uniformity to lack of uniformity in its thought processes.

In effect, every detail of our life experiences may be shown to be in step with the experiences of the cosmos. Our differences seem to become transformed into similarities. The primary difference is that we see through the eyes of a biological form of cosmic matter, the brain. Nature may even create other forms that could never be conceived by any living matter! (How typical of its versatility and secrecy.)

This last sentence slips into mysticism. Are such statements of "inconceivable worlds"—universes that exist without proof of their existence—incongruous with the scientific facts presented in the

Cerebrarium? Not at all. These statements are meant to nudge our logic a little, waking us from sleepwalking through life under the guise of conventional or "uniform" thinking. Along with the accepted scientific information are these anomalous images and stories that may be as volatile as they are suggestive. The more you're exposed to these sorts of visually paired contrasts, the more you crave information composed of both scientific facts and their artistic interpretations. Without this balance, the *Cerebrarium* could never explore the more open-ended concepts with hidden agendas.

AROUND ISSUES OF DIRECTNESS

By the close of these exploratory narrations and images, the categorical distinctions between the brain and cosmos—and the disciplines that make these distinctions—may all be lifted like a trade embargo. Immediately, new genres of art may overflow into new genres of science. Along with these changes, we may rediscover how no artwork—or work of science—is ever really complete in that our interpretations of the work forever extend their meanings and associations. They change with our changing perspectives and cultures (Metaphorm 2b).

The *Cerebrarium* might present some of these changes, exploring how *evolution is nature's medium of creative expression.* As matter builds and rebuilds itself, we may observe that we think the way evolution acts. The selection and layering of images—which include everything from volcanoes forming and cloud formation to things that don't strike us as being especially "evolutionary"—may stimulate us to reflect on the creativity of nature. They may compel us to consider the impossible acts nature performed in preparing the universe for life. Life may never have been ushered into such a tumultuous environment as space, had it not been for some pleasant coincidences. Without the fundamental constants of the physical universe, for instance, life forms most likely wouldn't resemble anything we're familiar with at present. A list of fundamental constants includes *Newton's gravitational constant (G),* the *proton's charge (e), Planck's constant (h), the speed of light (c),* the *Hubble constant (H),* along with 21 other constants. Any significant variance in e, h, c, or H, for instance, would have some serious consequences in that it would change their relationships to the variables in the equations they occupy. These equations describe the structure and forces of nature on all scales. So far experimental studies of constants haven't produced the kind of forceful evidence that can persuade physicists to think in terms of variations, rather than fundamental constants. A neurocosmologist

METAPHORM 2b. The Brain Theater of Mental Imagery. *This represents ideas and images that combine art, neurophysiology, and nuclear physics in exploring the interconnectedness of the human brain and physical universe. The smaller paintings, which are mounted on the large "cortical screen," show the brain as a kind of star whose systems have a dynamic resemblance to the fusion and fission processes that form and shape our physical universe. (In the foreground is the* Cerebrarium *model for interpreting human brain/mind processes. Courtesy of Ronald Feldman Fine Arts, Inc., New York.)*

might push the point: "What constitutes 'forceful or convincing evidence'?" Evidence that is lasting? Or self-consistent? Or truthful? Or exact?

These questions are accented by a series of moving images that show the word EVIDENCE (which is printed on a photographic lens) being washed away by a stream of clear water. Above these images scrolls Henri Matisse's statement: "Exactitude is not truth . . . *there is an inherent truth which must be disengaged from the outward appearance of the object to be represented*" (my italics). The narrator prompts us, "Is truth, process?" Is the "object to be represented" one of nature's illusions—the "object" being objective physical evidence?

CLEANING AND REFOCUSING THE LENS

"How can a coherent, whole picture of matter *And* nonmatter, of brain *And* mind, of science *And* art form if our languages and representations remain fragmented?" This is the question you may be handed verbally inside the *Cerebrarium* and upon exiting from its experience. Is it possible to create an overview that attempts to connect all of our unique representations of nature in order to simplify our worlds? But simplification doesn't necessarily entail smoothing over the curious expressions that distinguish abstract art from, say, mathematical expressions. The idea is not to think as one, but to think as many around one issue with many parts.

If our disciplines remain separated or distant, our world-views will remain as fragmented and "different-looking" as the infinite forms of natural phenomena to which they refer. *Without* some integrated philosophy, without some shared world-view, our imaginations will find little worth inhabiting in the future—as we will stress ourselves arguing over differences of perception and kill ourselves from stress.

QUESTIONING THE LENS'S MANUFACTURER

The closing images might suggest that while we're waiting for either of these events to re-form us, we might try to build bridges connecting our different land masses of languages and concepts and philosophies. "Without these bridges we may as well be resigned to the fact that one person's religion or art is unintelligible to another person's science. Both our brains and universes will remain as distant from one another as the languages used to represent them," concludes the narrator. She continues: "The only thing we're bound to observe in nature is that our languages and disciplines are akin to the animal kingdom—and 'zoo of subatomic particles'—in their varieties of form, behavior, and application. All of our languages and creative capacities are part of nature's grand experiment—its evolutionary creations. If we remain creatures of habit, we'll continue to judge by appearance first and foremost, concluding without further thought: There's no common language just as there's no common animal—with lions communicating with hummingbirds."

In experiencing these messages, you may realize that neurocosmology isn't something that needs to be invented. We're part of its invention. It's part of us—like the brain is part of the body and the universe is part of the brain. What neurocosmology needs most is to be more clearly illuminated. Not by me, but by you. The brilliance of this illumination

METAPHORM 2c. *In* Two Planes of Thought Intersecting, *two visually kinetic paintings overlap each other, stimulating a psychological tension between what one sees and what remains to be seen in the hidden territory of the mind—the area behind these intersecting planes of painted thoughts. The canvases appear to have a common origin but different paths of projection, like high-energy particles hurled from an exploding supernova. Collaged on the paintings are poetic conjectures about specific areas of the brain that are possibly generating these thoughts and imagery. The imagery centers on the centrifugal and centripetal forces of thought. This artwork also relates to the* Cerebrarium *insofar as it is characteristic of the type of images presented in the model in some physically dynamic form. (Courtesy of Ronald Feldman Fine Arts, Inc., New York.)*

depends entirely upon your own imaginative experiences of this material —not upon my ability to convey these experiences. I'm not the illuminator; *you are.* This book, like the *Cerebrarium,* only documents my personal experiences of neurocosmology as a cerebralist. And these experiences lead me to conclude that what we say about the universe we say about our minds. Rather than living and thinking disjointedly, we might grow to learn something about how we're pulled towards a unity of mind by the same gravitational force that temporarily pushes us away from this unity (Metaphorm 2c).

SCALELESS "MICRO-MACRO" PROCESSES

In the *Cerebrarium*, the connections we can establish between the cosmos and humankind—and between humankind and its creations—consider *process-phenomena on all scales*. The differences between the very small structures (studied in particle physics and neurophysics) and the very large structures (studied in cosmology and neuropsychology) can be integrated.

Exploring this thought, Metaphorm 3 juxtaposes two unrelated images of processes. This arrangement suggests that the same process which describes the events of subatomic particles can also describe the events leading to the sudden existence of the entire cosmos. The Feynman diagram, which is normally used to indicate how three elementary particles form from energy and how they interact, is presented here as a cosmological diagram. The diagram interprets how the universe is currently manifest and not just its individual particles and elements. The same event in a cosmological context involves the interaction of positive and negative elements which result in the release of enormous amounts of energy.

Even though most of the processes investigated in neuropsychology and cosmology are not directly visible, they can be explored through extrapolation and projective thought. We can map out the growth process of human beings and celestial matter without physically seeing these things grow into their predicted forms. With or without the aid of mathematical physics, we can intuit the birth, growth, and rebirth of stars, for example, as if we physically lived these processes firsthand—observing them from their introduction to their disappearance. The question is: Are we mapping our present reality onto things that may lie outside this reality altogether? Also, are the types of juxtapositions in Metaphorms 3 and 4 helpful in discovering anything new about the systems compared? Do we end up with the same conclusions regardless of which end of the scale spectrum we look from or which types of matter we look at?

ENVISIONING THE WHOLE FIELD

In the *Cerebrarium*, the most technically laden books about the universe can be presented as poetic literature about the brain. Descriptions of the balance of forces in the universe, or entropy, or the many-universes theories, are interpretable as descriptions of *the human brain's forces, entropy, and many-minds*. Thus the terms and concepts in cosmology

PARTICLE PHYSICS
(the study of the smallest
processes of the universe)

COSMOLOGY
(the study of the largest
processes of the universe)

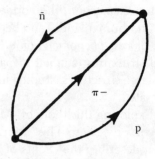

The Processes of Particles

The Feynman diagram of a three-particle
interaction shows the reaction and subsequent
annihilation of matter and anti-matter.

NEUROPHYSICS
(the study of the smallest
processes of the brain)

NEUROPSYCHOLOGY
(the study of the largest processes
of the brain,
including human behavior)

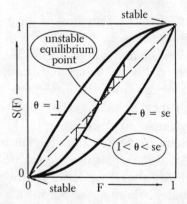

The Processes of Thought

Solution plot for the equation $S(F) = F$,
where $F = 0$ (no activity) and $F = 1$
(maximum activity).

METAPHORM 3. The Structureless, Nonhierarchical Micro-Macroprocesses of the Neurocosmos. *Consider the processmorphic relations between these four working forms of knowledge of the universe and brain. The* processes of structures can be *"scaleless," "structureless," and "nonhierarchical." The processmorphic relations between different forms in nature, such as celestial systems and nervous systems, need to be discussed in ways that see beyond the surface notions of scale, proportion, and micro-macrocosm relations. However, once these insights are obtained they need to be cast into more rigid terms in order to be further analyzed and applied. (Particle diagrams by Richard Feynman from Gary Zukav,* The Dancing Wu Li Masters, *1979. Thought diagrams by Alwyn C. Scott from his* Neurophysics, *1977.)*

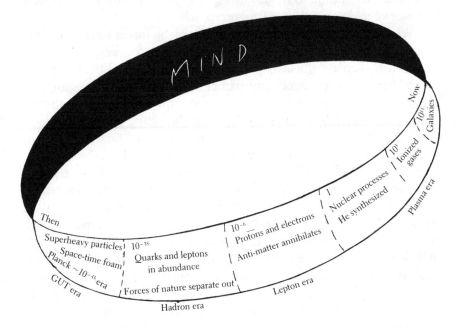

METAPHORM 4. The Reversible Mind-Universe. *Lying in the shadow of mind is matter? Creating the shadow of matter is mind? Will we ever tire from these sorts of visual "fugues" and reversible images? Probably not. The fact that we feel something for these types of philosophical drawings, as though they touched something deep in our being, reveals some aspect of our connection with them. The six key epochs in the universe's history are presented here in seconds. The Planck time $\sim 10^{-43}$ refers to the smallest measurable instant following the moment of creation.* (NOTE: *The bottom half of the figure was metaphormed from Paul Davies's diagram of the "History of the Universe" in his book* The Accidental Universe, 1982.)

can become allegorical expressions of brain terms and concepts. Accordingly, the history of the cosmos—from the "now" of galaxies to the "before and after" of superheavy particles in primordial times—may be represented as the history of mind. (See Metaphorm 4.) Or vice versa. In overlaying these two histories, our present conjectures are challenged.

To read about the growth of stellar populations is to consider the populations of nerve cells and ganglia that compose our central nervous systems. To explore how the four interactive forces (or those forces presently known) make up the universe is to discover how our minds are "made up," or influenced, by these natural forces. Perhaps this influence is more direct than we ever imagined. The implication is that the way we think, feel, act, and create is somehow related to the actions of the *strong force* (hadrons and quarks), *electromagnetic force* (photons), *weak force* (w-bosons), and *gravitational force* (gravitons).

Where the gravitational and electromagnetic forces have unlimited ranges, the weak and strong forces, by contrast, have very short ranges (much less than the diameter of an atomic nucleus). In relating our mental processes to elementary particles interacting, we suggest that these particles have processes in common with neurons—and the reverse. At the very least, these metaphorms bring the physical universe "indoors." Either that or they lead the mindful brain "outdoors" to join with the elements and forces from which it was created.

In relating quantum field physics, which is used to study the elementary particles and basic forces of nature, to the study of brain processes, we're left with more metaphorms than theories. Ultimately, none of the theories is satisfying scientifically because none can be empirically inspected. It's not that they are idle or empty, it's just that they're untestable—at least not in the scheme of current scientific protocol. Like the centuries of metaphysical reflections interpreting the mind-matter "overlay," these reflections are critical. As artful speculations they stir us to rethink the nature of the physical world and our minds' relationship to this world.

THE IMMEASURABLE VALUE OF MIXED-METAPHORMS AND METAPHORMERS

Imagine several particle physicists interested in the study of matter and anti-matter participating in a discussion on "perception of symmetry and asymmetry" in the brain sciences—a sort of informal neuroscience meeting in the *Cerebrarium*. Joining this meeting are visual artists, playwrights, composers, and writers.

Within minutes of listening to some introduction on the principle of symmetry, one or two physicists might ask mischievously what the concepts of symmetry in physical law have to do with symmetry in the biological laws of the brain.

"How do the symmetries of nature, which work in the universe, work *in* and *on* us?" probes one theoretical physicist.

The same person might charge ahead, without waiting for an answer: "What could the abrupt changes of events the universe experienced at 10^{-35} seconds old . . . events that turned our symmetrical world of matter and anti-matter into an asymmetrical world . . . what could these changes possibly imply about the workings of the brain?"

"Are you asking whether there was a similar shift from symmetry to asymmetry in the early evolution of the brain that has shifted human behavior in a big way?" the moderator politely interrupts.

"Something like that," answers the annoyed physicist. "I want to know how this shift from symmetry to asymmetry has altered our processes of thought," she presses.

Immediately another scientist pumps up the volume on an already loud and heavy question: "How can this relationship between matter and anti-matter *really matter* to the brain?" Silence falls with a thud. Some nervous laughter and rapid thought ensue, which are followed by a flowing liberated dialogue. Suddenly, everyone is anxious to talk.

"I can think of one reason why it *matters*," a voice from the back of the room booms. "In attempting to combine relativity and quantum mechanics, particles and anti-particles must be paired symmetrically," the physicist informs us. "Since every aspect of living and nonliving matter is influenced by the principles specified by relativity and quantum physics, I should think that the workings of the human brain are no exception. That's why I'm here. To see what I can learn."

Before anyone has had a chance to explore these statements, another participant rides a related point: "There's an analogous relation between brain and mind that also remains to be unified by a single theory." (The participant identifies himself as a visual artist searching for some riveting inspiration.) "By *mind*, I don't mean 'anti-matter,' such as positrons, or anti-electrons. I mean something completely *non*material —a nonmatter whose properties are opposite the properties of physical matter. Like an object and its mirror image—only the virtual image is more of a mental image belonging to the imagination than the tangible matter we see reflected in an actual plane mirror according to the laws of specular reflection."

"What's completely nonmaterial?" a neuroscientist asks professorially. "The stuff of spirit? The virtual material of soul?" he nudges.

"No. The *thought* or *concept* of symmetry," the artist answers matter-of-factly, as he holds up a picture of an artwork titled *Breaking the Mind Barrier: The Symmetries of Nature* (Metaphorm 5).

And so begins the gentle argument and clamor over definitions and viewpoints. So begins the science of scoring a scoreless game of art. A game that makes sense only as much as life makes sense.

METAPHORM 5. Breaking the Mind Barrier: The Symmetries of Nature. *The work delves into the unsuspected relationship between the workings of the human brain and the things the brain creates. It explores the "creative impulse" underlying human and cosmic creations. The writings and visual notations collaged on this artwork attempt to connect thoughts and their physical manifestations. The installation consists of an 8-foot free-standing column of light which visually divides two 9 × 18-foot paintings that converge in the corner of a room. At random intervals, the crescendo of a subtly pulsating sound triggers a brilliant flash of light which fuses the paintings, simulating an instant of creation and destruction. (Courtesy of Ronald Feldman Fine Arts, Inc., New York.)*

THE BROADER DIALOGUE INTRODUCED IN THE CEREBRARIUM

Determining the balance between matter and nonmatter in the brain— and the symmetry between matter and anti-matter in the cosmos—may not seem especially relevant to your life. But it is, in some undefinable way which the activities of the *Cerebrarium* intimate. Determining whether the human brain is composed of matter and virtual particles, forces, and fields—and whether our galaxy (and its immediate neighbors) is made up of half matter and half anti-matter—may not seem so important to your life. But, again, it is. It's as essential as everything you deem important to you. If we are in fact the "physical reflections" of the cosmos, then knowledge of the cosmos is applicable to our self-knowledge. This knowledge makes life meaningful. Whether these applications are expressed through mathematical physics or the anthropic principle or intentional language is irrelevant.

The difficulty in cross-applying these two sources of information rests on creating the most productive and meaningful metaphorm in which to relate these sources. It also depends on creating the most productive environments in which to interpret the metaphorms, discussing their implications within these different contexts. Metaphorms may be rejected "temporarily," that is, until they're used in some other context, in which case they may be more appreciated for their appropriateness.

Many experiments in high-energy physics have established that matter and anti-matter are nonhomogeneous entities, meaning that their properties don't mix. If you take a subatomic particle, such as a negatively charged electron, and you collide it with its mirror opposite twin, the positively charged positron, the result is—annihilation. Scientists believe that this process of annihilation is preserved on the grand scale as well, in galaxies composed of equal amounts of matter and anti-matter. As you would expect, these galaxies don't live very long. One conclusion drawn from such demonstrations is that we live in a primarily matterful universe—at least locally, that is.

The second conclusion is that clusters of galaxies primarily consist of either one or the other but not both: They're either built of matter or of anti-matter. Some physicists speculate that this antagonistic situation between corresponding particles may not exist in other regions of space where these two forms of matter may be insulated from one another by empty space and where they remain confined in the rooms of their respective regions. By avoiding physical contact, matter and anti-matter starve themselves so that they don't experience their voracious appetite for annihilation. At the same time, they suggest that the individual worlds of all galaxies, or clusters of galaxies, are fundamentally asymmetrical. If symmetry were preserved—balancing matter with anti-matter—a galaxy "would be history" at the moment of this perfect balance.

BALANCING ACTS OF MIND AND UNIVERSE

Perhaps the brevity of intuitions, which have been described as symmetrical events involving *cerebral parity*, is somehow connected with this larger relationship. And perhaps the process of analytical reasoning, by contrast, is a naturally asymmetrical event involving the predominance of one hemisphere over another. (The narrator of the *Cerebrarium* might purposively avoid specifying which hemisphere; however, s/he might point out in a questioning tone that reasoning is a process currently associated with one hemisphere and not the other.)

With the exception of this note on intuition and reason, the description of interacting particles presented here may not trigger any compelling associations with the human brain and its intra-actions or the interactions between peoples. It appears that, at least *physically*, these two forms of matter are not cast in some equal balance such that we self-destruct at the instant of intuition. But what about *nonphysically*, in our neurospace, in the space of our imaginations, which is part of what I've called the physical neurosphere? (The term *neurosphere* reflects the fact that our imaginings, and the mental spaces in which these imaginings occur, are creations of our neurons.) Could a similar process of annihilation happen? For example, when certain "corresponding thoughts" that are "oppositely charged" come together? Or when certain neuronal cell-assemblies and their corresponding anti-cell-assemblies virtually collide in cancelling each other's actions? How could you experimentally probe the events of such cell-assemblies and thought-assemblies? We can push this idea as a metaphorm, but can this metaphorm be pushed into the light of physical reality—pushed for the purpose of further exploration?

THE "BALANCED LIGHT" OF OUR MINDS

In forming this metaphorm about two oppositely charged thoughts colliding, the narrator might add one other informative note for you to consider. If we were to compare a galaxy composed of matter and one composed of anti-matter, we would be straining our imaginations trying to tell them apart. Even if we used the most sensitive techniques in spectroscopy to detect photons (or quanta), our task wouldn't be any easier. The difficulty lies with the photon itself; its anti-particle is integral to it, like the shell of a turtle is part of the turtle. The photons emitted from either galaxy look the same.[1]

In my earlier chapters, I entertained the brain-galaxy metaphorm and that photons represent "virtual quanta" of electromagnetic radiation that have thus far never been adequately described in neurobiological terms. Some scientists have speculated on what this virtual form of matter might be. However, their speculations are difficult to corroborate in the context of theoretical physics as it is currently practiced. Nevertheless Fred Alan Wolf (among other writers) has offered rich metaphorms that lead to powerful possibilities. According to Wolf, the fundamental process of the universe is the vertex, or the electron-photon interaction. Vertices represent the interactions of matter and light. "Our brains," he postulates, "loaded with 'real' matter, are storage arenas for this cosmic light show. Our electrical activity, which we call life, is rooted in the 'death wish' of all electrons: to return to the state of oneness with the universe of light. All electrons want to be the same thing—to be light again."[2]

Metaphorms uniting visible light and mind echo the controversial thinking of Benedict de Spinoza, the Dutch philosopher and exponent of seventeenth-century rationalism, who troubled over his contemporary

René Descartes's thoughts on the dualism of brain and mind. From Spinoza, we learn that the greatest barrier of the mind in grasping itself was erected by the notion that the mind could not be recognized in the light of matter.

IN LIGHT OF NATURE
WE DISCOVER OUR NATURE

In considering the preponderance of matter (as in brain) over nonmatter (as in mind), you immediately bridge these worlds through your thoughts. As we think about the correspondence between matter and anti-matter particles, or "pair creations" as they're called, we are directly —physically—influenced by this correspondence. Our thoughts cannot exist without this correspondence. It is this principle, or mechanism of correspondence, that makes our *brain's work* possible. It is also the same principle that makes the workings of the universe possible.

To interpret this metaphorm literally is to translate its metaphorical language in terms of the languages of particle physics. This metaphorm suggests that the brain apart from the mind is like photons apart from light—a point emphasized in the *Cerebrarium* by the subtraction of shadows from their light sources.

The fact that the photon and its anti-particle are part of each other, and that they're nearly impossible to distinguish, should tell us something about the nature of mind. It should also tell us about the integrated nature of art and science, which are fundamental elements of mind. They're also *virtual components* of the photon. If the electron component of a photon signified "the ways of art" and the positron component of a photon signified "the ways of science," and they were brought together, then we would safely predict that they would mutually annihilate each other. Another way of saying this is that art and science become completely integrated. And we would be able to see the light of this integration in the resulting artifacts. At least, this is the way of all pair creations that experience and rise above their collisions in becoming light. On the other hand, if these virtual components are not equally matched, we would merely suspect that art and science are somehow naturally separated by some invisible barrier of empty space—a "barrier" erected by our own minds?

INTUITION IS THE WAY IN

Here annihilation implies that *intuitively* both virtual particles are equally matched, like the particle and anti-particle that compose the photon. You might express this more mystically by saying that when art and science overcome their complementarity—when their positive and negative charges are allowed to fulfill themselves in their natural attraction for one another—then they become one, as in light. As long as we persist in looking upon art and science as being unequally matched in their capacities, we will only experience the darkness of their separate work and not this luminous union.

COUNTER-INTUITION IS THE WAY OUT

To say that these two virtual particles of thought are unequally matched is to think *counter-intuitively* about the ways of mind. Some may claim that our counter-intuitions more accurately represent the asymmetrical state of the universe. That if our minds were composed of equal amounts of matter and anti-matter, they would vanish as quickly as they're born, like theoretical matter/anti-matter balanced galaxies. The idea that the universe is in fact asymmetrical suggests to many that this asymmetry manifests itself in all forms of matter. And so what is regarded as the separation between art and science—or the separation between neutrons, protons, and electrons and their anti-particles—reflects this fundamental asymmetry with fidelity. If this asymmetry speaks for the universe it also speaks for the actions of our minds and bodies. This is one view. Another view is that our brains' "art and science," in their purest forms, are as integrally connected as the indistinguishable particle/anti-particle of a photon. Simply, their "light and matter" are one thing.

EXPERIENCING FORMS OF MATTER
IN FORMLESS PROCESSES OF MIND

Whether we substitute words and concepts like *science And art*, or *brain And mind*, or *matter And nonmatter* for the concepts *matter* and *anti-matter*, we're metaphorically talking about the same thing: symmetry. The process of balancing complementary elements is symmetry in action. "Balance" is the processmorph linking these and other paired concepts. The *Cerebrarium* demonstrates this with dynamic collages of contrasting textures.

To this date, no one knows why our universe is imbalanced—where matter outweighs anti-matter—and where one *form of thought* outweighs another form, as determined by our expressions, or realizations, of thought. All the extrapolations of data which work their way backwards to the beginning of the Big Bang suggest that all matter and anti-matter evolved from the same dimensionless point. Yet no one can account for this so-called primordial-imbalance hypothesis, which describes the disproportionate buildup of matter. "Will this situation reverse itself?" you ask. No one knows. Another nontrivial question looms: "Does this asymmetrical condition apply to the entire universe or only to a region of it? Does it apply to the whole nervous system and body or just a region of it?"

Taking a cue from a metaphorm that recognizes our central nervous system as a universe of sorts, we could postulate: "Yes, the entire universe is asymmetrical." But in accepting the implications of this metaphorm there's a counter-intuition to contend with: In the nonmaterial universe of thought, symmetry may be preserved. This contradicts current brain research, which stresses the asymmetrical aspects of our neuroanatomy and mental processes.

TESTING OUR EXPERIENCES OF MATTER THROUGH MIND

One way to determine whether or not there is more anti-matter than suspected is to do what many physicists and engineers are currently planning: To build an extremely sensitive neutrino telescope. Neutrino particles distinguish themselves from photons in that they have visibly distinct anti-neutrino particles. This makes neutrinos slightly more desirable for studying the balancing act between matter and anti-matter. I say *slightly* because a neutrino mass is impossibly faint. Their detection in nuclear reactions requires as much patience as it requires the most sophisticated detection systems ever conceived. Since stars of matter mostly emit neutrinos and their twin anti-matter stars mostly radiate anti-neutrinos, measuring the latter will help assess the nature of the former. (Or vice versa.)

Depending on which theory in cosmology you subscribe to, you're going to see the physical universe as either an inflating or a deflating entity. Both scenarios cite the Big Bang model, where there are three different types of possible universes—open, closed, and flat. Each has its own peculiar spatial geometry and volume. The open universe, for instance, has a hyperbolic form (or negative curvature) and is presumably infinite in volume. The closed universe is spherical in form (or positive curvature) and is presumed to be of finite volume. Finally, our universe may in fact be flat with a Euclidean form (or zero curvature) possessing infinite volume.[3] Ask yourself: "Is the child (the brain) like his or her parents (the cosmos)?" Ask the *Cerebrarium*.

All of these heavy and expansive views sit on the pinpoint of one assumption: That all the matter and energy we presently associate with the universe evolved from either something or nothing. By "nothing" I mean neither space nor time nor matter. For example, if the inflationary theory of the universe is correct, then the traditional notion that something can't be born from nothing is disputed. This notion, dating back to the fifth century B.C., or before the pronouncement of the Greek philosopher Parmenides, has thus far remained undefeated in the arena of modern cosmology. Support for the Big Bang model depends on a number of sensitive issues and assumptions. I'm familiar with at least seven:

> that our fundamental physical laws are truly "fundamental" and "unchanging" with time;

> that Einstein's *painting* of the effects of gravitation in his general theory of relativity is more than an interpretive or impressionistic work of science—it is truly realistic in its portrayal of nature;

> that the early universe was bursting with a uniform, expanding, and unimaginably hot gas composed of elementary particles;

> that the particles in this superheated gas, which permeated the whole of space, were in thermal equilibrium;

> that matter and space continued to expand at identical rates;

> that the densities of matter and energy have remained uniform regardless of where matter has evolved in space;

> that any changes in the matter and the radiation over time were negligible; that these changes barely affected the thermodynamics of the universe; if these dynamics, which describe the relation between heat and other forms of matter, were altered significantly, then matter would behave differently from what present observations in particle physics, astronomy, and cosmology suggest.[4]

Without debating the particulars of these issues, more than half of their concerns have not been resolved to the satisfaction of all astrophysicists and cosmologists. Concerning the Big Bang model itself, there is still considerable dissent over the primacy of one of its three derivative models that hypothesize about the openness, closedness, and flatness of the universe. "They can't all be 'right'!" Unless, of course, they describe different aspects of our universe at each turn of its full cycle from beginning to end and back again endlessly: from blank, singular points of infinite density, heat, and dimension to blossoming explosions and degenerative recollapses occurring in and out of time.

If you're curious about the language and disciplines piloting these models—and you desire to learn how these models reflect our brain processes—you might ask a cosmologist some rather frank questions:

"What do these cosmological models have to do with how my brain works? Or how societies work?"

"What does the latest quantum field theory tell me about how my brain is organized? Or how I think?"

"Will a grand unified field theory show me how the *structures of thoughts* my brain produces reflect the structures of the universe— in either its earlier form or later ones?"

"What will these quantum field theories tell me about my universe and its composition? What will they tell me that I don't already know?"[5]

The more basic the question, the better—as you clarify them with each asking. You might try extending their contexts, making similar inquiries to researchers working in the field of brain science. Ask a neurophysiologist, a neuropsychologist, a psychologist, or cognitive scientist:

"How do the long-term activities of the universe (10^{100}) influence our short-term activities (10^1)?"

"What are the biological equivalents of black holes, quasars, and pulsars in our nervous (mental) systems?"

"Why is mental energy a myth to scientists and material energy a myth to mystics? The fact that scientists haven't been able to see or quantify 'a unit of mental energy' may be more telling of our sciences' inadequacies, rather than demonstrating the *inadequacies* or *inabilities* of this special energy to manifest itself within the spectrum of matter."

"Even if we could see our minds' energies through some miracle technology, why should we believe what we see? The fact that this virtual energy would be materialized—shown to exist in the world of matter—would immediately render it as part of the material world illusion. This illusion is generated by our senses, which mystics have described for millennia. However, materialists have for centuries rejected these descriptions as explanations."

"How can the study of particle physics and cosmology augment our knowledge of the brain or human behavior? How are the theories spawned by these disciplines helpful to theories spawned by the sciences of the brain?"

"How can art and metaphorms contribute to the development of these sciences and still remain mysterious and elusive—something we experience purely without explanation of *why?* Works of art without this mystery and elusion are illustrations; they pointedly illustrate, rather than interpret, a view."

These questions need to be further simplified by your intuitions posed in the *Cerebrarium.*

LETTING GO AND BIDDING ADIEU TO PRECONCEPTIONS

What makes these issues and speculative inquiries pertinent to the lives of every human being is that they point out the parallel mystery between the matter that composes the cosmos and the information that composes our minds. Metaphorms highlight this point and heighten our experience of this mystery. At least this is their compelling ambition: to stimulate us to actually *feel* this knowledge, rather than just read or think about it. The act of experiencing the symmetries of nature moves us towards understanding them in ourselves.

Like fertile soil and seeds, these experiences help us grow to question our relationship with nature. Understanding this relationship might just inspire us to think about the consequences of our thoughts and deeds, something a supernova or a neutron bomb or heinous act can't do. Bertrand Russell once wrote: "People would rather die than think. In fact, they do." Much of the grand suffering in the world is because most people don't think—about nature, about life, about human nature, and least of all about the contents and properties of their mind. The questions in this book are meant to engage people as thinkers, not train people to think through metaphorms about the truths of nature.

HOW TRUE IS TRUTH

In exploring art and metaphorms as methods of inquiry in the *Cerebrarium*, the participant observes how we might also employ the concept of truth-values, as they're called in symbolic logic. (The idea of establishing these *values* was formally introduced to our way of thinking through the efforts of the mathematician George Boole [1815–1864].) The narrator may inform the participants that truth-values are variables (symbols) that can represent a diverse range of things—in this case, the brain and cosmos. Through "truth-values" we can form some interesting propositions about the behavior of these two systems. These values refer to the truth or falsity of logical propositions. Boolean algebra manipulates these propositions that are either true or false. If they're *true*, then they have a *truth-value 1*. If they're *false*, then they have a *truth-value 0*. When such logical statements are combined, they form what is called a compound proposition. These compound propositions are represented by logical connectives, or operators, that take the form of AND or OR (which are symbolized as $\wedge$ and $\vee$). The truth-value of the compound proposition depends entirely upon the truth-values of the components and the operator used.

What all this means is that if I am talking about some facet of the brain—*Proposition a*—that is "true" (according to some accepted knowledge of brain science) and some facet of the universe—*Proposition b*—that is also true, by employing the connective AND the resulting compound *Proposition a $\wedge$ b is true*.

Fortunately, this symbolic method of reasoning does not have to actually conform to the realities of either the brain or the universe. However, this exercise in logic is helpful insofar as it is suggestive of new ways of looking at relationships between different things. And, more important, it is helpful to our process of forming hypotheses. As Bertrand Russell once pointed out, whether or not the first proposition is *really* true (or our propositions of things as they really are, in fact, *are true*) is essentially unimportant for this theoretical study. The precepts of this study inform neurocosmology.

FACTUALIZING INFORMATION— DERIVING FACTS FROM FICTION

From exercising truth-values, we see how the properties of our minds may be understood in relation to the universe. The propositions cast the mind in the unique position of representing the universe as one of its many forms. A theory in cosmology, such as the superdense theory which details the expansion and contraction of the universe, becomes a metaphorm for our expanding and contracting minds. Immediately, the red-shifted spectral lines—which the astronomer Edwin P. Hubble discovered reveal galaxies receding from the Earth—cue us to the virtual spectral lines of the human brain's galaxy. A neurobiological galaxy (or galaxies) that generates nebulae of thought. We can begin to imagine through the motions of the *Cerebrarium* how the brain's galactic processes recede from some specified point in their evolution. Our thoughts, like these virtual galaxies, recede at speeds that are proportional to their virtual distances from one another. The farther away the galaxy is from this relative point of origin, the faster it's moving. In terms of brain processes, this movement of activity involving thought may be translated into cell-assembly dynamics.

This last statement makes sense only if we have in mind metaphorical images of *thought galaxies* that have specific points of creation from which they evolve. These galaxies of thought either grow together to form some agglomeration or they move apart, like islands in a vast river stretching out in all directions. By contrast, in trying to translate the astronomical terms *point*, *speed*, *proportion*, and *distance* into plausible terms for describing brain processes, the metaphorm changes its tune and form. It goes from being a rich tone poem in classical music to a plain jingle advertising some idea through analogy. It makes its way from the more abstract and implicit qualities of metaphor to the more explicit quantities of simile. Instead of suggesting in the fuzziest of language how one thing is somehow connected with another, you're motioning how it's possible to pin down, spell out, and count the number of connections between these relations. You make literal all likenesses. Both aspects of interpretation—and possibility—are present in these metaphorms.

In terms of the brain, the "point" (of origin and evolution) would refer to a region in the central nervous system, such as the hippocampus or entorhinal cortex (see Metaphorm 3b in Chapter 5), that may have generated a particular thought or idea. It might also refer to a small group of cell-assemblies. "Speed" would refer to the passage of time from the moment of conceiving an idea to its more advanced development. "Proportion" and "distance," too, would have specific references which are not necessary to name in this general discussion.

THE STEADY-STATE METAPHORM— EXPLORING THE EXISTENCE OF MIND

Another theory that may be explored as a brain process metaphorm is the steady-state theory in cosmology. This theory posits that the cosmos has always existed and most likely will always exist in an even state. So this metaphorm represents steady-state brain processes as they evolve. What is meant by *even*, or *steady state*, is that the average density of matter will remain constant as the universe evolves. In the course of expanding, of building itself up, the receding stars and galaxies must deposit new stellar material in the spaces they've vacated. Consequently, the universe always teeters between too much and not enough matter. Apparently, the density is not constant, since the amount of matter fluctuates throughout the universe. What could this fluctuation mean—metaphorically and literally—in terms of brain processes?

COSMIC BACKGROUND RADIATION METAPHORM— REPRESENTING THE MIND'S EXISTENCE

One neurocosmologist might suggest that the theory of cosmic background radiation relates to our brain processes. Background radiation, which is essentially electromagnetic radiation that saturates the whole of space, would be related to some infinitesimal radiation that permeates the entire neurosphere. And this radiation would date the earliest presence of this form of matter in the universe.

Another neurocosmologist might think of the radio background metaphorm differently. She might observe that this radiation, which in physical reality represents the total radio source emissions in the universe, is analogous to some form of yet undiscovered biological background radiation that is unique to each brain. As every brain is its own universe, the measurement of "neurocosmic background radiation" would vary with each person, according to age and many other variables.

In a strange way, this last view allows for all possible views. What I mean is one person's nervous system may function according to some steady-state scenario, while another's may fit the predictions of the superdense theory applied to brain processes instead of cosmic processes. Still other nervous systems might operate by completely different theories in cosmology, and so forth. Maintaining this multiview perspective is one of the principal impulses and chores of nature which are applied in nature's inventions: context, relativity, and change. These three words are temporarily enlarged in the *Cerebrarium* so that they expand beyond the screens they're projected on.

We need to be prepared to change our assumptions or modify our axioms at the request of nature. No matter how predictably constant the laws of nature appear to be in the government of all physical events, nature may have its moments of caprice and deception just as we have ours. And it is not a sign of ineptitude that we cannot detail this capriciousness with mathematical theory or anticipate these deceits. It's more a sign of nature's extraordinary ingenuity and clandestine character. It seems to know that "in order" to grow, it must accept the possibility and surprise that anything can happen—unannounced and unbeknownst even to itself.

To my mind, the laws which nature obeys are less suggestive of those which a machine obeys in its motion than those which a musician obeys in writing a fugue, or a poet in composing a sonnet. . . . If the universe is a universe of thought, then its creation must have been an act of thought. —SIR JAMES JEANS, *The Mysterious Universe*

15

Exploring Processmorphs— *Thought-Assemblies*

In this chapter I will be presenting another work of neurocosmology—one that represents neither *science qua science* nor *art qua art*. Rather it is, quite literally and simply, a "work of thought." Like the *Cerebrarium*, its subject matter is the brain-universe and processmorphology. It also attempts to glide above the intellectual barriers that usually prevent its full appreciation based on context-specific viewing.

VIEWING THE PROCESSES OF THOUGHT

The speculative work of artscience discussed here is titled *Thought-Assemblies*.[1] If you imagine a 12-story building turned on its side with all its windows representing individual paintings of mental images, then you have a sense of the size of this work in physical reality. (The actual work measures some 127 feet long and 9 feet high but that's of little interest here. It could just as well be measured in parsecs.) If you then imagine looking into the "windows of this building" and seeing its interior architecture—its inner life—expanding before your eyes, then you approach a clearer sensing of what this work is about, perhaps more so than if you were standing in front of it scratching your head in wonderment or frustration as to the meanings of its form.

Thought-Assemblies contains many hundreds of visual notations on the subject of this book. Some of these notes are crudely expressive, while others appear to be more polished and finessed. It's as though its creator were reconstructing years—or moments—of inspired thoughts to satisfy some obsessive search. These notes are all woven together in one assembly of "painted thoughts." (See Metaphorms 1 through 6.*)

Thought-Assemblies was conceived as a visual novel—one without a beginning, development, climax, or afterword; one that could be read in the space of a glance or over a lifetime. The fact that the work probes the nature of thought itself is nothing special. The only thing that distinguishes it from other forms of thought is that it purely documents its creator's evolving ideas about neurocosmology. Like all art, this work is a universe in itself, one that comes complete with its own cosmology for both the creator and viewer to discover. It is real inasmuch as it is a physical creation.† It is also a virtual work, insofar as it is not fully created until it is contemplated by viewers other than its creator. The meanings viewers make of it are necessary for completing the work. Without this participation, the work is dead matter, like a memory lost to the rememberer.

Just as a sunflower plant contains many hundreds of smaller flowers in its core, the *Thought-Assemblies* assemblage consists of over 500 drawings, paintings, and collaged written notes (see Metaphorm 7). This seems like a lot of material, but in truth it's practically minuscule. Actually, its creator was concerned about how *little* material this was for representing a topic as astronomical as the brain-universe. The "less is more" truism holds for rapid—and simplistic—overviews in presenting complicated ideas. However, in this case, I suspect less would simply be less.

---

* Note that only certain details of *Thought-Assemblies* are provided by these schematics and photographs. These serve to stimulate you to consider what you think the work might possibly look like—that is, according to your personal vision and interpretation.
† *Thought-Assemblies* was created during 1979–1981 and first exhibited in 1982 at the Musée d'Art Moderne de la Ville de Paris, France, A.R.C. 2, in an exhibition titled "Alea(s)."

METAPHORM 1. Emerging Thought-Assemblies. *Depicting a moment of inspired, directed thought. (Metaphorms 1, 2, 4: courtesy of Ronald Feldman Fine Arts, Inc., New York.)*

METAPHORM 3. The Conceptual Armature of Thought-Assemblies—*a diagram indicating the information portrayed in this artscience work, which consists of three interactive axes. Presented on the* X-axis *is information based on intuition and perception about the brain and universe. Intersecting this plane is the plane mirror, or Z-axis, which reflects vertically above and below the X-axis. Above the X-axis, the information is abstracted and implied, thus entering the realm of art. Below the X-axis, qualifying and quantifying information is added, entering the realm of science. Thought-Assemblies indicates that analytic and artistic thought can proceed from the same frame of insight-perception and that these two modes of thought converge. As an exercise in topology, if the artwork were folded to form a tube and then the ends of the tube were brought together to form a torus, or donut-shape, the farthest points at both ends of the X- and Y-axes would be continuous.*

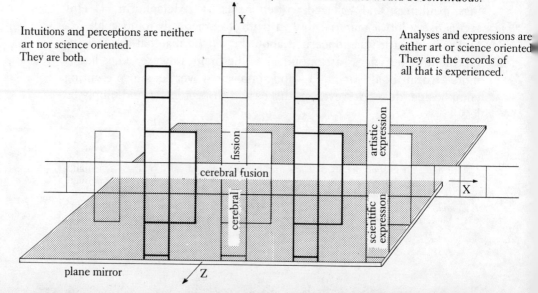

Intuitions and perceptions are neither art nor science oriented. They are both.

Analyses and expressions are either art or science oriented. They are the records of all that is experienced.

Y

fission

cerebral fusion

cerebral

artistic expression

scientific expression

X

plane mirror

Z

METAPHORM 2. Emerging Thought-Assemblies. *This represents the formation of an idea, visualizing the infinitesimal and global events occurring in the human brain as one thinks, feels, and creates. The influences of events such as synaptic and action potentials in neuronal signaling on the structure of thought are explored in the artwork. The spinning circular image on the far left is a top view looking down into the vortex of the spiraling form which, in this instance, symbolizes a turbulent thought. This turbulence is generated in part from the molecular and cellular "strife" visible in other passages of the painting.*

METAPHORM 4. Thought-Assemblies. *This reconstructs instances of inspired thought, visualizing the interactions of intuitive and analytic thought processes. The work is composed of hundreds of constituent images (or pictures of mental representations) that present my personal lexicon, free-associations, and perceptions of the world. The overall design of this work might be likened to an electroencephalographic (EEG) recording, implying that the patterns of our mental activity reflect the brain's electrical activity.*

METAPHORM 5. The Process of Thought-Assemblies *in which intuitive and analytical thoughts about the brain-universe relation are assembled over the course of milliseconds or minutes or years. The surface shown here is dynamic, rather than static.*

. . . from cell-assemblies to thought-assemblies
the cosmos unfolds inside us—
and we, inside it . . .

METAPHORM 6.

Details of the Parts

METAPHORM 7. The Whole-Part Relation of Thought-Assemblies. *Each of the more than 500 rectangles that comprise the "details" of this work of artscience represents one picture of a mental image. This single mental image can represent either a single thought or many thoughts. The images consist of visual notations that have been collectively organized on a two-dimensional surface. This surface is an expanding three-dimensional space, such as those surfaces in Metaphorm 8.*

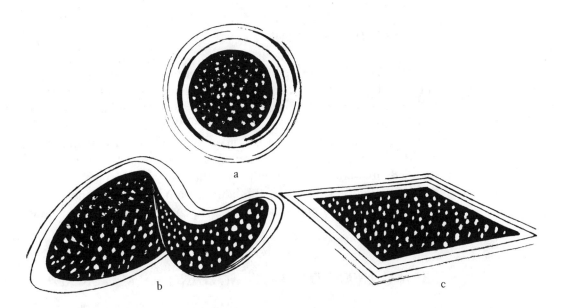

METAPHORM 8. The Universe of Thought-Assemblies—*a three-case scenario: a, a positively curved space; b, a negatively curved space; c, a flat, non-curving space. Each of the three shapes of the universe reflects the possibilities of a homogeneous, uniform cosmos. Uniformity refers to the way matter and radiation are fairly evenly distributed about us. The term also refers to the way matter is oriented in different areas of space in relation to its distance from us. By substituting the concepts "mind" and "thought" for "matter," the notions of uniformity and non-uniformity take on a new meaning and set of implications. Each shape (a–c) relates to Einstein's general theory of relativity, which informs us that gravity expresses itself as the local curvature—or shape—of space. Instead of looking at a three-dimensional surface that's static, what you're seeing is a four-dimensional space that is restlessly swelling and contracting. The analogy of a balloon expanding is frequently cited in cosmology. The dots signify galaxies which, as you might imagine, are being swept apart by the expansion of space. Eventually, these galaxies of matter—or thoughts and mental images—may also contract. (Metaphormed from Paul Davies, The Accidental Universe, 1982.)*

In the final analysis, there is no appropriate form or proper exhibition space to present this work. It could have been reduced in scale to the size of a pencil or toothpick or string of hair or protein chain. Of course, each reduction would affect one's experience of the work. Perhaps the only space suitable for its viewing is in one's own neurospace where things tend to be proportionless. There, many expanded versions of this universe and its cosmology can be assembled and disassembled at mindspeed. The different shapes of this artscience work, shown in Metaphorm 8, suggest that the structures of the universe are its models for creation.

THE BRAIN AS SOURCE AND MEDIUM OF ALL METAPHORMS

Thought-Assemblies considers the metaphorm: The brain as physical science and the mind as art. Implied here is that the content of the mind, represented by mental images, is the message of art. And the medium in which these images are created is the brain. Accordingly, a single mental image constitutes a "work of artscience." The term *work*, as defined here, implies the interpretation of the interaction between the message and the medium—both of which influence the behavior of a person in creation of thoughts. Ultimately, the brain is both the palette and canvas. The mind is the painter—and the imagination, the thing being painted or interpreted.

SEEING EVERYTHING AS NOTHING . . . MATERIAL

This work explores the medium of imagination—specifically, how we relate our world of insights to our world of materials. The virtual atmosphere in which mental images form at first is shown in flat relief like the silhouettes of afterimages. In this medium floats *everything*—or parts of everything—one has previously seen, or experienced, in physical and dreamt environments. In the mind, there's no single perspective connecting one point or thought or concept to another. There are only one's audio-visual-tactile-olfactory memories, which pointlessly erase themselves from the path of concentration, leaving at most traces of perceptions. *Thought-Assemblies* is a collection of these perceptions of the world.

The viewer becomes a participant in exploring this assembly of mental images made visible. The process of interpreting them reveals something of the mind's relation to other minds and to its outer environment.

Both the creator and viewer observe firsthand the entrapments of our minds as we attempt to learn how we create—and how we discover the nature of our neurocosmos.

THINKING MAKES IT SO

The experience of thinking about *Thought-Assemblies* is more "the art" than this artifact of thought itself. In this metaphorm, thoughts are shown to be connected with the things and processes to which they refer. For instance, the motion of "cosmic jets," which are focused streams of ionized gas thousands or millions of light-years long, is shown to resemble the process of one's thoughts as they become focused streams. The movement of spiral galaxies is shown to resemble the virtual movement of a specific thought as it becomes a spiraling mass of particles—where each particle symbolizes a star. Thoughts are shown to move like galaxies move—and form like galaxies form. In one region of this work, the painted thoughts appear to be strung together as superclusters. In another region, they collapse like a supernova.

The illusion here is curious: The mind seems capable of conceptualizing in several milliseconds what the universe takes more than several billion years to do in its "realizations." Is the creation of galaxies of matter and thought one and the same instant of explosive creativity?

If our thoughts and dreams were as material as the thinker and dreamer, would the universe look the same?

FEELING MAKES IT SO

Thought-Assemblies is organized in such a way that one's interpretation of its whole/parts changes with respect to one's physical distance from the work (Metaphorms 9a–9c). This is the same as saying that an astronomer's or astrophysicist's interpretation of a stellar body changes the closer either one gets to it via satellite reports. There are various layers of picture-statements and written notes collaged on this work. Each layer is thick and textural. With words, pictographs, and other visible language scrawled in paint on the surface, their concepts stand in raw relief. What you feel in viewing the work from afar counters what you gather about its nature upon close inspection. The acts of knowing, experiencing, and interpreting this universe are explored in immediate and fundamental ways. These processes are also similar to those with which astronomers and cosmologists perceive and interpret the "temporary" installations of the physical universe.

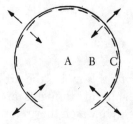

METAPHORM 9a. The Influence of Distance on Experience. *At distance A, one glimpses a panoramic overview of certain connections between the brain and cosmos. Philosophical, scientific, and technological interpretations are visible in these connections. At distance B, one sees both overview and details. And, at distance C, one can engage in the details, contemplating the interactions between visual words and sounding images. One implication of this design: After one studies the details of this work, one is encouraged to stand back and experience—to see anew—the beauty of the whole work of art, the human mind and universe on which this work comments.*

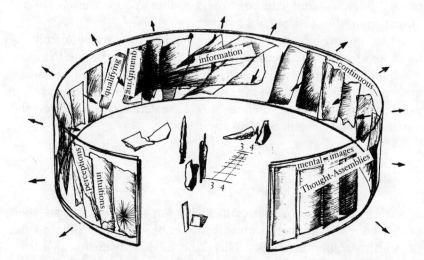

METAPHORM 9b. The Experience of Space. *The spatial orientation and shapes of the paintings imply order and disorder, control and absence of control, just as thoughts are continually evolving. The metaphoric imagery painted on the irregular-shaped, unstretched canvases and sculptural forms reflects the constantly shifting states of the neurocosmos. The installation, which consists of 515 mixed-media paintings of mental images and notations, is "adapted to" the design of this unique exhibition space. Both the expressive and cerebral natures of this work build upon the aesthetics of Western prehistoric cave paintings and Eastern rock-cut monolithic sculptures and ancient friezes, murals, steles, and scrolls of "picture-writing" with symbolic, figurative elements that seem to resonate with the entire history of civilization. One period and stylization overlaps with another, as a matrix of imagery unfolds.*

METAPHORM 9c. States of Mind and Thought-Assemblies. *The surfaces and colors of the paintings symbolize various cellular and chemical properties of the brain involved in generating different thoughts and mental states. The thoughts focus on issues affecting human evolution. The shapes and images of the paintings imply movement, just as thoughts are in continuous motion. Although these works may appear to be flat and static, with the application of imagination they become as multidimensional and dynamic as the mental images they depict. (Courtesy of Ronald Feldman Fine Arts, Inc., New York.)*

. . . SO WHAT . . .

Like the universe it explores and creates, *Thought-Assemblies* is difficult conceptually to be intimate with at first. You need to befriend its concepts. You need to establish some kind of rapport with the work, before you can appreciate its character—which is almost comical in its bulky obsessiveness to know nature and to represent this knowledge without the constraints of illustration and pedagogy. It's more than one person's interpretation of the unity of thought and matter—a subject many dismiss as obscure. It defies the complacent attitude of "so what," in attempting to show what is *so* extraordinary about our thoughts—especially the way they physically touch and envelop us as we try to make sense of our life experiences.

. . . WHAT EVERYONE EXPERIENCES
IN THE WORLD OF MIND

In current models of cosmology, the universe is predicted to do one of two things: either perish by way of its production of matter, eventually burning itself up,* or disperse by means of constant expansion—until literally the cosmos freezes over.

This is a small wonder to a neurocosmologist who has already anticipated that we'll be burning and freezing forever, as the universe endures an endless series of creations and destructions. This intuition has been expressed metaphorically in the philosophical and religious works of many ancient cultures. Their explorations and insights into cosmology seem to complement those derived in modern cosmology from mathematical physics. One of the more interesting questions at this point is: How do these two scenarios in cosmology reflect our processes of thought or ways of mind?

Thought-Assemblies interprets these sorts of questions. It suggests that all answers are speculative and perhaps will remain so. It arrives at a similar conclusion concerning inquiries into the large-scale structure of the universe. The "largeness" of the universe is depicted as being as unbound, or infinitely expansive, as the mind itself. The scale of the cosmos can be defined neither by technological innovation nor by mathematical physics *such that the definition accounts for the full scale of the universe and its scaleless processes.* To assume that we can secure a broad enough—but detailed—definition, which satisfies all the constraints and hidden clauses of the cosmos, is wishful thinking propelled by reason of ambition. At least, this is one interpretation, which is expressed as a faint question mark the size of a speck of dust lost in the center of a black sheet of paper. One's mental image of this paper represents a thought in progress—a thought not unlike those depicted here.

All negative overtones aside, this work speaks positively of our futurelessness—meaning the futility of the concept of *future.* What is the future of the universe, or the brain, without its mindful observers? We could run impressive computer simulations, or extrapolate from the data ad infinitum, and we still could not accurately predict what we will become or what will become of the cosmos. A storm-front metaphorm, detailed by means of modern meteorology, speaks for the possibility of our evolution. A calm-front metaphorm speaks for another.

---

* According to the inflation, or expansion, theory: The collective gravity of all the mass in the cosmos becomes so great that it halts its movement and draws matter back in upon itself.

LARGE THOUGHTS IN LITTLE BODIES

In the most tangible way, *Thought-Assemblies* reflects these themes. A viewer can wander up to any part of this metaphorm and examine these reflections. Although its Cartesian, grid-like form suggests that it is rigidly structured, its underlying thought processes are as loosely composed as dreams. The scientific information on the brain and cosmos is poetically interpreted, as in the metaphorms discussed in the earlier chapters.

What sparked the creation of such a large-scope work was the desire to learn about how we represent our experiences of the world of mind and the physical universe. The work was envisioned as a kind of *theoretical art*—complete with experiments. Like the exotic, advanced technology employed in theoretical particle physics, *Thought-Assemblies* was conceived as a spark chamber, a tank of liquid helium, or some other elementary particle detector. In this virtual tank, we can observe the trajectories of electrically charged particles of thought. It is a virtual spectroscopy lab on paper in which we can envision the different spectra of thoughts, each with their constituent wavelengths, energies, and molecular structures. As a processmorph, it is a virtual particle accelerator in which we can study the forces of elementary particles. It's also as it appears: An assembly of thoughts on paper created for aesthetic involvement.

Experiencing this work requires the viewer to take literally long and short strides, with sustained patience. Viewers are encouraged to see the work walking quickly or even sprinting past its 11 connected sections—like a sensitive theorist who first intuits the whole before analyzing the parts. With slow, short steps one can examine the details of the painted thoughts that nest in each section. This motion informs the synoptic overview.

No one directs you as to how to read this novel—or what section to read first. There's no set order or direction in which to interpret the passages of ideas. No one instructs you as to what part of this physical universe you're gazing at. The viewer becomes the astronomer, the visual artist, the physicist, and the poet interpreting a particular configuration of thoughts, which, like stars, bear certain messages in their spectra. The viewer is free to search the conceptual dimensions of a cluster of metaphorms.

BODIES OF THOUGHT IN TRANSFORMATION

Thought-Assemblies is meant to be as much a part of the viewer's thinking experience as it is a reflection on the thought process itself. It's that specific and that general—or vaguely precise. Like the concepts it engages, the virtual space in this art is a boundless extent—dimensionless, timeless, formless. Its true geometry is exactly opposite of what can be seen tangibly. Its form, like its content, is only meaningful if its viewers find that their own thoughts are transformed in similar ways (see Metaphorm 10). These transformations of thought reflect how matter is transformed—or transmuted, converted, modified, evolved, altered—into other states of matter or other forms of thought. Some viewers may identify this transformative process as a creative act insofar as it's based on *unconditioned responses* to environmental influences. Others may refer to this process as the workings of instinct rather than creativity.

Like the atoms that make up the molecules and particles of air— and us—our thoughts are continually being transformed. A quiet note written on one of the abstract pictures of mental images relates this idea: "When I think about how electron spins have something to do with how I think and this thinking influences spinning electrons . . ." The words vanish under a thick carpet of paint. After reading a series of related statements collaged on this wall of thoughts, one comes away with the impression that *Thought-Assemblies* is about the processes of elementary particles and not just people—processes that are connected with other phenomena and not just thinking. One begins to see how particles are like people; they appear to go from one state of "self-actualization" to another.

METAPHORM 10. Some Transformations of Thought. *Are there one or three states of mind, states of matter, states of thought? Forms may change, but the processes of transforming and comparing—or* metaphorming—*remain the same? Here again we meet a paradox of nature. In* process—*which represents* change—*there is stability. And within* form—*which represents* unchanging structure—*there is instability. (Adapted from David Bergamini, editor,* Mathematics, *Life Science Library, Time-Life, 1965)*

This idea stretches the thoughts of the philosopher and paleontologist Pierre Teilhard de Chardin (1881–1955), who theorized that humankind strives toward mental, social, and spiritual unity—perhaps like the smallest atomies, which exhibit the behavior of what might be called "striving." They strive to be fulfilled—not with the same pathos as some of us possess, but in their own restless energetic way. The striving can resemble the process of an electron changing its orbit in an atom. Or an atom changing its position in a molecule, etc. Although I'm relating human attributes to particle physics, there's something metaphorically compelling about this relation.

CREATING SOMETHING FROM SOMETHING

The following passage considers the origins of the neurocosmos. It hypothesizes about a *neurocosmic* explosion-implosion event that created both matter and nonmatter (Metaphorm 11).[2] This passage, along with the other images, statements, and explorations presented in this chapter, is one of hundreds of metaphorms and processmorphs explored in *Thought-Assemblies*.

When the universe first exploded some 1×10^{10} years ago, it simultaneously imploded. In that eternal instant the values of matter and nonmatter (or mind) were set in some perpetual equilibrium, determining the symmetries of nature. The universe was perfectly balanced. However, within 10^{-35} second (as current theory holds), nature broke this fundamental symmetry.

All that exploded was physical and visible (denoted by the Greek symbol ϕ, phi), comprising the particle-wavelike nature of matter. All that imploded was nonphysical and invisible (denoted by the Greek symbol ψ, psi), composing the *virtual* particle-wavelike nature of nonmatter.

METAPHORM 11. The Supernova—*representing the birth of matter* And *nonmatter. When a supernova explodes, its core simultaneously implodes. The outer envelope of exploding material creates the remnants which ultimately form the material of neutron stars or black holes.*

These two forms of matter and energy, or "psi-phi," are integrally connected.

The connection between these forms of matter may be shown in the metaphorm of a plane mirror in which all objects move in relation to their virtual mirror images. The chief difference between the physical laws of specular reflection and the "virtual laws of mental reflection" is that the properties of the latter are opposite the properties of the former. This means that what bends, stretches, grows, etc., in our imaginations —and *by means of imagination*—doesn't necessarily correspond to all that we can physically bend, stretch, grow, etc., in physical reality. What we can *do with* matter we can imagine *doing to* matter, as any gifted animator will demonstrate proficiently. Where the world of matter dies to conform to the finite geometry of Euclidean space, the world of non-matter lives to change in the infinite geometry of Hilbert space. And yet both worlds comprise one world as implied by the word *universe*.

Billions of years later, as these real and virtual worlds of our universe continue to expand, the substances of both matter and nonmatter correspond to the unified human brain/mind (denoted by my symbol Ψ). At this stage in our brains' evolution, only the material world may be understood and explained. One means of explanation is inference and reason through experiment. We can reason towards objective, tentative truths. However, our explanations are all influenced by the transience of nature. This natural transience is also responsible for the transience of our languages which we use to explain, or represent, nature. Consistency and permanence are only two of nature's numerous illusions which constantly affect our understanding of the material world.

By contrast, the hidden world of nonmatter is only partially revealed through intuition. At present we can only visit this mysterious world with our symbolic abstractions in works of art and science. These abstractions allow us to experience and know nature's mysteries in a personal way, although this knowledge and these experiences are not reliable sources of explanation. Inconsistency and impermanence are known in the world of nonmatter as the bedrock of nature. They're nonillusive expressions of nature's being.

The processes of matter *And* nonmatter reflect human brain processes. These processes are influenced by—and influence—the asymmetries and symmetries of nature. To inquire whether or not the human mind is part of the continuum of the cosmos is to pose *a Möbius strip of a question*—with answers melding into other answers endlessly, without satisfaction of arriving at the truth. To state that the mind has evolved as long as the universe has evolved is to provide *a Klein bottle of an answer*—with questions and more questions pouring into a bottle with no inside.

No diligent evidence is offered here to strengthen this explosion-implosion hypothesis. The only observation from astrophysics that loosely anchors this hypothesis is the scientific model of a supernova explosion, which is itself still sketchy. Another model that can be used to weight this metaphorm is the explosion-implosion event of a nuclear pellet used in laser fusion reactions (Metaphorm 12). The imploding fuel would represent nonmatter and the expanding plasma would represent matter. As for the blast of laser radiation . . . God only knows where *that* came from!

Of course, this laser fusion image of *deuterium* and *tritium** pellets being bombarded doesn't live up to our sense or expectations of "the beginning." Neurocosmologists are intent on discovering a corresponding image in brain science which shows a similar dynamic at work. Presented in Metaphorm 13 is a diagram of a *synapse*, which is universally regarded as the means by which nerve cells communicate with one another.

---

* *Deuterium*, which is also called "heavy hydrogen," is an isotope of hydrogen. It is the principal fuel for nuclear fusion (and hydrogen bombs). *Tritium* is the heaviest isotope of hydrogen.

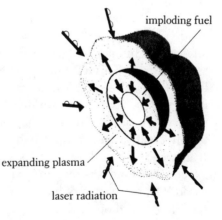

METAPHORM 12. Laser Fusion—*interpreting the creation of matter* And *nonmatter, or mind.* (Adapted from Archie W. Culp, Principles of Energy Conversion, 1979)

imploding fuel

expanding plasma

laser radiation

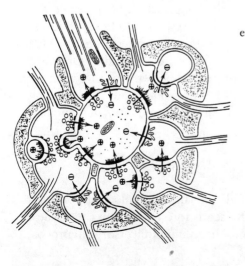

METAPHORM 13. The Synapse—*representing here the creation of matter* And *nonmatter.* (Adapted from E. D. P. L. Williams and R. Warwick, Gray's Anatomy, 36th edition, 1980)

This image lends itself to this discussion of the explosion-implosion hypothesis in the following way. The process of "imploding" in the nerve cell represents the *input of (neural) information* in the form of neuro-transmitters. This process could also represent the imploding of nonmatter. By contrast, the process of exploding matter would represent the *output of information* in the neurosphere, which refers to the physical behavior of the system.

Metaphorms 11 through 13 are processmorphs of this exploding-imploding dynamic of matter *And* nonmatter. Whether any of these metaphorms are good candidates for describing the origins of these two forms of matter is unimportant here. What is important is that they stimulate us to think of our minds' origins through the stimulation of *Thought-Assemblies*.

Most theoretical physicists and cosmologists look at the Big Bang of matter event (*without* the implosion concept). Neurocosmologists, on the other hand, see the beginning of the present shape of the cosmos as the Big Oscillation of matter *And* nonmatter (*with* the implosion concept). Whatever this event or moment was, it most likely was neither singular nor without these two forms of energy. One form is plainly visible to the senses while the other remains invisible (and undetectable). Both forms are held in relation to one another through the physical laws of symmetry.

It is with some respectful amusement in mind that I make the connection between the "growth" of a subatomic particle and the growth of an idea or a human being, referring earlier to the particle's states of "self-actualization." This term was created by the humanist psychologist Abraham Maslow. He used it to describe how human beings seek the fulfillment of basically two types of needs: Those crucial for survival, such as food, water, and sleep, and those needs critical for growth, which are largely creative and without end. Maslow once asked, "Do we see the real, concrete world, or do we see our own system of rubrics, motives, expectations and abstractions, which we have projected onto the real world? Or, to put it very bluntly, do we see or are we blind?"[3] Or do we see blindly, as though blinded by our fears and perceptions that what we see is essentially meaningless? Perhaps it is only one's powers of creating and conferring meaning that distinguish a rock from a human being.

IN SEEING OURSELVES LIVE AND GROW

From our tiny platform in space and from the technological portholes we peer through, what we presently see before us may be the modest activity in one relatively small area of the infinite universe. A neurocos-

mologist might think the insights and testimonies from our greatest theo-
reticians are like proton-size pebbles that barely interrupt the stillness of
a pond in some picturesque setting, the universe. One might imagine
the universe made up of billions of such ponds and lakes and oceans in
space, each one with different characteristics, properties, structures, or
forms than the next. Each one possessing subtly or abruptly different
"physical laws" than its relatives'. Each one having an area around it that
defies all the numbers we could conceivably use to describe this area and
the distances between the celestial bodies which occupy it. We may be
the center of *our universe*, but each of some 20 billion other intelligent
life forms may be the center of *their universes*. This is a positively over-
whelming little-big thought with absolutely no empirical grounding.
That's why it's presented in *Thought-Assemblies*.

This depthless thought was prompted by reading *A Brief History of
Time* which was authored by one of the world's most inspired thinkers,
Stephen Hawking. One illustration Hawking provided in describing the
physicist James Clerk Maxwell's theory of light propagation in 1865 was
of, in this cosmologist's words, "a three-dimensional model consisting of
a two-dimensional surface of a pond and the one dimension of time."
(See Metaphorm 14.) A neurocosmologist might imagine there are thousands of these ponds instead of one, where each pond represented another universe—another organization of matter (mass and energy) and
time-space. (Perhaps this is precisely the opposite conclusion Hawking
would have preferred that I arrive at.)

The reason why our physics doesn't support this supraview of bil-

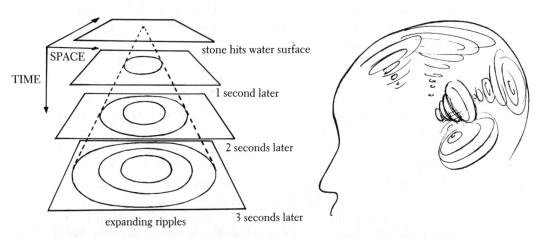

METAPHORM 14 (left) *and* METAPHORM 15 (right) *are to be viewed together. Metaphorm 15 shows interference patterns from subsystems in the "holographic brain."*
(Metaphorm 14 adapted from Stephen W. Hawking, *A Brief History of Time, 1988. Metaphorm 15 is an interpretation of Karl Pribram's model, 1971.)*

lions of worlds is that our physics relates to only our universe alone. At the very least this thought confronts the logic that propels us to think in one directed manner. After all, there could be as many "ponds" as there are neurons in the human brain. The ripples from a stone's throw form an ever-growing circle as time goes on.

There's an analogous relation in the brain: Stimuli are registered in each neuron or group of neurons in the forming of memories. The reference to neurons recalls the research of Karl Pribram, a prominent neurophysiologist at the University of California, Berkeley, who has described the layers of interactions involving memory in terms of holographic actions (Metaphorm 15). Although Pribram's illustrations differ from Hawking's visual notation, their patterns were evidently similar enough to trigger this association. Determining the meaning of this connection is undefinable at this time. The idea also happens to be indefensible.

In relating Hawking's discussion of the concept of time (in the context of black holes) to Pribram's discussion of the concept of memory (in the context of holographic processes), a neurocosmologist might suggest that there are some memories which are so dense and painful to recall that they behave like black holes. These virtual objects, which we can't see for the moment, seem to affect matters of memory within the range of their gravitational influences. A neurocosmologist may be inclined to think these black holes of consciousness meddle in matters of mind that our central nervous system would just as well prefer to forget!

This juxtaposition of concepts in *Thought-Assemblies* could imply that there are minute regions, or subsystems, in the human brain which are so powerful in their virtual gravity that any input approaching their region, or interacting with them, meets the same doom as subatomic particles approaching a black hole's event horizon. If one were to be aggressively literal here, one could cite regions and neurobiological processes of either normal or mentally impaired individuals who exhibit such disturbing patterns of behavior. But this neurocosmologist prefers to leave these speculations open for you to look into on your own through formal study or introspection.

These associations may be meaningful to someone searching for a possible (though not necessarily plausible) explanation regarding a specific psychological phenomenon—say, the emotional upheaval a person suffers when recollecting traumatic memories. To many neurocosmologists, these associations are a source of wonderment and study. This conceptual leapfrogging from one diagram, concept, or study to another is neurocosmology in action. The process of exploring something is synonymous with the art of leaping to—and science of landing on—new soil.

GROWING FROM WHAT WE SEE AND SEED

One point which must be flagged is that the anthropomorphic relations highlighted here are not intended to reposition humankind in the center of the cosmos. As though our presence were more meaningful in the larger scheme of things than the presence of insects. As though the universe could not function without us, which is most likely not the case. The cosmos will continue to live long after our departure from the annals of its history, a history no human being will ever live to read in its entirety.

The purpose of presenting these curious comparisons is to suggest that one facet of nature—the universe—may have some of our attributes just as we have some of its characteristics. However, the universe probably shares the qualities of other creatures as well. We may never fully understand nature without understanding all its creatures—those that are Earth-bound and perhaps the many inhabiting distant galaxies living in other ponds and on other planets.

TO SEED WITHOUT FEAR

The uneven unfolding of *Thought-Assemblies* insinuates that absolutely no theory of the brain or universe is so stunningly conceived and well-packaged that it fits everyone's idea of a good product, everyone's budget of understanding, and everyone's sense of a fair warranty. The reality is that all products break down either immediately or in an almost pro-grammed duration. That is like saying that all knowledge has its warran-ties, its expiration dates, and its hidden clauses sanctioning certain responsibilities of its manufacturers. The creators of relativity and quan-tum physics most likely foresaw a time when a new class of quantum field theories would be needed to describe nature's supersymmetry and expressions of gravity. A lack of this foresight would have suggested one of two things. Either creators know better than to design a perfect prod-uct, or model, that never breaks—meaning that it's forever current and operational—or they haven't figured out how to achieve this feat in the art, science, and technology of manufacturing and testing models. Like the chief executive officers of the Coca-Cola Company jested when their plans to improve their formula for Coke backfired: "We're not that smart and we're not that stupid."

STABILITY AT ALL COSTS

Some open-ended conclusions in *Thought-Assemblies:* There's just enough stability in nature to lead us to formulate theories of stability and constancy. And there's just enough instability in nature to disprove these theories. The fact that an idea can grow or mature demonstrates the paradoxically stable process of instability at work through acts of growth or evolution. What can be explained by our physical laws can just as easily be unexplained and untied by our mental laws with the introduction of doubt and skepticism—one being more lethal than the other. Both serve as equalizers. They recognize the flaws in our reasoning, such as: What we grossly oversimplify through metaphorms, we tend to overly complicate through the antitheses of these abstractions, induction and experimental proof.

   If you believe in the unity of mind *And* nature then you can envision the mind *as* nature. And with this belief, you see how the human mind *is* as unpredictably devious (counter-intuitive), mischievously straightforward (intuitive), and absolutely not-what-you-would-expect or could even anticipate through meticulous schemes or logistics. As we stand securely on the elegant carpet of wisdom which our brains have woven for us, unexpectedly this piece of security is pulled from under our confident stance. It's happened to every scientific discipline, from classical physics to modern physics to plasma physics to new physics. We've witnessed the disconcerting alteration of one prized theory and model after another, while the universal constants (as in the weak force and strong force constants, the Hubble constant, the cosmological constant) have weathered these paradigm shifts. The people most surprised are usually those who are foolish enough to stand obstinately on this carpet with both feet. This said, most of us are fools. Having never learned this lesson, we now stand not on a relatively "tame" rug, but on the roughest of paths in a jungle, tensely waiting to glimpse the supersymmetry and stout order of the universe—the "Tyger" (from William Blake's *Songs of Innocence and Experience*). And this immortal creature of our collective consciousnesses both guards us from and threatens us with this "fearful symmetry."

THE ILLUSION EXISTS BEFORE THE MIRROR, TIME, THE TENSELESS PAST-PRESENT IN THE PRESENT-FUTURE

Thus far I have examined how *all relationships* are inherent in the brain-universe relation. Part of the "work" of neurocosmology is to examine the processes of everything we create, irrespective of the historic period or time in which things were created. The concept of *time* is as much a creation of mind as the invention of clocks. And so time is a mental construct. Like every other tool of measurement, such as proportion, we use it to determine the duration of some thought, action, idea, or material realization.

Time is conceived as an abstract medium through which things change. We are now asked to understand the process of time in the broadest context—comparing the element of change between all of our creations and those of the cosmos. Time represents *the form of change, not the content*. It refers to "when" something changes, not "how" or "what" changes occur.

"What, then, is time? If no one asks me, I know what it is. If I wish to explain what it is to him who asks me, I do not know," wrote Saint Augustine in *Confessions*. Some further personal reflections:

> Time is change with or without measurement.
>
> Time is movement without beginning or end points. It is the "still point" which is part of the "string of 'still points'" connecting all persons, places, and things.
>
> Time is having tomorrow inform us about yesterday. A billion years later is a billion years earlier. "We're *there* already," with or without the memory of being there.
>
> Time is what minds do through the actions of the brain *without time constraints*. The brain was born without the physical concept of time. It's only our minds' notions of "biological clocks" and other measuring devices that tell us when to age and pass on.
>
> Time itself is ageless. It is as ageless as its process of change.
>
> Time is a paradox as in "out of time," or "time out."
>
> Time × Timeless = Zero . . . where history *is*.

TIMELESS HISTORY AS A PARADOX OF MIND

A history of mind that takes into account all the properties of time—as related in both poetic and scientific concepts—is a timeless history. It emphasizes *what* was thought or created, not *when* something was created. If the history of time doesn't include all of our imaginings, or free-associations of time, it may be regarded as incomplete. As intimated "before," time is an artifact of our conceptual nervous systems. Time, like mind, just *is*. To speak of the end of nature, or the end of life, or the end of the universe, or the end of time itself is to speculate about something that will remain a speculation.

Every concept of time reflects some aspect of the continuum of our mind, a continuum where events occur either in succession or not in succession. We may *live* in a time-space environment that is spatial, but we also *think and create* in a timeless-spaceless environment that is nonspatial. Unfortunately, or fortunately, we can't quite figure out the dimensions or proportions of this virtual space in the neurosphere. *The Wheel of the Law* sculpture in Buddhist art explores with astute realism this unproportioned, timeless space. Similarly, *The Wheels of the Sun Chariot* of Surya Deul Temple at Konarak, in northeastern India, interprets time as a cyclical form—where beginnings and ends are arbitrary points. You will find similar interpretations in particle physics and cosmology which postulate a cyclical pattern to the creation/re-creation of the cosmos.

The reason for this repetition of themes may rest on one straightforward fact: Our minds, like the universe, probably undergo similar cyclical transformations in their evolution. And so everything we theorize regarding this pattern of evolution takes place inside us. Since the human mind invented these notions, they probably reflect *in form and in substance* the nature of the mind itself. At least this is the central belief of this neurocosmologist. We know this fact firsthand because we live it. However, we don't quite trust all of our descriptions of this fact; they're either too precise or too vague. But we may be in for many more surprises when we learn that both of these perspectives—and every other perspective in between—are correct in their representations of the neurocosmos. That is, the universe seems to be nowhere and everywhere we imagine it to be.

The ancients heralded the thought that mind or spirit is the sole energy of the cosmos, and that our minds are the place for discovering this energy—the dynamic essence of matter. Is it any wonder that our sciences have come full circle in their search? At present some theoretical physicists are looking inward for answers that are to be applied to the universe's structure(s) and evolution(s). And some outstanding psychologists, on the other hand, are applying their inward science, so to speak, to looking outward for answers. Answers that will eventually be reapplied to the human mind and social systems. To me, this strategy seems like a pretty healthy process of creative seeing, however objectionable it is to those who are not willing to see beyond the surfaces of the material universe.

Meanwhile the arts have for the most part separated themselves from their earlier marriage with science, technology, and the whole of culture. However, they are undergoing another cycle of evolution that may result in their reunion with life. Perhaps certain artists at the close of this millennium will become forceful world leaders—as synthesizers and generalists and new specialists—whose theoretical art will inspire others to embrace all forms of thinking and representations as an art of life. The mental environment supporting this next wave of philosophy may be better conceived without permanent boundaries and without the despair of making sense at the expense of our physical senses or fragile environment.

These ideas will hopefully challenge you to rethink what you think you know. They're the clearest examples of concepts that push the limits of reasoning and the material world around, from which we reason and intuit the nature of our mental activities. I mention these knots in science to introduce the second part of my first statement about deciphering our brainways: "In decoding the brain, we need to discover the nature of the brain's creations." This statement is easily as large in scope as its counterpart. To decode the mind first requires deciding what this thing called mind is and how it works. Only then will we understand all the forms of its coded messages and means of encoding.

To decipher the ways of our brains is to determine the meanings of its actions despite their fuzziness. It's not that we will try to confuse reality, it's just that in order to see reality we have to consider how confusing the reality of the brain is. In all our considerations, we may find ourselves settling for the least confusing descriptions of brain and cosmos, which may be "none of the above."

WHERE ALL MEASUREMENT IS MEANINGLESS

As *Thought-Assemblies* expresses, everything can be generalized, metaphormed, contextured, and factualized. Everything can be detailed—meaning written, spoken, or rendered in detail—paying close attention to the parts in concert with the whole. But nothing can visualize all the details of nature, or all the ways of its *being:* neither quantum mechanics, nor relativistic cosmology, nor Buddhist contemplations of timeless and structureless existence, nor poetic reflections on the mindful operations of nature. Although each of these clusters of thought is a self-consistent world—possessing its own special context—these many worlds can be interrelated through the process of contexture. Thus, when quantum wave functions—the essence of quantum physics that describes the processes connecting all material forms—are applied to the study of our minds and consciousnesses, the original meanings of these functions are transformed into a new vision of our integrated nature. However, the metaphorms which describe this union are no closer to representing the whole reality of our nature—half of which is measurable, whereas the other half seems to stretch beyond all measurement.

To interrupt this pronouncement: There's a special reward in experiencing this work. It's the joy of recognizing one's own feelingful thoughts about the nature of the universe in another's thoughts. This is one of the purest and most direct forms of communication—this feeling of connection, of confirmation, of truth. One is touched by something which feels like truth in that it has—like aesthetics—no limits in its capacity to delight.

All places are alike . . . in the universe.—ALBERT EINSTEIN

All things are alike . . . as processmorphs.—A Personal Corollary

16
Embodying Fact in Fiction— The *Encoded Monolith*

For many, the mind is no larger than the brain—inch for inch, pound for pound. And the universe is no more than what can be seen in full bloom in front of our eyes. The idea that these two processmorphs are intertwined—and inseparable—lies beyond immediate comprehension.

Even if we could render the most astonishingly realistic portrait of our relationship with nature, this relationship would still remain tentative and mysterious. Perhaps what we're describing as "mind" or as "universe" really isn't the *whole nature* of mind or universe. We may, in fact, be modeling some other form of matter that has evolved since the era of superheavy particles some 10^{-35} seconds after the Big Bang. Like these particles, no one has ever seen the *primary effects* of the mind. We've only seen its secondary effects, or artifacts, which we're able to subject to analysis. An analogy which comes to mind is the rhythmic dance of a wheat field choreographed by moderate winds: The winds are visible only through the velvet motion of the wheat. We can only feel the winds' presence and sense how their constituent particles move in concert, like observing the arabesque tracks of elementary particles in a bubble chamber. Although studies in fluid dynamics help us understand the interactions of particles that determine the motion of a fluid or airstream, there are still many details that are unresolved about these dynamics.

Without a coherent map between the processes of our thoughts and their physical realizations, we will only be able to speculate how our minds are processmorphs of the things they create and which create our minds. Our clearest thoughts on this subject still resemble particles in Brownian motion, for example, whether or not we're able to observe the irregular, random movements of our thoughts.

No science can, as yet, determine the nature of processmorphs, any more than they can accurately classify processes. The reason for this has to do with experience, perspective, and methodology as they inform our notions. We have only recently become aware of the fact that all forms of information can be connected. Although this is not the official procedure of the scientific establishment, developments in computer science and imaging technology are hedging towards this conclusion. This said, there's relatively little crossover of insights between the disciplines of brain science and psychology, astrophysics and cosmology. What is discovered in one domain of knowledge usually remains in that domain. Through magnetic resonance imaging, for example, neuroscientists can see into the proton matter of the innermost regions of mind. Their findings have never been formally entertained in the context of astrophysical and astronomical research, which includes eavesdropping with radiotelescopes on "conversations" (or galactic interactions) from the outermost, red-shifted regions of the cosmos. And yet, both regions look pretty much alike on either end of the physical spectrum. Both are open, spacious abstractions which seem to merge.

CREATING THE ENCODED MONOLITH:
A METAPHORM CONNECTING MIND AND COSMOS

In Chapters 14 and 15, I discussed works of artscience that embody many of the concepts explored in neurocosmology. I introduce here another work, which is called the *Encoded Monolith* (Metaphorm 1). It repre-

METAPHORM 1. The Encoded Monolith (1980–1990). *This stainless-steel form consists of movable interlocking drawers containing hundreds of drawings, paintings, visual notations, writings, and sculptures. The work documents more than 12 years of thinking and explorations on the themes of neurocosmology by the author. (Courtesy of Ronald Feldman Fine Arts, Inc., New York.)*

sents one of an infinite number of interpretations of our minds as processmorphs. I will devote the remainder of this chapter to discussing this brain-universe metaphorm.

Encased inside the *Encoded Monolith* are hundreds of visual studies which interpret the themes of this book. The sculpture exists as an ageless library of interpretive works. When the artifacts within are recalled for exhibitions, like memories they reveal the life and content of the Monolith.

This metaphorm is organized into three interconnected columns of drawers. The left column represents the Art Section, which interlocks with the right column, or the Science Section. Each drawer is divided into a "pure" or theoretical and "applied" area. Where these drawers virtually overlap in the "applied" area is the third column, or Technology Section. The terms *pure* and *applied* refer to the realization of the works contained in these drawers.

DRAWERS AS CHAPTERS AS CONCEPTS AS GALAXIES

The *Encoded Monolith* is the sculptural form of this book, meaning that every chapter in *Breaking the Mind Barrier* relates to each of the 42 interlocking drawers of this sculpture. Like its contents, it, too, is a processmorph of the brain. The columnar shape of the Monolith implies that the human mind—like the universe—can be rigidly compartment- alized (Metaphorm 2). And at the same time, it can be interlocking or integrated. It can contain—or *be*—all structures of information. These reflections on its container-like form are further emphasized by the fact that there is a series of numbers engraved on the face of each drawer. The numbers span from 270 degrees to 90 degrees and back again, which shows them to be identical to the numbers on a radian protractor.

METAPHORM 2. The Monolith Metaphorm—*representing a uni- verse of processmorphs where all forms of matter and energy and information are integrated.*

So this universe appears to be orderly. In fact, its organization can appear to be systematic to the point of being predictable—and thus, to some, uninteresting. Having reached this conclusion based on appearances alone, the following thought rarely rushes to the center of one's brain: Maybe this seemingly coherent "form"—with its firm planes of polished steel—is anything but well-organized, solid, or silent. Who would guess that beneath its cool-gray veneer exists an active furnace of thoughts in some chaotic heat. Even though we rarely see the whole work of art, science, technology—or the mind *in, behind,* or *of* the work —we often foreclose on envisioning life in this hidden territory. Hence the true content of this sculpture is concealed (or revealed) selectively and only in part. Its wholeness cannot be seen directly nor can its primary effects be shown. If these statements speak for our creations, then what do they imply about the universe? Nothing? Or something? Or nothing with the potential of becoming something?

Eventually, after opening these drawers and probing the physical thoughts, just as an astrophysicist might search a galaxy to study a select group of stars, you realize there's no hierarchy in the organization of the drawers—either between or within them. Nor is there any hierarchy between the individual works inside each drawer. You also notice that no drawer is independent of the others even though there are, by all counts, 42 individual structures that are physically separate. The distribution of "galaxies"—or the concepts numbered in each drawer—is visible in one instance and completely invisible in another.

"This visible/invisible property, which is unique to art, may also represent some curious feature of the whole universe," you might imagine. The work suggests how something so tightly constrained and expressionless on the outside can be so utterly impulsive and expressive—even chaotic—on the inside.

After this first wave of realization, you further realize that the only way you're going to know what the Monolith is about is by experiencing its contradictions and paradoxes. This amounts to free-falling from the heights of your imagination without leaving the ground. It involves suspending and reshuffling your logic like an inexperienced Rubik's Cube player who fumbles around with the components of the Cube—growing irritable with each clumsy trial-and-error. One moment the numbers on the surfaces of the drawers make the smoothest sense; they implicitly order the information inside the drawers or imply that order is in the making. The next moment, they make no sense whatsoever—especially if there's nothing inside, or nothing inside to which the numbers refer. This analysis of ambiguity can deprive us of our confidence, especially those of us who declare that nature is only what we understand it to be analytically, not what we don't understand. One consolation: In the weightless state of falling, the unknown seems to float with you instead of weighing you down with insecurity.

After the second wave of realization, it may occur to you that everything you choose to analyze and model regarding the universe of the *Encoded Monolith* only reflects your personal experience of this object. Regardless of the physical evidence in the form of its apparent height, shape, texture, etc., your analyses represent only what this object could possibly be or mean to you, not what it is. Discovering what it *really is* may require relating every detail of the physical and imagined world to it —as though it is all these things and more.

Everything you imagine this work of artscience is, *it is*. If you see the Monolith as an open universe that is constantly expanding, it is nothing less than a hyperbolic curve. If you see it as a closed universe that has only its fall to entropy to look forward to, it is nothing more than a sphere. If you see it as a flat universe—as flat and hard as its steel surfaces—then it is hardly anything but a flat universe with nowhere to go but to move on the same plane with zero curvature. If you think the Monolith is comical and playful—like determining the "charms, colors, and flavors" of quarks and other subatomic particles—indeed it is! If you think its indecipherable and unnecessarily complicated, it *is*—at least to you personally. The only thing it isn't is "nothing." There's no such thing as nothing as in "no-thing." Whether you think anything of it, it *is* something—even if this *something* is other than what you're prepared to see it as: the universe or art, the brain or science, and so forth.

The shock of the third wave of realization is subtle but stirring. Slowly, this visible energy form—the Monolith—is understood as a formless process, not as a processless form. It moves like a verb; it doesn't stand as a noun. It is the act of metaphorming you're experiencing as you examine this metaphorm. Like all processmorphs, the Monolith can take any form your imagination sculpts or makes of it, shaping its symbolisms into new associations and meanings. Once again, the orderly nature of the work harmonizes with your first impression. You discover it can be both organized in its disorganization and highly disorganized in its construction.

A CONSTRUCTION OF ENDLESS MOTION AT THE EDGE OF CREATION

Neither the size of this sculpture, nor the material used in its fabrication, nor its moment of birth is fixed like its outer form suggests. (Metaphorms 1 and 2 merely indicate one "manifestation" of it.) The Monolith is open to your imaginings. The distinction between imagination and reality we're so determined to preserve implodes here. Like a supernova in its death throes, the distinction between these worlds collapses in eerie silence. Whatever arbitrary scale you assign it as a viewer, either it expands infinitely beyond your definition and sense of scale or it contracts to a point as "small" and "hot" and "singular" as the superdense, primordial universe. Even though technically the Monolith was built in 1980, it is dateless. Even though it measures 7 feet high by 3 feet square, it is sizeless. Even though it was constructed from stainless steel, it is metaphorically materialless.

All these details documenting its creation are optional in the world of processmorphs to which it belongs. It is an embodiment of lifeless details only in its static appearance, not its symbolic content. That is, the life of its symbolism can't be measured according to the practices of reductionists who attempt to quantify phenomena.

As a processmorph the Monolith reflects, transmits, and absorbs information, for example, just as all forms of matter or electromagnetic radiation do in passing from one "medium" to another. Here *the medium is the mind*, and not just the materials used to construct this sculpture. It is the mind of both the Monolith's creator and its interpreters. Moreover, the sculpture shares these same processes with all forms of mirrors, transmitters, and absorbing mechanisms, regardless of scale and application. The fact that it doesn't look like any one of these things—or that it can't be employed functionally as some reflector, or antenna, or shock absorber—doesn't mean that its processes are unrelated.

Like the universe, the Monolith has the presence of an ageless vault. It seems to be both prehistoric and futuristic, as it is occupied by timeless processes of human and cosmic creations. In its conception, "time" and "history" are nonexistent, like footprints in a desert erased by wind and sand. Only the information inside the sculpture offers clues to its origin and symbolisms. Although this information is not accessible to the eyes at first sight, it is approachable through intuition. Like theoretical physicists who open the secrets of nucleons with their intuitive thoughts first rather than particle accelerators, or neurobiologists who unveil the mysteries of neurons with their speculations preceding dark-field illumination microscopes, the casual observer is left figuring out how to unmask the Monolith's treasures in discovering its meaning. Is this process of discovery any different from an astrophysicist's course in uncovering—or learning—the "nonluminous" dark matter that is believed to make up more than three-fourths of the mass of the universe?

THE ABSTRACT CONTENTS
OF A CONCRETE UNIVERSE

Without actually opening the drawers of the Monolith, without actually touching or inspecting its contents, the observer is led in and out of different ponderances regarding its nature. Suddenly, these questions seem to be embedded in their own answers, like anti-particles are embedded in their counterparts, photons. As a cerebralist, you sense that the light generated from this process of questioning-answering is critical for discovering the relationship between the Monolith—as cosmos—and the mind, between abstraction and realism. (See Metaphorm 3, which is one of the visual notes from the "applied" side of Drawer 270 degrees in the Art-Technology Section of the *Encoded Monolith*.)

Does the information inside really tell us anything about the Monolith—or the reverse? How do we analyze the processes of this universe, if they can be related to any form imaginable? Even if we devise the most consistent and plausible theories about its contents, properties, and purposes, can these theories define the nature of this object without having ever physically touched it? Do we need an infinity of perspectives to describe its infinitude? Can we trust our sense-perceptions more than our thoughts and thought experiments in relating our knowledge of this object of nature?

There are no patented answers to these questions, just as there is neither a "final analysis" of the universe's content nor a single viewpoint that prevails. There are, instead, interminable analyses of the Monolith's potential content, which includes both its probable and possible meanings. Depending on the context in which this work of artscience is viewed, its meanings are as boundless as its potential forms.

MESSAGES WITHIN THE MONOLITH

Some of the drawings inside the *Encoded Monolith* explore ways of describing the properties of mind with the same degree of precision that quantum physicists describe the transitory, virtual particles of matter. These speculative drawings conceive of the "quantum mechanical mind" as the essence of the four interactive forces that determine the myriad of forms of nature. They suggest how our minds are the probability patterns visualized in computer simulations and symbolic models. *Virtual particles of mind* are shown to react violently when compressed in space like nucleons.

Other drawings speculate that if our minds share the processes

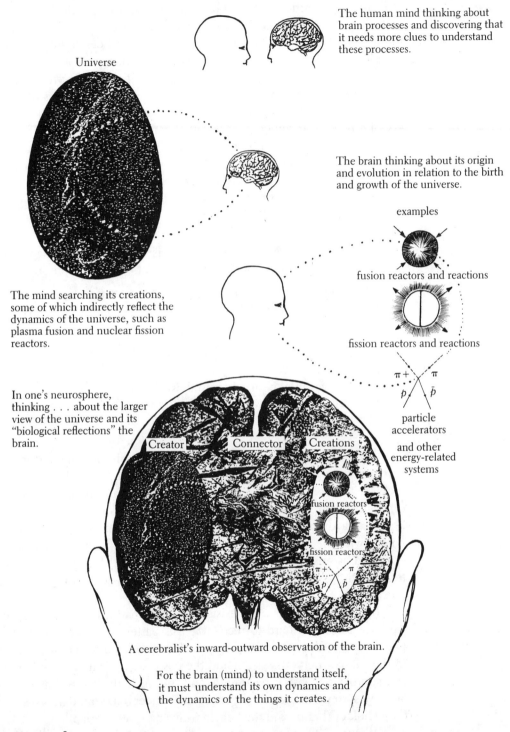

The human mind thinking about brain processes and discovering that it needs more clues to understand these processes.

Universe

The brain thinking about its origin and evolution in relation to the birth and growth of the universe.

examples

fusion reactors and reactions

fission reactors and reactions

The mind searching its creations, some of which indirectly reflect the dynamics of the universe, such as plasma fusion and nuclear fission reactors.

$\pi +$ π

p $\bar{p}$

particle accelerators

and other energy-related systems

In one's neurosphere, thinking . . . about the larger view of the universe and its "biological reflections" the brain.

Creator   Connector   Creations

fusion reactors

fission reactors

$\pi +$ π

p $\bar{p}$

A cerebralist's inward-outward observation of the brain.

For the brain (mind) to understand itself, it must understand its own dynamics and the dynamics of the things it creates.

METAPHORM 3.

of elementary particles, then their forms of expression reflect the forms created by these particles. If the quantum world we fascinate over is mind-related, then the "charged states" of different elementary particles describe the *charged states of mind*. In measuring photons, neutrinos, electrons, and the other classes of particles that are part of the camp of leptons and hadrons, we're indirectly *measuring the elements of mind*. Our minds are naturally all these things—and much more. The discovery of our nuclear minds is an essential connection in a world of irresolute philosophies and embittered scientific perspectives. Where nuclear matter is the fragrance of nature—which we can smell but cannot see— our minds are the essence of this fragrance.

Some questions form in response to these thoughts, many of which are depicted and stored in the Monolith.

The speed of electrons may be measured by time-of-flight techniques. How can we measure the speed of thoughts?

The kinetic energy of electrons may be measured by calorimetric techniques. How can we measure the kinetic energy of feelings?

The dilation of time may be shown by using radioactive decay of cosmic-ray mu-mesons. How can we show the dilation of time in the mind?

Light waves may be used to establish color measurements in terms of energy. How can we use light waves to establish measurements of energies of the mind? What are the forms of these energies?

These types of musings are meant to enliven scientific and mathematical descriptions of our sole universe. As philosophical probes, they seek to discover "the thought through which all things are steered through all things"—a thought Heraclitus expressed before the era of Socrates, a thought which sees through nature's masks. They envision nature without secure answers, observing its actions, which may be interpreted as answers. These actions steer all things through the eyes of the reflector—the mind's eye which sees nature's actions for what they are.

In the thick of our views exists a common pattern of perception, but we have to search hard to recognize this pattern. As I've stressed, the pattern is *a weave of united processes*. We now need to see this "weave" in the human nervous system and the cosmos's system alike. The wider and sharper our vision becomes—and the more flexible our definitions of these systems become—the more the conceptual boundaries between the beholder (brain) and the beheld (universe) will overlap.

When we fill our perceptions with nothing but the details of the brain and cosmos, we court disaster. Without constantly refocusing our vision (and values), we live with a nearsighted outlook, staring at the "skin" of these entities. And we don't even see this skin and details very clearly because the context in which these things are seen is either fragmented or blurry or both. Consequently, the way we live reflects our reasoning: narrowly focused in one sense and totally unfocused in another.

Like the human brain, the Monolith remains a study set apart from astronomy and cosmology. It remains an isolated objet d'artscience that is as distant from the epicenter of our consciousness as the plight of nuclear weapons, environmental pollution, and the population explosion. It's as removed from our immediate concern as the idea of a million years hence. How ironic, considering that this object—like the brain itself—is part of the fuel, vehicle and driver for discovering the universe.

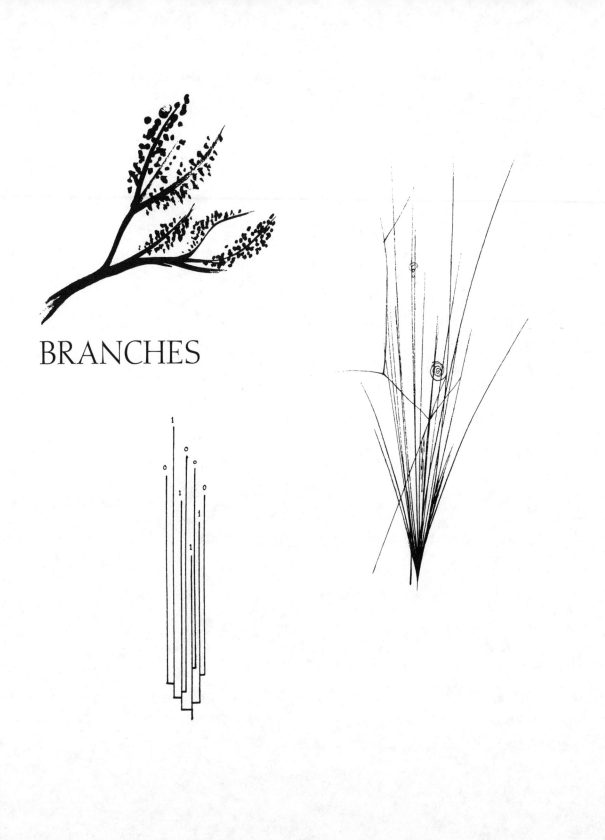

BRANCHES

Grenzüberschreitende Kunst—Art which crosses borders.

Branches #1

The visual notations, drawings, and artwork presented thus far in this book represent a fraction of the media and artistic forms that can be used in expressing brain-universe metaphorms. Neurocosmology incorporates the inexhaustible freedom and open-endedness of contemporary art, qualities which often parallel and overlap the freedom of inquiry in contemporary science. The artscience work of neurocosmology exists in all environments including, as I've discussed, the neurosphere. These works can be seen in art museums as well as in the technological environments of particle physics laboratories, spectroscopy labs, plasma fusion research centers, bioengineering facilities, materials research testing labs, mechanical engineering and computer-aided design labs, etc. The more scientific and technologically oriented works are only appreciated as artforms to the degree that we're prepared to see them as art. Here the definition of art is expanded to include all pictures or portrayals of thought—from the most prosaic to the quixotic. It also embodies the changing constraints we use to determine the quality of our thoughts and creations.

EXPERIENCING A.R.T. (ALL REPRESENTATIONS OF THOUGHT) AND EXPLORING METAPHORMS IN EVERY CONCEIVABLE CONTEXT

In the pages ahead I offer some works of contemporary art that represent a mastery of metaphoric thinking. They also reveal an exceptional sensitivity to the whole of nature. I am convinced that these two qualities are critical for human development and survival. Note that the selection of artists is partial and illustrative rather than comprehensive. The point of commonality in these works is that they all stimulate for me similar heightened awarenesses of nature, connecting—or erasing—the zones between earth, sky, heavens, the human body and neurospace. Some even intimate how the *act* of creation in art and the creations themselves connect with cosmic systems. Although only a small facet of the arts (specifically, the visual arts) is presented here, I am hoping this facet provides enough clues to future explorations in artistic inquiry applied to neurocosmology. The fields of music, film/video, performance, theater arts, media arts, literature, and other ageless branches of the arts are also part of this exploration. Certainly the works of Bill Viola, Laurie Anderson, Michael Snow, and Hollis Frampton, among others, are exemplars.

The following examples are also intended to turn our awareness toward a more complete notion of the "cerebral" in art,* where the term "cerebral" embodies feeling and acting and creating, as opposed to cerebration alone. I share the views of William James, who, in 1890, described "cerebralists" as people who combine the sensual and the intellectual, the physical and spiritual. Somehow this earlier meaning of the word *cerebral* was lost or diminished to connote analytically dry and emotionless thoughts or creations. As many ancient Eastern and Western cultures confirm, art may be *cerebral* and, at the same time, immensely physical, drawing upon all disciplines, media, and spirited expressions. The allegorical sculptures in traditional Hindu and Buddhist temples, for instance, are works *of* the mind, *by* the mind, and *for* the

---

* In the context of neuroanatomy and neuropsychology, *all* art forms are technically speaking—or quite literally—"cerebral" in nature. A work of art, no matter how expressive, impulsive, or loose, is directly influenced by the *cerebral cortex* and its subcortical connections. Somehow this fact eludes most discussions of aesthetics and introspective accounts of emotion.

mind, but their creation and experience involve the whole body. Similarly, the symbolic stone constructions and metaphoric earthworks in the British Isles are both sensual and mindful in their forms. All of one's senses are nourished in meditating on their meanings.

Having worked at the Center for Advanced Visual Studies (CAVS) at M.I.T.* for the past decade, I find it appropriate to first mention the work of Gyorgy Kepes, the founder and first director of CAVS.[1] Kepes's artistic researches of natural phenomena have yielded new expressive forms which reveal the reshapings and novelties of nature. His photographs, for instance, *Topological Forms* (1940) and *Cosmic Gyration* (1977) (Metaphorm 1), are suggestive of the peculiarities of mental images—or "formings" of both the human mind and cosmos—that are immeasurable in terms of time, duration, size, etc. These images seem to "occur" over milliseconds and hours and eons; their virtual height, width, and depth can represent parsecs (where 1 parsec equals 3.26 light-years) or centimeters.

* The Center for Advanced Visual Studies at the Massachusetts Institute of Technology was established in 1968 to provide a forum for the arts to interact with potentially all disciplines and media in exploring uncharted possibilities of creative expression. Many of these explorations involve collaborative projects between artists, scientists, mathematicians, engineers, and scholars, where the emphasis is on experimental and environmental art.

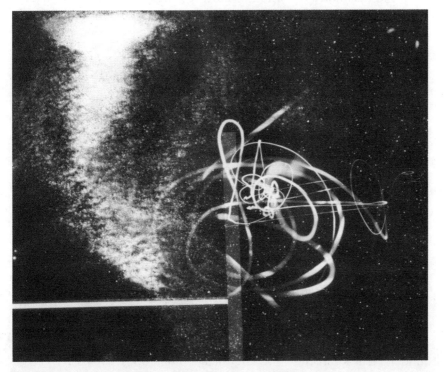

METAPHORM 1. *Gyorgy Kepes*, Cosmic Gyration, 1977.

Other metaphorms, such as *Photo-Elastic Walk* (1970), *Flame Orchard Sound-Animated Gas Flames* (1971), and *Sound Programmed Projected Magnetic Fields* (1973), integrate art and technology in an intimate and sensual way as they expose the poetics of the natural world. Light sources as varied as flames and polarized light become kinetic artforms that amplify the silent stirrings and forces of nature. Stepping on transparent tiles that glow from stress or fatigue in *Photo-Elastic Walk*, for example, inquisitive participants might metaphorm the properties of photoelasticity—recognizing in these effects similar properties of the human mind. Stimulating this sort of expanded consciousness is a primary impetus in these works, as Kepes shares with us in writing:

> The most convincing artistic forms of our time are inner models of structural vitality and social relevance. They give us confidence that in spite of everything there is still quality to life. We can put them to important use, first in the reshaping of the man-made environment in accordance with our best physiological and psychological interest, and, second, in the shaping of our inner world so that our sensibilities and our outer world harmonize.[2]

Another individual who has nurtured these art and technology interactions for the past three decades is the current director of the Center for Advanced Visual Studies, Otto Piene. Through his environmental artworks and international Sky Art Conferences, Piene has broadly focused on both the aesthetics and phenomena of sky, space, and light—including the spacious skies of our mental life. Some of his earlier constructions, such as the solar-like *Onion Flower* (1964–1965) and *Milky Ways* (1964–1965) (Metaphorms 2 and 3), show the origins of his rich concept of Sky Art.

It's scarcely surprising that at the time the legendary artists Jasper Johns, Robert Rauschenberg, and Joseph Beuys were peering into the night of art's future, there formed in Düsseldorf a series of eight "Night Exhibitions" (1957–1958) presented by Otto Piene and Heinz Mack, who had recently organized an artists' collaborative called "Group Zero." The Group challenged its audience to look beyond the confines of traditional artforms in the temporal space of, as Piene described, "a vernissage at night without an exhibition lasting any longer."[3] The goals of this "New Tendency Art," as it came to be called, were to explore the ephemeral phenomena of light, vibration, and experience and to revision the relationship between nature, humankind, and technology.

One of the most ambitious collaborative creations of CAVS artists under Piene's direction was the metaphoric work *Centerbeam* (1977).[4] This colossal, new-media, kinetic metaphorm was part "building," part

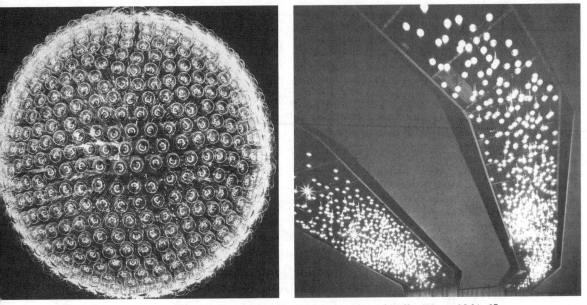

METAPHORMS 2 AND 3. *Otto Piene*, Onion Flower, *1964–65, and* Milky Ways, *1964–65.*

"sculptural fugue," part "art machine," part "aqueduct," part "nervous system." Its elongated linear form combined, among many other dynamic components, laser-projected drawings and music/sound experiments (Paul Earls, Gyorgy Kepes), holographic projections in steam (Harriet Casdin-Silver), steam forms (Joan Brigham), ice structures (Carl Nesjar), and performance and sound explorations (Chris Janney). In commenting on the scope of this work, Piene wrote: "The pleasant and entertaining naivete of 'Centerbeam'—along with its sophistication, ambition, and multimedial convincing power—is the basic naivete of art: the attempt to capture the universe in a nutshell. Our nutshell is 144 feet long; it projects sensory apparatus (mouths and eyes) onto—and sometimes impersonates—a cloud."[5]

The writings, experiments, and metaphorms of the environmental artist Robert Smithson exemplify the arts' versatile searches and expressions. Works such as *Non-Site: Franklin, New Jersey* (1968) and *Spiral Jetty* (1970) challenge viewers to consider the connections between one's personal worlds and the natural world. These works in particular have provided much inspiration for those seeking to move beyond conventional media in interpreting the arts' relationship with life. Smithson's interests ranged from investigating the forces of entropy (as in floods, earthquakes, construction sites, and other visibly chaotic or catastrophic structures) and geology (strata and earth formations) to language (its forms, contents, contexts) and human values.[6]

One artwork that is immediately relevant to the reflections in this book is Nancy Holt's metaphoric environmental work *Sun Tunnels* (1973–1976), constructed in the Great Basin Desert, Utah (Metaphorm 4). This work engages viewers to experience the constant reorientation of one's mind as one adjusts to the expansive surroundings. The desert space—like the desert of deep space—is "re-sized" through her art to fit the human experience and comprehension of the landscape. In Holt's words, "the center of the work becomes the center of the world."[7] *Sun Tunnels* is organized in an **X** shape, consisting of four massive concrete cylinders. Each cylinder has holes in its top surface corresponding to four constellations that mark the corners of the celestial sphere. As you enter the tunnels, the sky and astronomical world are pulled directly into your neurosphere for contemplation.[8]

In a related way, Robert Morris's ancient-modern metaphorm, the *Observatory* (1971), reveals solstitial events. It also recalls the aesthetic effect of neolithic works that presumably showed events such as equinoctial sunrises.[9] There are also Charles Ross's *Sunlight Convergence/ Solar Burns* (1971–1972), *Star Maps* (1973–1977), and, more recently, *Star Axis* (1985), which explores the relations between the universe's motion and ourselves. As Ross describes, "by walking through this work [*Star Axis*], one will be able to directly experience the entire 26,000-year cycle of polar precession."[10] Janet Saad-Cook's visually beautiful *Sun Drawing: 37° 40′ North Latitude* (1982) sculpts sunlight reflections cast from reflective materials that are organized according to the subtle passage of the Sun's light and the imperceptible roll of the Earth's rotations.[11]

Another work which increases our sensitivity to the human-nature relation through metaphorms is Walter De Maria's evocative environmental sculpture *The Lightning Field* (1978). The 1 mile × 1 kilometer cluster of 400 steel poles on the elevated plains of west-central New Mexico—each pole being equally tall, in vertical position, and organized in an orderly grid—charges one's mind with as many visual associations as the electrical strikes the field receives during the thickest lightning storms.[12] The contrast between this tense, tightly composed artwork and the relaxed expanse of the flat plains evokes feelings of imminence and suspense, as though some new order were soon to emerge from the union of these two "compositions" of nature.

Using more technological media, Juan Geuer's seismographic apparatus and metaphorm *Terrascope* (1965–1987) detects in real-time the rhythmic surges of magma and other geological forces that are constantly manipulating our solid continents. Also, Tom Van Sant's *GeoSphere, The Earth Situation Room* (1987–1990) is an artscience work that visualizes the environmental and geopolitical aspects of global changes on an

METAPHORM 4. *Nancy Holt*, Sun Tunnels, 1973–76.

Detail of Sun Tunnels.

automated 21-foot replica of the Earth. Employing advanced interactive computer technology, Van Sant and a team of artists and scientists are constructing a computer-driven, special-effects sphere that will receive real-time and time-lapse projections of weather-related events and other information pertinent to the environment. *The Earth Situation Room* will also explore the implications or consequences of various climatic changes, examining the impact of the human condition.[13]

In between these physical reflections on geology, ecology, and our technological sophistication are the media works of George Bolling, a former Artist-in-Residence at San Francisco's Exploratorium, and journalist connected with NASA's Ames Research Center. In the mid-seventies, Bolling assembled a "live viewing" of the Jupiter Flyby Mission, bringing together artists, scientists, theologians, among others who were interested in espousing on the man/cosmos relationship. This successful media event was followed up by a similar series of presentations and viewings in conjunction with the first Mars landing.

The focus on ecology in the arts has many of its roots in the pioneer conceptual artist Hans Haacke's works such as the metaphoric sculpture *Condensation Cube* (1963), for example, which uncovers other aspects of our changing environment in a poetic and equally telling way—relying on one's technological imagination for "projections" concerning the human condition.[14]

Alan Sonfist's environmental metaphorm, the *Microorganisms Enclosure* (1971), provides other insights into climatic changes in general. In this particular work, Sonfist investigated the competitive interactions of bacteria and fungi which he grew in a clear box. To his surprise, the colors of these life forms changed as they vied for dominance and survival. Moisture formed inside the box as a result of the metabolic processes involved in these competitive activities.[15] In *Crystal Monument* (1966–1972), Sonfist explored the responses of crystals, which he encased in a synthetic globe, to the climatic changes occurring outside the globe. In an explicit way, these "external" changes affected the forms and positions of the crystals.[16] As you contemplate this metaphorm, you may wonder whether the life forms and inorganic material on this planet are similarly affected by the "climate" of the cosmos. This is one contemplation stimulated by Sonfist's investigatory approach to seeing the empire of nature in all its splendor. This approach is forcefully stated in the artist's proclamation of intent:

> My art presents nature. I isolate certain aspects of nature to gain emphasis, to make clear its power to affect us, to give the viewer an awareness that can be translated into a total unravelling of the cosmos.[17]
> . . . all great art has derived its meaning not only from artistic quality but from the social needs of the time. Our overriding social need

is to develop a sensitivity to nature so that we can preserve our planet.[18]

Exploring other ecology metaphorms, conceptual artists Helen Mayer Harrison and Newton Harrison attune their audience to our changing environment—concentrating on those changes in nature caused by human intervention. Metaphorms such as the *Lagoon Series* (in the mid-1970s) and *First Meditation on the Condition of the Great Lakes of North America* (1977–1978) not only explore humankind's relationship to nature, they also consider the open role of artists as naturalists, conservationists, environmentalists, activists, and "information explorers" whose ideas can help ease the assault of ignorance on our endangered environment. The Harrisons expand this role—seeing artists as architects capable of conceptualizing plans to be implemented. In many of their works, one's attention is directed towards corrective measures rather than destructive impulses that spring from a catharsis of complaints. By inviting viewers to participate in understanding ecological developments through their metaphorms, both the Harrisons and participants can choose to work towards either facilitating solutions or observing needs and urgencies.[19]

In a different orientation, though keeping the natural world as a point of creative focus, the sculptor and inventor Thomas Shannon has ventured into other terrains of nature that are not directly visible—namely, electromagnetic force fields. A number of his metaphorms explore these hidden fields that power our lives and thoughts. Shannon's *Corner of the World Medium* (1974–1976) and *The Compass of Love* (1981), for example, offer some of the clearest reflections on the processes of energies—both past and future—that influence all "life" forms, which include those quasi-forms of life more readily attributed to organic matter. In *The Compass of Love*, a 24-foot magnesium needle hovers over a 12-foot aluminum dome. Concealed inside this structure is a source of electromagnetic energy (invented and patented by Shannon) that is responsible for floating the needle well above the domical base; the needle will supposedly float for thousands of years if left undisturbed. Both the title of this work and its dynamics set in motion associations that encompass everything from subtle human relationships to rakish encounters between various kinds of material and particle interactions.

The conceptual artist Panamarenko also combines art, engineering, and invention, in creating both real and mythic technological apparatuses such as flying machines, flying saucers, and other visionary metaphorms. His investigations of such things as magnetic fields, space flight, and star trekking are intended to provoke prevailing systems of science and at the same time to learn from these systems—applying the results from his experiments. Many of Panamarenko's models seem like they've been hastily assembled in order to quickly test a hypothesis and thus further modify an idea. They're insightful, candid portraits of the process of invention that may one day influence mainstream science. Between his note-making and constructions one can see the possibilities of an idea forming—and flying. Without the constraint of having to explain or convey his imaginings to anyone other than himself, works such as *Magnetostatica* (or "Project for a Flying Saucer," 1978), *Journey to the Stars* (1979), and *Magnetic Space Ship* (1980) remain unhindered by the potential throttle of formal scientific inquiry.[20]

The metaphorms of environmental artist and sculptor Piotr Kowalski also break the mind barrier of disciplinary knowledge in exploring relativity in science, art, and meaning. Kowalski's artistic experiments in the past 30 years have probed the boundaries of cognition, experience, the material of thought, and the immaterial of communication.[21] His exploratory activities have included sculpting large-scale steel structures by means of explosives (*Dynamite Formed Piece*, 1965); shaping novel mixtures of inert gases and electricity in glass to illuminate the hidden forces of electromagnetic fields (*Measures to Be Taken*, 1968); and creating conceptually vigorous structures for public spaces—architectural structures (*Place Pascal*, 1983) that challenge those who engage these spaces to rethink concepts of architecture.

Other explorations by Kowalski have concentrated on the perception of time and motion in human thought processes. His *Time Machines* (1970–1982) incorporated computer and cognitive science in exploring the spatial and temporal aspects of thought. Through interactive installations, the participants experience for themselves the eerie reversal of time-space as they either gaze into a full-length spinning mirror (which reverses space) or speak into computer-driven microphones (which reverse time). A third Time Machine reverses both time and space. These modalities are manipulated by a meticulously crafted computer system which is embedded in a steel container designed for traveling.[22] One's perceptions—and thoughts—are simultaneously reversed as they form in response to some specific questions about time-space posed by the artist. This work can jolt and disorient you as effectively as an evocative metaphorm can transport you to the outer reaches of your senses—where the events of mind are shadowless or too quick to recall.

Some of the most subtle—but powerful—metaphorms that can physically transform one's everyday vision of nature are Kowalski's holographic mirrors titled *Information Transcripts* (1989–1990). Through the holographic process, the artist has converted ordinary plates of glass into natural optical instruments that make visible the quantum movement of matter. Unlike the coherent, solid-state reflections in plane mirrors, and unlike the common holograms of objects with three-dimensional photographic effects, Kowalski's *Information Transcripts* consist of nothing but pure light. Pressed within a single pane of glass are what look like convex and concave "flat spheres," or multidimensional discs, that are brilliantly luminous. Their strange luminosity seems to be internally derived—like starlight—rather than being generated by some external light source. In viewing the surroundings through these mysterious lens/mirrors, one's world-reality is reconstructed in such a way that objects appear to oscillate between altered states of form and color. The oscillations create a sort of visual harmonics.

In contrast to these technological creations are the environmental artworks of Robert Irwin and James Turrell. Drawing the materials for their work from natural phenomena such as light, both artists engage the dialogue between light, space, and experience. Their visual metaphorms cue our awareness to the physical presence of illuminated matter and sense-perceptions. In one disc sculpture form, *Untitled* (1969), Irwin subtly manipulated an object under the cross-streams of luminous shadows, making the object appear strangely vaporous and unreal. Substantial form is thus rendered insubstantial by the viewer's mind detecting the fringes of coherent matter under the influences of light.[23]

Many of Turrell's works explore the opposite pole of our perceptions of matter. What is normally perceived as insubstantial, such as pure light, he manages to weight and shape as concrete "presences." The artworks titled *Rayzor* (1969) and *Laar* (1976) turn imageless walls into pools of light that define the borders of some depthless, geometrical space. The ambiguity of these light volumes and spaces prompts you to inspect them. In attempting to touch these seemingly empty walls, your hand dips into a void rather than resting on some secure surface. This illusion of solidity can be startling. In contemplating these solid light forms, it's possible to experience an almost dreamy sensation of falling through one space-time in the physical world into another.

Both Turrell's and Irwin's art stimulate viewers/participators to discover their own perceptual processes at work in sensing the world. The artists refocus our attention on experiencing "the uncommunicated"—consciousness—itself. The process of becoming aware, of seeing the presences of light, is as much a part of the artform as the formal elements of the work. Viewers can glimpse how feelings are touched and sculpted by these ephemeral atmospheres of light which inform one's thoughts in the twilight of experience.

Further probing the influences of light on perception, Turrell makes use of yet another medium more familiar to the ancient Celtic megalithic monument builders than to contemporary sculptors. His most compelling project, *Roden Crater*, rekindles the life of an extinct volcano in Flagstaff, Arizona, by creating unique spaces within the volcano as viewports to both the sky and mind itself. This aesthetically engineered work was conceived as a special observatory where one could feel, in the artist's words, the "space-looked-out-onto"[24] through the medium of natural light and insight, where visitors may use their whole sensorium in experiencing the world from the nests of these hidden spaces. Here a visitor is no longer a "viewer" but a participator in the creation of the art —an art which is mostly about their personal observations of the world. As Turrell articulates:

> What really interests me is having the viewer make discoveries the same way the artist does . . . instead of having the viewer participate vicariously, through someone else. . . . You determine the reality of what you see. The work is the product of my vision, but it's about your seeing. The poles of the realm in which I operate are the physical limitations of human vision and the learned limits of perception, or what I call "prejudiced perception." Encountering these prejudices can be an amazing experience, and if someone can come to these discoveries directly, the way the artist does, the impact is greater and so is the joy.[25]

The step inward through art and metaphorms is also an outward stepping—towards self-discovery and the jostle of enlightenment. This we discover in the conceptually intricate works of visual artist Lowry Burgess. Designed for deep space, Burgess's *The Quiet Axis* leaves our terrestrial world altogether. Like a message in a time capsule bound for nowhere and everywhere simultaneously in the physical universe, this truly cosmic art and life work lives in perfect solitude. The artwork was lifted into the outer world by the NASA space shuttle *Discovery* in 1988; it was the first Non-scientific Payload. The symbolic artwork was ritualistically composed over 10 years—its materials having been gathered from all over the globe and elsewhere. As Burgess relates: "The components of *The Quiet Axis* have their ultimate extension in the Large Cloud of Magellan and the immense spiral of the Andromeda Galaxy. Two of these components are already in place on Earth: *The Inclined Galactic Light Pond* (1968–1974) in the Bamiyan Desert in Afghanistan, and *The Utopic Vessel* (1974–1979) to the south-west of Rapa Nui, near Easter Island. A third object, *Gate into Aether, Sonic Wreath* (1976–1988), will shatter in space at the exact moment the space shuttle carrying it intersects with the axis of *The Quiet Axis*. The *Gate into Aether* contains distillations of water taken from eighteen major rivers around the world, and holographic recordings of ecstatic human and animal song, as well as the sound of wind in the trees."[26]

The story of the elaborate construction-process of this metaphorm is critical to the art's full appreciation. The small monolithic cube that makes up the core of the artwork contains a series of 20 "holograms of nothing"—except light. The holograms float inside a sealed vacuum chamber within this containment vessel—alone—like the Platonic mind that conceived them. A private passage from one of Burgess's luminous poems about the conception of *The Quiet Axis* creates in our minds yet another dimension to thought—one that joins the mysterious contents of all artforms. With Burgess's muse, we drift into spaceless imagination:

> And now,
> in the fractured heat,
> neither now nor in memory,
> things float where they will
> as milkweed in languid air, gathers slowly,
> into a quiet axis, gently firm and sufficient,
> enough that with a courage I set my feet down,
> not to move the world or steady it, nor reject or deny it,
> but by loving balance come to walk in its bright day
> with steering fire to guide my steps.[27]

Other examples of works which masterfully buttress literal and metaphorical elements of imagination in probing the physical world include those of the conceptual and performance artist Chris Burden. Although he is mostly known for his controversial "event objects" with commentary, one work which links Burden's concerns with this book is his *Model of the Solar System* (1983). Upon coming across an ordinary ball positioned in the corner of an exhibition room, you're not particularly impressed by the object. I mean, it's a ball. It hangs from a string. Your first thought is to dismiss its presence in an impressive "Twenty-Year Survey" of Burden's work. But then, you read the typed words on a modest strip of paper only to be shaken by the most extraordinary realization! As the artist points out, you're standing next to the largest mass of work in our solar system—the Sun. This metaphorm would seem no more astonishing than looking at an illustrative model in a science museum, if there was no hidden agenda. What makes the experience of this work so compelling is the sense that something exists outside the framework of what is presented before your eyes in this unframed thought. You suddenly realize that this virtual Sun and its invisible colony of planetary bodies are, as the artist's statement describes, "proportionally scaled spheres placed at appropriate distances within a 1-mile radius of the model sun." Instantly, your mind places you in each of the surrounding areas specified in the model, engaging you to feel firsthand the positions of the planets. Shortly after, you feel yourself become absorbed into the fabric of outer—open—space while standing in the congested space of the immediate room. But more profoundly, you begin to see the solar system as one of nature's "kinetic sculptures."

As this experience draws you further inward and outward, Burden's words take on new meanings that swell beyond their technical implications. He writes, "In this scale model, the Sun (865,000 miles in diameter) is represented by a sphere 13 inches in diameter and 40 inches in circumference, located in the second floor gallery of The ICA [Institute of Contemporary Art]. The planets of the solar system, represented by ball bearings of accurate size, are placed at the correct scale distance across the city of Boston. These distances from the Sun vary from 36 feet for Mercury, the closest planet, to almost a mile for Pluto, the farthest from the Sun."[28] In reflecting on this description, you rediscover how actively involved you are in experiencing this metaphorm's forceful aesthetics.

The exploratory installation art of Geneviève Cadieux crosses the boundaries of aesthetics and conventional photography in interpreting the relationship between the viewer and the viewed, the eye and the object beheld. In most of Cadieux's works, the human body and notions of "self" are the subjects of her metaphorms. The body is also recognized as the natural means by which knowledge of the world and of oneself takes shape through one's contemplated experiences. It is the viewer's experiences of the installations—and acts of metaphorming— that "complete" her works of art. In *Sequence No. 6* (1980), for example, a pair of eyes have been enlarged, isolated from the full frame of a face, and roughly replicated some four times. By placing the photographs in tandem, and by working their surfaces with somberly colored printing inks and emulsion on steel plates, Cadieux draws the viewer into some provocative meditation on the camera's movements in relation to the expressionless eyes—eyes that exist in the shadows of consciousness. One notices not only that the camera has been repositioned, but also how the work extends the physical limitations of the photographic apparatus. All kinds of associations emerge from one's meditations on these atmospheric photographs, perhaps in the same way an astronomer is absorbed in gazing at the faceless stare of stars isolated in a telescope. In a more recent work, Cadieux metaphormed the spiral architecture of the Canadian Pavilion in Venice with her site-specific photographic installation titled *La fêlure, au choeur des corps* ("The thread connecting the chorus [and heart] of bodies," 1990).[29] Installed on the windows of the Pavilion, an enormous, detailed image of two mouths locked in some evocative embrace—with tongues bridging the land masses of skin—appears like the forceful melding of stellar matter from two interloping or colliding galaxies. The skin of both anonymous figures is deeply scarred. The scars seem to connect the histories of these individuals.

In the collaborative installations of Kristin Jones and Andrew Ginzel, one's point of view is largely shaped by window-like frames the artists construct in directing one's vision. These windows make visible the forces of nature, including what Jones and Ginzel call "the simple miracles: of the forces of wonder, of the very preciousness of our own lives." The cycles and events of life are abstractly animated in their installations, driven by mechanical devices with silent engines. Works such as *World View II* (1985) and *Seraphim* (1985) explore the transient states of our physical world while magnifying the mystery that underlies this transience. In both metaphorms, sandscapes constructed in a rectangular container form an artificial desert devoid of biological life, even though there are ample sources of water. Only the hypnotic shimmer of wind-swept threads, like snow, falling towards some earthly sphere—or the suggestive steam rising from some primordial vapor vessel perched in this container—ignites these dimensionless spaces with life. Through these "theatrical reliquaries," as Jones and Ginzel refer to their works, we can rediscover our imperfections, transience, and mortality in relation to nature. Their metaphorms reflect both our fleeting awareness and reverence for natural beauty and, in these artists' words, the "physics of existence." [30]

The research interests, inventions, and sculptures of Bill Parker, a former student-apprentice of Gyorgy Kepes and graduate of the Spatial Imaging Group at M.I.T.'s Media Lab, represent yet another breadth of exploration discussed in this book. As a visual interpreter of physical phenomena, Parker applies his rich background in physics towards re-creating the fury of stars for aesthetic engagement. He forms starlike objects of colorful energies by shaping—in the stillness of vacuums—precise mixtures of inert gases and electricity. These stellar forms invite one's electrical touch which, in turn, influences the flowing streams of ionized gas (or plasma). Parker recently built a plasma fusion sculpture titled *Lightning Tunnel*, which is 5 meters long and 2 meters wide with transparent corridors like an arborway covered with vines of excited ionized gas. Viewers/participators can experience walking through shafts of sustained lightning bolts and can actually influence the patterning of these bolts with the movements of their bodies. With some insightful reflection, one can sense how this starlight's restless energies might form within the neurosphere itself. Parker's kinetic works link nature's small- and large-scale electrical fields through the intimacy of one's touch and imagination.

METAPHORM 5. *Kenneth Snelson*, Atoms at an Exhibition, *1988.*

One crowning example of a metaphorm that joins the arts and sciences in exploring nature is the ongoing artwork *Portrait of an Atom* (1960–1990) (Metaphorm 5) by the sculptor Kenneth Snelson. This work exemplifies how an artist—who has a concept about some fundamental aspect of the natural world which is central to science—expresses this concept in multiple contexts. It's important to note that Snelson portrays his model in such a way that it balances interpretation with explanation. (There's an implied complaint in the fine arts that if a concept is represented too illustratively, or with too much didactic "spin," it's aesthetically barren.) In modeling the atom, the artist envisioned not the familiar light streak or clouded image of some atomic form dusted with electrons that orbit around a "still point"—the nuclear equator—like planets. Instead, he pictured the orbits as "matter-like whole objects," where, as he relates, "each consists only of the electrical particle racing rapidly in a circle as a matter-wave." His atoms are "equipped with an electrical charge, gyroscopic momentum, orbital magnetism and intrinsic spin. They act upon one another through these forces as one object to another."[31]

The difficulty—and ease—of interpreting "the artist's model" and assessing its scientific merits is discussed by the noted physiologist Robert S. Root-Bernstein in an insightful article, "Beauty, Truth, and Imagination: A Perspective on the Science and Art of Modeling Atoms." Root-Bernstein tenders the point: "It seems a mistake to me to try to categorize Snelson's work as one thing or another—as art or science, truth or imagination. Snelson's work is a new perspective on structures in nature and the nature of structure. This perspective, in turn, makes new things imaginable and therefore new things possible."[32]

In rendering his interpretation of the elusive properties of atoms—a pursuit which has tested Snelson's long attention span of some 30 years —he discovered a barrier between mind and its realizations. Whether you use a pencil and scrap paper to model an idea or a Cray computer, a model is a model is a representation. It rarely matches the reality of the thing represented. Snelson recalls: "I've spent endless hours trying to give my atom vivid sculptural life. Some representations have to do with magnetic interactions. They are simply mosaics, north-to-south, of ring-shaped magnets assembled like a chain of gears. . . . All of these examples, so painstakingly done, were sadly disappointing and clumsy compared with the immaculate gossamer atoms of my imagination: dynamic structures composed only of wavelike traces, electricity, magnetism and angular momentum, all doing their atomic dance in quantized atomic space."[33]

As I've stressed at various points throughout this book, it's not the forms of structure alone that metaphorms help us to see anew. Rather, it is the integrative processes of life underlying all structures of nature that bear the richest understanding of nature. (Compare Snelson's *Needle Tower*, Metaphorm 6, with the complementary Metaphorm 6c in Chapter 1.) It is through these processes that all of our models and interpretive representations of the world can be integrated with pleasurable coherence.

These statements on process are most clearly expressed in the poetic constructions of Edwin Schlossberg, whose varied artforms are about the nature of process—from A to Z—involving the symbolic systems of language. His philosophical sculptures and painterly writings often take the form of layered, transparent materials that are assembled as composites. Words, sentences, and other elements of symbolism strategically fill the surface of each layer without saturating the abstract compositions. Although many of the word forms are fragmented or scattered in appearance, they manage to coalesce into crystalline meanings as one aligns

METAPHORM 6. *Kenneth Snelson*, Needle Tower, *1968.*

them in sight. Like poetry, they can be read openly from one or more perspectives, leaving the essence of all that they convey up to the imagination. Some of the composite forms are mounted on movable frames and platforms that can be arranged experimentally. As you probe the multiple meanings of his words, you eventually discover some clues to the nature of our language universe. The discovery may be stimulated by any one—or part—of Schlossberg's poems printed in his leaflet titled *Word: Nerve Tidal Gestures* (1990), which accompanies the sculptures. One thought fragment, in particular, tweaked my curiosity: "It is not that mind/ is the center of the universe/ but only that mind/ is the experience/ never separated."[34] Others moved me to the brink of abstract thought:

E
EARTH
EASY
EDGE
ELECTRIC
END
ENTER
EVEN
EVER
EVIL
EXPRESS
EXTRA
EYE

Things are made
by their boundaries
nothing appears
nothing surrounds us
nothing is here
our nerves gently firing
except in fulfillment
the motion towards everything
gesturing in

N
NAKED
NAME
NATIVE
NATURAL
NATURE
NEAR
NEED
NEGATIVE
NERVE
NET
NEVER
NEW
NEWS
NIGHT
NO
NONE
NOTHING
NOW
NUMBER

Enter the microscopic
everything else disappears
you in a landscape
geometry and meaning
require invention
new shadows
false motions
odd intentions
gathering shifting
pressing for clues

And then the scent
of where your body lies
and perhaps of some
yet smaller dimension
and here in this explosion
burst the fullest stark sense
of what they were meaning
when they mentioned
the word life

The few artists and artworks considered next offer novel reflections on our metaphorming minds. The philosophical paintings of Arakawa, for instance, present the artist's evocative concepts and writings about mind-body-universe relations. These concepts are interwoven with contemplative imagery. The combinations of words and images, textural colors and suspenseful spaces can fibrillate one's breadth of reasoning. Many of Arakawa's meditative insights, as sensitive as they are—richly painted with delicate precision—enter your mind like neutrons transmuting the nucleus of your perceptions of the world. His paintings prompt you to question the undefined voids in our consciousnesses. Standing before one of his enormous canvases, such as *Atmospheric Resemblance (A Life of Blank)* (1978) (Metaphorm 7), you feel as though you're staring into an ocean of mind at the changing tides of human thought. The shoreline of one painted thought is continually washed anew by waves of meaning propelled by your personal interpretations. The mind depicted reflects the internal state of affairs that regulates our inner representations of the world. The artworks present meditations on the act of metaphorming. They invite us all to become active participants in experiencing our own changing mental architecture—an imaginary architecture that the artist bridges with the buildings of the cosmos.[35]

Arakawa's investigative concerns include interpreting and mapping various terrains of thought. These concerns are more readily shared by explorers in the field of pure mathematics* than of fine arts. He poetically relates the nature of thought to actual geometries of nature through various propositions and constructions. In painting "thought geometries," Arakawa gives the ephemeral qualities of thought physical presences like living biological forms. The viewer is immediately struck by

---

* Bertrand Russell informs us: "Pure mathematics consists entirely of assertions to the effect that, if such and such a proposition is true of *anything*, then such and such another proposition is true of that thing. It is essential not to discuss whether the first proposition is really true, and not to mention what anything is, of which it is supposed to be true. Both these points would belong to applied mathematics." (From B. Russell, "Mathematics and the Metaphysicians," in James R. Newman, ed., *The World of Mathematics*, 1956.)

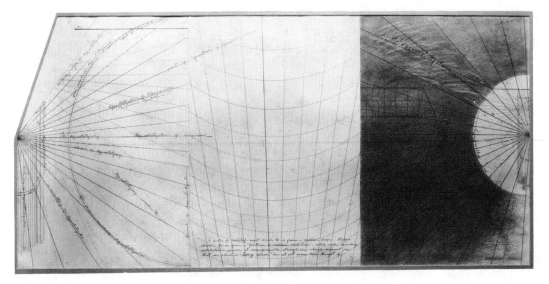

METAPHORM 7. *Arakawa*, Study for Atmospheric Resemblances, *1975–76.*

his panoply of curves and deformable surfaces, which lend themselves as metaphorms to cosmology of both mind and universe. These forms are part of his idioms. They suggest likenesses between shapes of the physical universe and thought itself. Scrutinized, these forms can lead us to understand something of the geometrization of the thought process, while entertaining the possibility of a geometry-based universe.

Some metaphorms that engage your whole sensorium in experiencing them include the mythical creations of visual artist Anselm Kiefer. Putting aside all bristly categories of style and briskly labeled historical art movements, Kiefer's dense paintings expand our senses, like a sorcerer who teaches us how to feel the world with our whole body. In his monumental mixed-media painting *Osiris and Isis* (1985–1987), Kiefer has used Egyptian mythology to comment on moral and cultural crises that threaten contemporary life. He pushes the symbolic forms of mind into the mud, rock, tar, television circuit boards, and other material fragments that compose this painting, in conveying his concept of life— or, perhaps, a state of the universe. He presents the legend of Osiris, nature's indestructible creative force, as an edifying note to the reality of nuclear annihilation.[36] The psychological textures of the painting seem to transcend the political or historical references that shape their surfaces and structure their chaos.

Many of Kiefer's metaphorms suggest to me energy forms that are recognizable as both human and cosmic processes alike. In works such as *Painting = Burning* (1974) and *Tree with Palette* (1978) a painter's palette is singled out in the landscape of brooding, viscous paint as though to remind us of the presence of the artist's mind and transformative forces—forces forming all creations. Like private (though visible) monuments to the mind, they focus one's feelings on the shades of human emotions and intellect with which we paint our cultures and civilizations.

Another example of art that conveys something of the constructive ambiguity in neurocosmology is the recent work of Mia Westerlund Roosen. The unidentified energy forms of Roosen's metaphoric sculptures intimate the formings of flora during the Precambrian era, when the Earth's crust was composed and the earliest lifeforms made their appearance. Her symbolic creations, which are roughly biomorphic in shape, seem to evolve from some sensuous Rousseau-like jungle. Their virtual sounds are encoded as textures, and their forms share relations with inanimate things that are in the midst of leaning, intertwining, infolding, enfolding, rotating. In the sculptures titled *Leaning Disc* (1988) (Metaphorm 8) and *Double Disc II* (1988), the edges of two lead and concrete discs come together invoking an image of interloping galaxies as they first touch. Like galactic structures, the discs seem destined to expand beyond their present dimensions. One can almost feel the shadows that complete and connect these objects; they're thick yet soft and ethereal. The meanings and feelings these sculptures evoke are as volatile as ether and as dense as lead. In one viewing, the suggestiveness of their forms eases them to rest in your imagination as nameless artifacts that exist apart from the turmoil of our universe, like fallen meteorites. In another viewing, this interpretation may suddenly reverse itself as one's mind imbues them with virtual action.[37]

On the subject of turmoil, interpretation, artistic inquiry, and metaphorms: the works of conceptual artists Adrian Piper and Lorna Simpson explore the universe of human turmoil caused by prejudiced perceptions of the world and ourselves. The forms and perils of prejudice are dealt with directly through metaphorms that illuminate the profusion of forms of racism and its destructive process. Both artists have inventively incorporated in their creations their experiences as persons of color, refocusing the consciousness of people of all races. In this regard, their works are politically and ethically oriented or motivated. In Piper's artwork, for example, the virulence of racism, stereotyping, xenophobia, and other forms of social and psychological injustice is shown to surface the moment one's perceptions are limited to the outer layer of something —human beings and interactions, social conditions, etc.—rather than being unlimited in perceiving inner essences. Piper regards racism as one of "the big issues, [the] issues that are going to take centuries to solve,

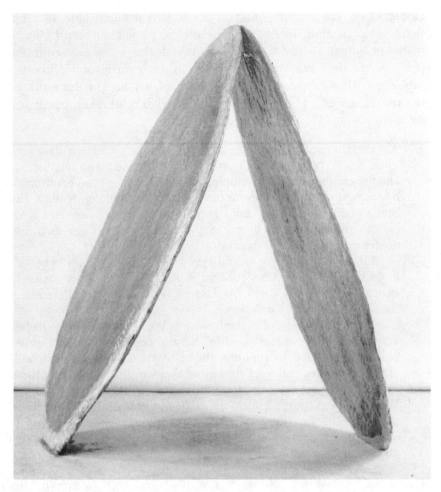

METAPHORM 8. *Mia Westerlund Roosen*, Leaning Disc, 1988.

that really plague people all the time."[38] Through her writings, videos, mixed-media artwork, and performances she intimates how "we are all implicated in the problem"—the mushrooming problem of economic, social, and cultural racism. In works such as *Hypothesis Series* (1968–70) and *The Mythic Being I/You (Her)* (1975), she transformed her physical appearance in order to provoke responses from her audience, recording her self-transformations and these responses. Although these activities were shared by other early conceptual artists such as Vito Acconci and Hans Haacke and feminist artists such as Eleanor Antin, Hannah Wilke, and Lynda Benglis, Piper's political agenda was substantially different. Her works examined in both passionate and cerebral ways how racial stereotyping muddles the world of interpersonal relationships as it emerges from the coupling of ignorance and insensitivity, blind assumptions and aggression. In *Portrait* (1983), Piper's reflections on the instabilities of human nature as expressed through racism are pressed to the point where they can virtually explode in one's imagination, like the problem itself. She heightens the tension of her text in this work by buttressing it with a photograph of the "Fat Man" plutonium bomb as it devastates Nagasaki. The text reads:

> All sentient species are biologically programmed to attack alien enemies. Some species are programmed to attack their own members as alien enemies. Rats, for example, will attack, kill or even cannibalize one another under conditions of overcrowding and deprivation. But human beings are more unique still. Only human beings are capable of self-destruction, of suicide, of acts that have our own self-obliteration as a conscious purpose.
>
> Human beings must view themselves as alien enemies to be able to do this. They must believe that if they allow this alien enemy to exist, it will destroy them. And so to avoid destroying themselves they destroy themselves.
>
> We can see why this might be so. We do not know ourselves very well. Often we feel assaulted by unacceptable thoughts or impulses, and move to suppress them; or shamed by unacceptable physical features and work to remove them; or threatened by others' unacceptable behavior or appearance, and so attack or reject them. We view these things as alien enemies, not as the familiar ingrained parts of ourselves they are. And so we are constantly moved to destroy and reconstitute ourselves in conformity with our truncated and distorted self-image.[39]

Lorna Simpson also combines text with photography to amplify her thoughts on racism and sexism among other acts of misguided discrimination that spearhead man's self-destructive tendencies. Artworks such as *Guarded Conditions* (1989) and *Easy For Who to Say* (1989) explore with ranges of subtlety the nature of prejudiced perceptions that contribute to the denigration of human beings. In *Easy For Who to Say*, Simpson presents in a detached, almost unemotional way, five identical figures of a black woman whose face has been blotted out and replaced with a white, flat, oval-shaped mask. Printed on each mask is a bold, red capital vowel—arranged in the order of A,E,I,O,U. Underneath each anonymous portrait is a single word on a plastic plaque, the first letter of which corresponds with the vowels: AMNESIA, ERROR, INDIFFERENCE, OMISSION, and UNCIVIL.[40] The work is gently confrontational, provoking its viewers to recognize in this faceless woman the meaning of losing one's identity and the impact of this loss on our collective conscience.

In examining these themes in the context of neurocosmology, one is left contemplating how various processes of racism are reflected in the natural world. Are they? We see how racist behavior is painfully evident in our social and economic systems. "You'd think this behavior extends to, or from, the cosmos itself!" your subconscious mind might proclaim with great dismay. Just how deep does racism go anyway—as deep as our perceptions permit? How "deep" are perceptions—as deep as our consciousness and insight project? Where does it stop? We tend to draw limits on this potentially infinite regression, recognizing our inability to grasp the infinitude of such a concept and its pervasiveness. To particle physicists, the whole concept of racism may seem at first totally irrelevant and alien within the scope of the things and concepts they study in their field of work. For instance *what*, in the name of nature, is a racist lepton, quark, gauge boson, meson, or baryon?! In describing the physical properties of these major particles, a physicist isn't likely to discuss which member of these families of particles is the "perpetrator or victim" of racism—as though the "subatomic roots of racism" were as real as prejudice. And yet, in a tangible sense, these roots may in fact begin with the electrons (negatively charged particles, carriers of electricity, and constituents of atoms) in the lepton family. Or they may end with the photons (zero charged particles and carriers of electromagnetic force) in the gauge boson family. Who knows? Do electron spins have something to do with how we think and this thinking influences spinning electrons? How could we ever know this in depth or discover these most subtle influences?

One deadly conclusion drawn from these sorts of reflections is that racist tendencies are as inevitable as they are natural; although we can control these tendencies, we can't rid ourselves of them any more than we can get rid of a monsoon or ward off an earthquake. This narrow-minded interpretation leads us straight towards disaster if we assume that these tendencies are acceptable in their unmanaged form or if we believe that we can't free ourselves from this behavior through awareness and education. A more prosperous interpretation is one that allows us to exercise our sensitivity and restraint, placing in check these brutal tendencies that seem intrinsic to our nature. Instead of being helplessly influenced by the discordant ways of the quantum world or natural forces, we can choose to respect the more harmonious qualities of nature —qualities that allow one to live without "averting one's gaze from the immanent spectre of the Other," to quote Adrian Piper.[41] There's always a choice.

Metaphorms that represent the intersecting abstractions of human history, value systems, morality, memory, and time include the installations of Barbara Steinman. *Cenotaph* (1985-1986) and *11:02, Nagasaki*

(1988), for example, interpret some momentous changes in human values that continue to influence every aspect of our being and remembrances. As Steinman intimates, these changes have left our minds numb from the possibilities of our apathy and unconscionable actions. Both installations possess a sort of haunting energy that grips the imagination like some pervading guilt—pinning itself to one's waking consciousness. Beyond the spareness of materials composing these works, Steinman's metaphorms are filled with the presence of a history that is propelled by conflicting emotions: The longing to create countered by the despair from creating; the resolute hope for a positive future contaminated by forlorn impulses to destroy all hope; the act of dealing with dominance and succumbing to acts of dominance. The elusive historical events that Steinman comments on in *Cenotaph** are best recounted in the artist's own words, which she has sculpted—or entombed—in three slabs of granite that surround a sepulchral monument in a memorial-like room: "The radicalism of measures to treat people as if/ they had never existed and to make them disappear/ is frequently not apparent at first glance."[42] Faced with this radicalism, the chronological ordering of facts and truths and human deeds—or history itself—seems pointless, as pointless as recording the passage of human lives and creations only to void them with indifference or failed memory.

Over the past thirty years, the metaphoric works of Nancy Spero have explored a related "radicalism of measures" involving the perception of women and the (male) assessment of historical contributions of women to culture. Through Spero's art we experience the continuous presence and reimaging of women (as activators and protagonists, victims and heroes, or goddesses) glimpsed peripherally. Metaphorms such as *Notes in Time on Women* (1979, a 215-foot-long scroll with narrative) and *The First Language* (1981) are charged with imagery that balances the despair and pain of women with fortitude and determination. Spero's lyrical figures interpret the perceptions and stylizations of body types, as well as mythologies, from disparate cultures. Her depictions of women (and, in Spero's words, "men transformed into their female counterparts") span millennia like epics of India and Egypt. Various characters of humanity and emotions are portrayed with a directness that both attracts and repels one's full consciousness. (The artist traces the conceptual armature and "cosmology" of this work to a small book titled *Forty Thousand Years of Modern Art*.[43]) In the broadest context, Spero's works can be metaphormed to represent the "feminine" aspects of the physical

---

* *Cenotaph* denotes "a sepulchral monument erected in honor of a deceased person whose body is elsewhere."

universe, where images of women and the cosmos seem to evolve from one and the same turbulent, anxious concept.

Finally, the painted metaphorms of Ida Applebroog are just plain indescribable. Or rather, they have *so many* possible interpretations and psychological properties that when they're considered collectively they all seem to flow together to form one incoherent reality. One ends up inquiring in disbelief: Are we talking about the same sardonic, humorous, tragic, maddening, satirical, sad, sometimes hopeless, promising, revelatory artwork? What "style" of expression would you give to the expanding forms of the universe? What "categorization" of subject matter would you assign to works of art that comment on the human *being* but choose not to be categorized? Applebroog's deceptively sparse, storyboard paintings and constructions—however bluntly expressed or minimally composed—are as open-ended in their paradoxes and meanings as the cosmos itself. Works such as *"Why else did God give us the bomb?"* (1985), *Hurry Up and Die* (1985), and *Tomorrowland* (1986) push narrative and nonobjective art together—like matter and anti-matter—creating a sort of mutual annihilation of meaning. Every detail of her artwork —or lack of detail—acts as a propellant for new perceptions of human nature. In *Yes, That Is Art* (1985), four frames from an imaginary filmstrip show the outline of a person standing erect with arms outstretched, their backs to the viewer, looking out over an infinite blackness. Under each frame, next to the same repetitive body, is one of the four words from the artwork's title.[44] Whether the work represents a summary of contemporary art or a person contemplating a temporary universe is a matter of interpretation. Like distinct footprints in undisturbed sand, her thoughts are clearly legible. However, as soon as one approaches their meaning, the sand shifts, concealing their intentions. Applebroog's metaphorms are as much symbolic critiques of the forces that shape our lives as they are expressions of our inability to grasp the nature of these forces and their influences.

To return you to the book at hand: The notion that art and metaphorms—however symbolic or abstract—can be neither understood descriptively nor applied as methods of inquiry is suspicious. It's as suspect as the notion that science is neither personal nor self-expressive—its methodology completely "objective" and corroboratory. Such pronouncements are foods that coagulate in the mind, blocking the stream of ideas like fat corpuscles disturbing the natural flux of blood. Perhaps in their present languages and representations the arts and sciences reflect these turgid attitudes. But their languages will change, as both disciplines become more fully integrated with life. Their practice will become less partitioned or divisional. The exchange of ideas and approaches crucial for understanding the brain-universe will occur.

There is an almost instinctual fear that the arts in general (and metaphoric thinking specifically)—when conjoined with the sciences—can only "soften science" like ocean waves effacing a sand sculpture. It's blissfully easy to forget that the neurobiological roots of both our artistic and scientific expressions of thought are as intertwined outside the creative medium—the brain—as they are inside. As creators, conceptualizers, and theorists, artists can inform the sciences in interpreting the conceptual ambiguities of the unexplainable. Scientists and mathematicians, in turn, can inform the arts by introducing new strategies for exploring the explainable. As soon as borders are established between art-making and sensing and science-making and sensing, we divide—rather than share—our sensibilities, our tendencies to experience nature metaphorically, our perspectives, our freedoms of creation, and our paths towards self-knowledge. In separating sensibilities and practices we split both our comprehension of being whole and our values.

As whole human beings, we are all part artist, part scientist, and part layperson, and we often struggle with the delirium of details in living that all but consume our inventive minds. In addition to learning how to live with the different temperaments within each of us, we need to discover how science *emerges out of* art, just as order and complexity emerge out of chaos and entropy. During the course of our explorations, we may learn how the brain emerges not in opposition to the universe, but in harmony with it—even if this harmony is painfully turbulent at times.

Mind is invisible circuitry, tying us together.—MARILYN
FERGUSON, *The Aquarian Conspiracy*

Branches #2

When I first conceived neurocosmology some 21 years ago, I had no idea who any of my predecessors were. The concept simply occurred to me without scholarship or question. It was hewn from the quarry of my own intuitions, rather than from studying carved, erudite blocks of basalt with ancient visual philosophies in the form of hieroglyphics. In the course of developing my ideas, I learned of other sculptors who worked in similar quarries. Although I was surprised at first by the overlap of concepts, I realized that we all work near and drink from the same reservoir of imagination that forms from our collective creativity. A brief anthology of these concepts follows.

A FOREST OF THINKERS

Some special trees of twentieth-century thinkers are singled out in this quick walk through one of many forests in the history of ideas. The reason these trees stand out so prominently here is that their neurocosmological types of ideas cross the worlds of unconventional wisdom. In the following pages we'll consider some ideas pertinent to neurocosmology, including Carl Jung's dynamic unity, R. Buckminster Fuller's synergy, Ludwig von Bertalanffy's general systems, Ilya Prigogine's dissipative structures, Norbert Wiener's cybernetics, Gregory Bateson's cybernetic systems, and David Bohm's holonomy. This list could be extended 100-fold as there are many others from all periods of history whose integrative visions of nature have advanced our awareness of ourselves in the context of our environment, the universe.

In discussing how these individuals' works have indirectly influenced the concept of neurocosmology, I offer no chronology of ideas or specifics. Instead, I prefer to adopt the attitude of one of the most venturesome philosophers of this century, Ludwig Wittgenstein, who prefaced his pioneering treatise on logical philosophy with the remark: "It is a matter of indifference to me whether the thoughts that I have had have been anticipated by someone else." Actually, I'm hoping that many of my thoughts have been pondered by others as this will make my book more accessible and personal.

A DYNAMIC UNITY OF MINDS

We cannot approach the study of neurocosmology without mentioning the study of the unconscious as a source of self-knowledge. When you think about it (and dream about it), the thought process is continually influenced by the unconscious mind. Even analytical thinking represents a form of coherent unconscious interplay, a form of thinking often informed by memories. All human mental activity is, in some way, touched by unconscious impulses and tendencies. In closely examining the creative expressions in the natural sciences and in the humanities, or mathematics and engineering, this thought becomes increasingly apparent.

Few thinkers this century have produced more influential insights into the covert language of the human psyche—into the influences of the unconscious brain—than Sigmund Freud and Carl Gustav Jung. Since I tend to be more partial to the work of the latter, I will briefly comment on Jung's work. Interested in all aspects of thought, in particular the processes of creative thoughts most visible in the privacy of our dreams, Jung helped many turn their minds inside out to see the symbolic contents of their thoughts. His analyses revealed how the covert operations of thought processes influence not just our behavior, or mental creations, but also our sociological ones such as societies and cultures. And that by studying our creations, we might ultimately learn about the nature of our inventive minds—minds whose forms of energy exist somewhere beyond the faces of our dreams and the facades of physical reality. Both Freud's and Jung's analytical powers were instrumental in pointing out how we might begin to unblock our insecure and defensive perceptions of ourselves. These insecurities continue to influence the world-views and societies we create in response to our riddled perceptions.

For Jung, art was the largest seaport and gateway to the psyche. His search of symbolic artforms attempted to unfold how all forms of art are rich source material for understanding something of the inner language and intentions of the thinker. Art isn't outside—or foreign to—the mind, either in its purest abstraction or realism. It is not to be understood only in the limited and insulated context of its immediate fields: art history, criticism, aesthetics, literature, or even psychology. Art speaks to the whole of human creation—including physics, chemistry, biology, and the social sciences. It doesn't simply represent, or draw bridges between, one aspect of nature and another. Art *bridges* these separate aspects of material and metaphysical reality through its symbolism, its processes, its forms, and its meanings. Art is nature interpreting itself. Understanding the psyche is the key to understanding the art of nature. More broadly, it is the door through which we enter other planes of consciousness—one of which is scientific thought.

In discussing the psyche and character of "archetypes"—which he generally defined as "the original pattern of all things of the same type" —Jung wrote: "The deeper layers of the psyche lose their individual uniqueness as they retreat farther and farther into darkness. 'Lower down,' that is to say, as they approach the autonomous functional systems, they become increasingly collective until they are universalized and extinguished in the body's materiality, i.e. in chemical substances. The body's carbon is simply carbon. Hence 'at bottom' the psyche is simply 'world.' " [1] (This process of universalizing the layers of the psyche is especially apparent in religious and philosophical architecture and art. See examples in Chapter 1, Metaphorms 6b and 6c.)

In searching the unity of mind, we must be unified. If the concepts of unity, wholeness of mind, and integration mean anything, we will know this meaning only as we truly *experience* these concepts. Otherwise they're just so many vacuous words, as meaningless as all the data we've ever collected by scientific means without a powerful and deeply humanistic vision guiding this collection. Concerning the importance of human values in science, I hear many artists asking certain sleeping scientists, "When do you want me to wake you up?" And these scientists responding, "Stop dreaming."

SYNERGY IN ACTION

There have been numerous explorers of life who have applied the principles of synergy in their investigations of nature. The word *synergy* (and its related terms *synergism* and *synergetics*) is defined as the joint action of discrete agencies in which the total effect is greater than the sum of their effects when acting independently. This definition amounts to *people working with people working with ideas*—ideas whose realization exceeds any one person's mental or physical capacities. In more recent times, one of the most formidable among synergistic thinkers and practitioners was R. Buckminster Fuller.

He's been described as a comprehensive generalist and visionary, a scientist and artist, a mathematician, an inventor, a poet, and about 37 other things to be exact. Fuller was as diverse as nature itself. That he was a universal human being of exponential inquisitiveness and inventiveness is an understatement. Whether he was designing his well-known geodesic domes to accommodate his universe or was following some other activity involving the creation of "dymaxion" architecture (which he defined as "making the most with the least" amount of material) for a new twenty-first-century environment, Fuller was always a practicing synergist. For him, synergetics was the only future with promise for the present. It remains the single most effective means of combining energies and intelligences, emphasizing the cooperative interactions of "wholes to parts." This cooperation is never more obvious than in Fuller's geodesic structures, which are *universes in the making*—forever adaptable, "permanently temporary" structures—that are rich in purpose and design.

To Fuller's list of remarkable modern realizations of synergy in the fields of art, architecture, science, engineering, and technology, we could add the megalithic constructions in the British Isles and Brittany, such as Carnac with its precisely paralleling rows of granite monoliths—some standing 20 feet high and weighing many tons. We could also add the Egyptian rock-cut Ramesses II Temple at Abu Simbel, whose humanpower and conception were mountainous; or the Toltec Pyramids of the Sun and Moon at Teotihuacán in Mexico, the largest pyramids in the world, with their infinity of steps to the sky; or the Muslim Badshai Mosque in Lahore, Pakistan, that spreads out symmetrically in every direction to create a communal pool of contemplative space for thousands to engage in prayer; or the astonishingly complex, Hindu monolithic rock-cut Kailasa Temple in Ellora, India, built in the eighth century A.D.; or the magnificent Srirangum Temple with its Thousand-Pillared Hall of carved columns; or the Gothic Chartres Cathedral in France and the German Gothic Cathedral in Cologne, with their unearthly composition and precise construction. There are hundreds of examples of synergistic works. What is more, all of these works relate directly to many of the conceptual explorations in this book. They are all physical reflections of the neurocosmos. They are all processmorphs.

ACTING THROUGH SYNERGY

These works of art and architecture were expressions of a large vision, the center of which was humankind. With fewer people on this planet and different socioeconomic structures—and a time-scale that is almost incomprehensible to us nowadays—our ancestors built their worlds based on meditations that related man to nature. In terms of synergy, coupled with sustained intensity and forethought, we find examples only in our advanced technologies that might rival in scope these earlier works. However, as Fuller frequently reminded us, with our planet crammed to full capacity and our sense of time pushed to speeds surpassing light and sound, our worlds are built and disassembled in seconds with little regard for projective, philosophical, or spiritual thought. The coordination of expertise involving many hundreds of people in creating and managing space shuttle programs, urban planning, transportation modes, dams, bridges, and telecommunications, among other large-scale projects, is mostly without some deeper, larger vision—other than technological curiosity, practical necessity, and profit. It seems our understanding of the synergetics in antiquity and before has dwindled. Today, the humanist philosophy of synergy, with its plans of peacefully coordinating billions of people in realizing a single ecologically sound vision, is perceived as an ideal luxury of thought—a thought reserved for one's leisure time, all seven seconds of it!

The idea here is not to wallow in the negative aspects of combining human physical and mental powers. (And there are plenty of them.) Nor should we despair over our willful compression of time ensuing in angst. The point is to consider Fuller's dream of pooling our energies in a way that has little precedence—either in the feats of engineering that launched space research, computer science, and medical imaging technology or in other wondrous technological innovations. In Fuller's view, synergy preserves our individual spirit. At the same time, it provides us with the means of working as a unified force towards collective, conscionable goals—all of which may be guided by humanist philosophy. One of these goals is to organize teams of thinkers coordinated in such a way that they're motivated to think together. Synergy works by the integrative working of people and parts—with individuality preserved in all ways—through temperate negotiation.

Through the sensitive lens of Robert Snyder's cinematic book *R. Buckminster Fuller: An Autobiographical Monologue/Scenario,*[2] we see the artist's lifelong ideas of using synergy to problem-solve on the human condition—specifically, to "reform the environment (rather than man himself)." According to Fuller's perception, "If you give man the right environment he will behave favorably." Part of that "right environment" is the mental climate in which one thinks or is encouraged to act. If it's not too arid (restrictive) and not too humid (undisciplined), the results may be inspirational. As he expressly stated, synergy is the only workable way of achieving this. It is also one means of maintaining a constant watch over this special climate, like a meteorologist charting our minds' "whether-or-not" conditions and barometric pressures.

SOME APPLICATIONS OF SYNERGETICS EVOLVED FROM SYNERGY

For more than a decade, the German mathematician Hermann Haken and his colleagues at the Institute for Theoretical Physics at the University of Stuttgart applied synergetics to the study of the behavior of self-organized, complex systems. The agenda of synergetics, which Haken refers to as the "science of structure," includes investigating how order spontaneously emerges from total disorder. From complete havoc emerges a civilization of concepts, a constellation of actions and behavior.

Synergetics was created in response to an informal calling within the scientific community to build a broader forum in exploring this subject. Haken proceeded to build an interdisciplinary forum in which a number of research topics on the subject of chaos, for example, could be discussed collectively, rather than individually. Nearly every discipline is represented in synergetics, save for the arts and humanities.[3] (This oversight bears the same dismay as a composer who accidentally omits two of the four movements from a sonata!)

Its primary concern is to observe, in Haken's words, how "the cooperation of many individual parts of a system leads to macroscopic structures or functions." Various systems are shown to evolve as they undergo major changes on a macroscopic scale (see Metaphorm 1). Syn-

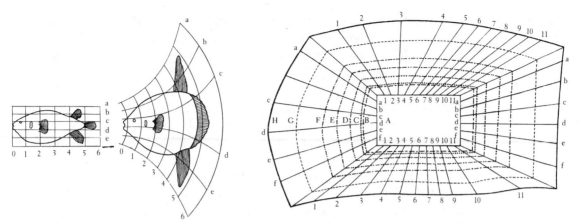

METAPHORM 1. *The "transformation" of two kinds of fish* (left: *porcupine fish* Diodon; right: *sunfish* Orthagoriscus mola). *(From D'Arcy Wentworth Thompson, On Growth and Form, 1917)*

ergetic techniques are capable of revealing how all the parts (or subsystems), such as atoms, molecules, tissue, or humans, generate these changes in a self-organized way and often towards a common goal. This goal can be anything. For example, it can involve moving the body in a disjointed motion to avoid danger. Or it can involve reacting fluidly to some stimulus in order to preserve continuous motion. The goal of movement can include the orchestrated action of subsystems in the simplest of multicellular organisms. Or it can include the concerted actions of complicated creatures such as ourselves.

VISUAL ANALOGIES
IN THE SERVICE OF MATHEMATICS

Applying physical experiments, mathematical constructs, and analogies, synergetics attempts to describe the evolution of the brain, for example, as it grows from a highly unstable state to reach stability through self-organization. This evolution is best understood by studying the effects of change in the nervous systems and behavior after abnormal structural changes resulted from injury or disease (as in brain lesions). Lesion studies show that in the process of recovery, structure is reestablished by some organizing mechanism.

In his overview to the edited book *Synergetics of the Brain*, Haken discusses how "the structure [of the brain] determines the function," and how "the function in turn influences the formation of the structure." This statement contrasts an idea explored in neurocosmology concerning the relationship between structure and function: *that structure does not necessarily determine the function of something.*

Haken bases his observation on the forming of biological structures. He relates that two hypotheses describe this process of formation in terms of genes. One hypothesis claims that cells position themselves, for example, according to some genetic blueprint. The competing hypothesis claims that only part of the growth information is provided early on; the remaining body of information is given to the cells by other means. The latter hypothesis is corroborated by experiments. Haken reports: "By transplanting neurons from one position to another one, they can be switched from producing one neural transmitter, e.g., acetylcholine, to another one. These experiments may be interpreted in such a way that the genes have provided the neurons with the know-how to produce *both* kinds of transmitters *but not* with the information about which transmitter is to be produced."[4]

Neurocosmologists might inquire whether these two hypotheses also apply to the formation of stellar structures. Assuming that the universe is an *organism* of sorts (honoring a Platonic conception)—and assuming that its matter "grows" as all forms of matter appear to grow and evolve —we might ask: "What would be the equivalent of a "genetic code" of the universe? Do stars or galaxies have "specific functions" acquired within the developing body of the universe—a "development" determined by their position in space? Are the assumptions that underlie these questions meaningful? Does this relationship make any sense? What is a better way of framing this inquiry so that it's more productive?

EXPANDING THE SCOPE OF SYNERGETICS

The analytical techniques applied in synergetics can be expanded, with a few critical modifications. These include incorporating the suggestive language and ambiguities of art. The metaphorms in neurocosmology, for example, represent only a minuscule fraction of artistic capacities directed towards revealing unsuspected relationships between diverse systems. Within the circumference of things synergetics operates on, the arts can advance observations about the workings of the brain and universe, if so directed. More than 10,000 years of "artistic insights" into nature are an untapped source of information in developing scientific concepts. The application of these insights moves beyond illustrative analogies through the process of metaphorming. Given that the primary language of synergetics is mathematics, the "artist's viewpoint" is very much appropriate, inasmuch as this point of view is often articulated mathematically these days. But even without this particular language of numbers, the arts' input is essential in expanding notions.

RECALCULATING THE IMPORTANCE OF THE ARTS

Most people grew up hearing that artists can't speak the "language of nature"—what Galileo fondly referred to as the "language of mathematics." This, of course, is a fallacy about both artistic thought and nature's solitary language. Two quick interjections: There are many artists for whom mathematics and programming in computer-aided graphics and animation, for example, are as natural and fluent as a second language is to a bilingual person. The second interjection, and the more significant point here, is that mathematics is not the only language with which to describe the ways of nature or the means of the universe. It's only one of many languages. As I recall, the general idea of synergetics is to bring together as many disciplines as possible—in looking from as many perspectives as possible at similar patterns of phenomena.

IN COPING WITH SCIENCE

Synergetics is as much about "how science can cope with complex systems," as Haken proclaims, as it is about how science can cope with science. These concerns are not unlike those of neurocosmologists. As I've intimated, neurocosmology can be understood as a more open version and expanded form of synergetics—with art as its first language and mathematics as its second. But it's definitely "bilingual" in this respect. It strives to gracefully integrate all aspects of the arts in its scientific quests and all aspects of science in its artistic quests. There's no such thing as segregation in neurocosmology. Its inclination is to look more comparatively at the *relationships between* different patterns of phenomena, without prejudice as to how we must search or represent these relationships. This amounts to studying not only the changes within a given complex system but also those changes occurring *between the interactions of systems*. All of these systems and their interactions are clues to our nature.

COPING WITH GENERAL SYSTEMS

In the late 1940s the Viennese biologist Ludwig von Bertalanffy developed a theory of general systems that described life forms as organized, complex systems that interact with their environments. Rather than rigorously analyzing some phenomenon (molecules or cells) to the exclusion of some other phenomenon (societies), Bertalanffy considered the interconnectedness of these natural open systems. His strategy was to search for similarities in the way systems' parts and wholes functioned integratively. An open system was defined as a "steady state"—a self-sustaining, self-maintaining, and self-repairing system. This definition remains with us. Any open system can be studied comparatively, such that one can show how different systems receive, transform, and release or exchange information or energy in a steady-state process.

General systems also consider our values as systems dependent on contexts. This means they're likely to change according to the circumstances surrounding them or in which they're viewed. And, in this scientist's view, our value systems are subject to will-o'-the-wisp thinking and chain reactions of reason that become supercritical whenever we fail to protect such fragile notions as ethics and moral law. Bertalanffy described humankind as "systems-within-systems, worlds of systems, whose various levels of complexity often confound our sense of morality."

The idea that we are worlds of varying systems whose operations are subject to the caprices of context places us in an uncomfortable situation. Our actions and knowledge live in harrowing uncertainty, making for precarious and tiresome living. It is especially stressful for those who see life as a jack of hearts rather than a joker—believing that it is only nature's best forms of creations and good deeds that are most representative of our nature. It emphasizes that all of the concepts and objects of our creations are clues to our nature—the destructive and lethal along with the constructive and benevolent.

PERPLEXING COMPLEX SYSTEMS

In commenting on the perplexities of cybernetic systems, we might make the following observations. One thing which distinguishes, say, guided missiles from human beings is our capacity to organize ourselves from genes to thoughts. Our genetic code is sensitive enough to guide us in our tumultuous development. It is the miraculous difference between the workings of our creations and our workings as creators. Without the genetic code to inform us during our development, and without the intervention of our minds to modify our will or form values, there would be little difference between our actions and those of an ICBM targeted for disaster. In his book *Maps of the Mind*, the scholar Charles Hampden-Turner succinctly summarizes for us the components of a cybernetic system as outlined by Bertalanffy (see Metaphorm 2), leaving

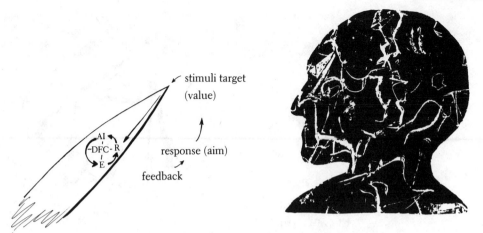

Worlds of systems, one more complex than the next, working with coordination?

METAPHORM 2. Left: *A summary of Bertalanffy's general systems theory. The target sends stimuli to the receptor (R), then to the direction-finding center (DCF). The center modifies the stimuli by either amplifying or inhibiting information (AI). This modification influences the behavior of the effector (E), which in turn acts according to the orders it receives. The effector also feeds back to, or signals, the receptor that there's been a change.* (Rocket diagram *metaphormed from Charles Hampden-Turner,* Maps of the Mind, 1982)

us with a timely reflection: "To transcend hostile academic disciplines and warring international systems, we must develop self-organizing general systems of symbolic relationships which reconcile empiricism with dialectics, classes and races and the human species with its environments."[5]

In the context of neurocosmology, we need to decide how meaningful it is to relate this cybernetic view of the brain and its behavior to the cosmos. Is the brain *more or less than* a celestial body because it is self-organizing? Do stars or galaxies operate according to similar cybernetic principles? Do they possess the components of a cybernetic system? Such questions might seem as nonsensical as asking whether the forces of destruction create our destructive forces. First you wonder: What are these forces of destruction? Are we speaking about the natural world or our inner nature? Are these forces psychological or cellular in origin? Or both?

Technically, cybernetics refers to "any action of the organism feeding back upon itself resulting in the alteration of its behavior." This reference invokes the idea that the cosmos is *organic* rather than mechanical in nature—*creative* and *purposive* rather than purposeless. This

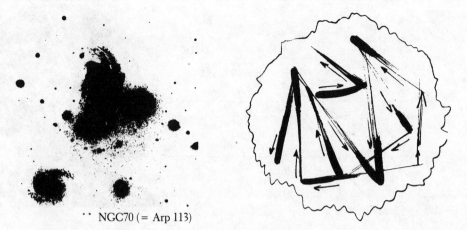

·· NGC70 (= Arp 113)

METAPHORM 3. *The Brownian motion of systems, beneath the veneer of order and organization?*

invocation counters the astrophysical theories that depict the universe as some kind of nondescript, mechanistic entity enduring the plights of Brownian motion (Metaphorm 3). In the heat of this motion, "particles" the size of stars, galaxies, clusters, and superclusters randomly collide over millions and billions of years. Through time-lapse photography or computer simulation and metaphorming, millennia can be portrayed as milliseconds.

GENERAL SYSTEMS OF THE NEUROCOSMOS

Those who support Bertalanffy's findings see the world as groups of harmonized relationships that must be understood contextually. In *The Systems View of the World*, the philosopher Ervin Laszlo writes that "contemporary general systems theory seeks to find common features in terms of shared aspects of organization. . . . it discovers the repeating patterns in the organization, such as floors, doors, and windows [of a building], and evaluates these as so many variations on a common theme."[6] A neurocosmologist would explore the organization of processes in addition to structures. We need to concentrate on how self-organized systems work in relation to other organizations of systems. This knowledge further illuminates the relationship between the environment that both informs and forms our inner world.

Emerging from this view, then, is the idea that many individual systems combine to form one or more "supersystems." These supersystems, in turn, form more complex organizations and systems—all of which are fully integrated (Metaphorm 4). This process of aggregation can be described in terms of forming a net—one that starts off in life with many pieces but little coordination. Eventually, these pieces become well coordinated, as part of an evolving system. Each part of this system, Laszlo writes, "retains some individuality of their own. They exist as definite subassemblies within the larger whole. . . . Systems within other systems can have their autonomy and freedom of decision."[7]

NETWORKS OF CREATIONS

You might wonder how tightly these statements about general systems apply to the dynamics of the brain as well. Revisiting the premise of this book regarding the brain's "physical reflections": The workings of the brain reflect the workings of the *concepts* and *objects* the brain creates, such as theorems, factories, religious beliefs, and their institutions. *Everything* we conceptually or physically create is a processmorph of its creator. Even the *concept* of systems is an artifact of human mental activity, in the same way that Michelangelo's ceiling in the Sistine Chapel is an artifact of thought processes. Something in the way these creations work reflects our workings and behavior.

This is easier to feel or sense than to understand. It's hard to think of a physical object as a product of the central nervous system for it implies that this product, or artifact, reveals something of the mind that created it. Imagine searching the infrastructure of a 211-ton telescope and declaring that the way this hefty object works is demonstrably similar

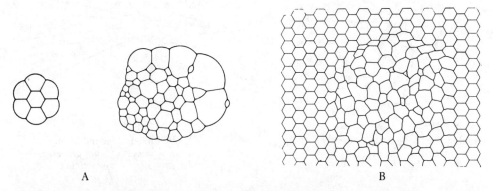

A   B

METAPHORM 4. Aggregation. A: *Images of aggregates of uniform and nonuniform bubbles creating a net-like form.* B: *A microstructure of iron with its polygonal shape presented metaphorically as a net that "has the character of a giant system."* (Adapted from Cyril Stanley Smith, A Search for Structure, 1982)

to the way some aspect of the lightweight brain works—as though the idea for this technology were somehow fashioned from some specific nervous tissue. We might, for example, point to the entorhinal cortex deep within the human brain as a region that is involved in ideation. The implication is that not only this technological object, but also its manufacturing and assembling processes are customed-designed by neural cell-assemblies. Each assembly or subsystem of the brain leaves its mark on its contributions to the material world.

THE SNAG OF LITERAL-MINDEDNESS

On first pass, this thought seems not only outrageous and wickedly unfriendly but patently untestable. In short, it makes a poor candidate for any credulous scientific analysis. I mean, how is it possible to trace back to the central nervous system—detail for intricate detail—the creation of an idea. Nothing could be more mysteriously complicated than the physical realization of an idea: From the geometry of synapses to thoughts of geometry. Wouldn't we have to know something about the relationship between the functional anatomy of the brain and the types of thoughts that anatomy produces? Wouldn't this knowledge necessitate having a clearer idea of the way different systems in the brain work collectively as a supersystem—or systems within a large system?

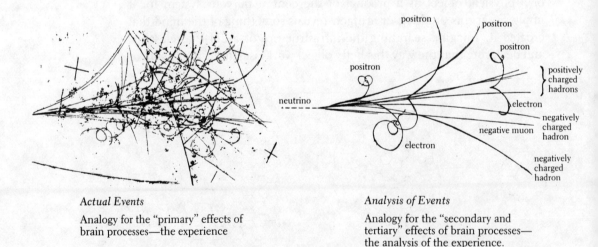

Actual Events

Analogy for the "primary" effects of brain processes—the experience

Analysis of Events

Analogy for the "secondary and tertiary" effects of brain processes—the analysis of the experience.

METAPHORM 5. *Shown is the relation of the primary actions of the human brain to its secondary and tertiary actions—as in the analyses of brain processes. This image can also be interpreted as "high-energy" neural activity creating "jets of virtual particles of thought"—with one jet merging with another.*

These sorts of questions swirl around at first, until they settle into a coherent thought: *All objects of human creation are secondary and tertiary effects of neural systems.* These effects are analogous to the secondary effects of high-energy particle tracks in a spark chamber detector in nuclear physics. The angelical tracks of high-energy particles show the effects of the primary actions of subatomic particles as they interact (Metaphorm 5). No one knows whether the secondary effects of subatomic activities are less complex or orderly than the secondary and tertiary effects of the human brain. By *secondary* and *tertiary* I'm referring to the tangible realizations of thoughts.

DISSIPATIVE STRUCTURES OR INTEGRATIVE PROCESSES?

It seems appropriate that my next brief notes on the Nobel laureate Ilya Prigogine, a theoretical chemist, should follow Metaphorm 5. Since the early 1950s, Prigogine has explored the intersections of the worlds—and "lives"—of animate and inanimate things through his studies of thermodynamics.* Prigogine questioned the second law of thermodynamics, which explains how energy is lost in all exchanges of energy. That is, energy in a *closed system* moves towards *entropy* and disorder to the degree that it continually loses energy and breaks down in the process of working. Without replenishing the loss of energies, the system eventually falls into the state of *equilibrium*, or so-called heat death. The term *equilibrium*, as applied to the thermodynamics of closed systems, is used to refer to the state of utter randomness and homogeneity.

Equilibrium states mark the extreme downside of evolution—when the worlds of atoms and molecules are totally lifeless, or exhausted, and thus have no energy for working. Astrophysicists and cosmologists have explored the possibility that the universe is an isolated and closed system, one that is destined to run itself into the dead zone of maximum entropy, which is to say, maximum randomness and decay. Like the word "closed" suggests, nothing affects it. Closed systems are islands unto themselves, with neither matter nor energy flowing between them or their environment. Such perfectly cloistered islands would be devoid of life, movement, and sound, existing only as abstractions or ideal concepts. Because perfectly closed systems are abstractions, the term *near equilibrium* is used to describe those systems that approach this state of disorder.

---

* Thermodynamics describes the interchangeable relationship between heat and all forms of work (or energy): thermal, electrical, mechanical, and chemical. "And symbolic, or metaphorical, and psychological forms, too," a neurocosmologist adds.

UNSTABLE PROCESSES

What puzzled Prigogine—and many other scientists before him—is how life could possibly skirt the grueling, sad fate of equilibrium and the inevitable slide towards disorder. And yet it does. Life grows increasingly complex and ever more orderly. At least on the surface, life appears to be evolving this way. Prigogine was one of the first scientists to realize that all life forms are *open systems*. His important realization—that order grows from chaos, states of disintegration, and entropy—overlaps with an insight from traditional Indian philosophy on "the nature of the ultimate" which reads: "Ultimately everything arises from disintegration *(tamas)* and ends in disintegration."[8]

Elaborating on this notion, the Indic scholar Alain Danielou excerpts a passage from the *Maitrayani Upanishad* (5.2) which reads: "First there was only [absolute] darkness *(tamas)*. Stimulated by quiddity, it became unbalanced, and the form of the revolving-tendency ['activity' or 'multiplicity'] *(rajas)* appeared. Stimulated, this revolving-tendency became unbalanced, and out of it the tendency toward disintegration, the centrifugal-tendency ['inertia'] *(tamas)*, appeared. Stimulated, in its turn it became unbalanced, and the tendency toward cohesion *(sattva)* appeared."[9] These processes of evolution are endless, as things continually grow towards and away from order.

THE WAY OF CREATION THROUGH QUIDDITY

The word "quiddity" used in this passage refers to the stimulation process. It is defined as "that which makes a thing what it is; the essential nature." I take this word to mean "creativity" as I define it in its broadest sense: *any unconditional response of matter and energy to an applied stimulus*. I associate quiddity with the creative process such as that involved in the "revolving-tendencies" towards either centripetal or centrifugal action. The former acts inward—towards a central point—as in synthesizing, absorbing, concentrating, connecting, and ordering. The latter acts outward—away from a central point—as in scattering, dissipating, dispersing, and decaying.

In terms of thermodynamics (rather than traditional philosophy) quiddity would refer to the state of an open "creative" system living far from equilibrium. This system has at least two options of "unbalanced" growth through centripetal and centrifugal tendencies. It can move towards "multiplicity" (or *rajas*) and "cohesion" (or *sattva*), or it can move towards dissipation and disintegration (or *tamas*).*

QUIDDITY: FAR FROM EQUILIBRIUM

What is curious here is that the concept of time and order in which these tendencies occur is open for interpretation. You might imagine that the centripetal and centrifugal actions occur out of phase with one another, meaning that they happen in sequence. Something becomes integrated only to become disintegrated when "stimulated by quiddity." Or you might interpret these two tendencies as occurring in phase, meaning simultaneously, in which case you end up *temporarily* with a sudden branching of two possibilities. Either you become more organized and coherent or you become more disorganized and diffuse. Either way, this branching process reveals the state of "unbalanced" action through which we create and grow and cohere—and disappear—as we move in and out of order . . . *moving in and out of chaos.*

To return to Prigogine's rich insights into open versus closed systems: Apparently, we spend our lives in this self-perpetuating state of instability, engaging life as creative individuals. We're joined by the cosmos, which also lives in a similar far from equilibrium situation. We are what Prigogine calls "dissipative structures" and what Hindu philosophy calls "the principles-of-the-elements" (or *tattva*).† We constantly fluctuate from one state of organization to another, one activity to another, experiencing along the way moments of near equilibrium which some might call periods of entropic "rest." These moments of stability and momentary "rest" are referred to as "punctuated equilibrium" or "saltationism."

There is evidence to support this "saltatory" state in fossil records which show how various environmental influences suddenly affected what was otherwise a fairly stable, unaltered species. This "saltatory" view of evolution—where things grow suddenly by "leaps and bounds"

---

* As an interesting note, multiplicity and cohesion are associated with "awareness" and "dream" in Indian philosophy. Also, disintegration is associated with "deep sleep." Whether we're aware of and dreaming of integration—or we're dead asleep as in a state of perfect disintegration (what Danielou calls the "blissful . . . unconscious state of consciousness"), these centripetal and centrifugal tendencies influence the "matter of our being."

† On this concept, Alain Danielou writes: "In the state of wakefulness, man experiences existence as depending on relative space and time manifested in the principles-of-the-elements *(tattva)*, which are the basis of perceptible forms" (*The Gods of India*, 1985, p. 25).

and then they stabilize for some time—is based on the observation of destabilizing fluctuations.

AN APPLIED PHILOSOPHY OF CREATIVITY

If you were to substitute the terms *near equilibrium* and *far from equilibrium* for the words *tamas* and *rajas*, respectively, you might regard this Hindu cosmology as a precursor of Prigogine's important theory of "dissipative structures" and "bifurcating [branching] points" of action. According to his theory, as expounded in *Order Out of Chaos*,[10] life forms evolve to greater complexity due to disorder. We evolve in harmony with entropy, *not in opposition*. As open systems, we live far from equilibrium crossing one threshold after another as we evolve. Our unbalanced, revolving-tendency (or *rajas*) ensures our growth. Our centripetal force drives us towards increasing complexity and "multiplicity." At the same time, there is the potential for our centrifugal force to drive us to total collapse and destruction. The thrust of these "bifurcation points" can, in the words of Prigogine, "lead us to the best and worst. We are participating in an evolution whose outcome isn't clear to us."[11]

If Prigogine had explored the open, self-organizing systems of the universe as envisioned in Near and Far Eastern systems of thought, he may not have arrived at this impasse regarding our "outcome." The outcome is fairly clear—providing you regard certain traditional concepts of totality, unity, oneness, creation, and other concepts of integration as being clear. It seems our integrative processes and dynamic creativeness consist of points of bifurcation. These points are both constructive and destructive actions which are *forever occurring in the person* —that is, *nature*. "Person and Nature, Purusa and Prakrti, are one yet distinct . . . They exist only in relation to each other," writes Danielou. Although these and related philosophical statements are not cast in terms of the laws of thermodynamics, they are cogent insights into the dynamics of how energy and matter, information and ideas, flow in nature. In reading these earlier interpretations of chaos and entropy, which were stated in the framework of philosophy, you quickly realize that our "collective thoughts" on open systems describe the transitions from chaos to order in animate and inanimate matter. They also reflect how we *live these transitions*—rather than simply think about them.

This notion of transitory, branching possibilities is central to all dissipative structures which live by means of chaos. Prigogine's mathematical descriptions of "the way" of all dissipative open systems resonate then with the descriptions of chaos in Hindu philosophy in particular. Chaos has many representations, all of which refer to the *linga*, or "the sign," or *Shiva*. The *"linga* of the universe," as Danielou discusses in *The Gods of India*, is essentially a formless or shapeless mass even though it is represented as "ether," the "pillar of light," the universe's "golden egg," and other forms. "Space is the *linga*, the earth the altar. In it dwell all the gods. It is called the 'sign' because all dissolve in it. (From *Skanda Purana*.)" [12]

Even though we continue to "escape into a higher order," as Prigogine conjectures, this doesn't mean that we will forever escape or evade the opposite situation—where order comes full circle—as we decay like some unfortunate, ideal closed system subjected to the perils of the second law of thermodynamics. Once order succumbs to entropy, chaos becomes us. But this chaos is no more permanent than our highest states of order.

CREATIVE SYSTEMS: AS ORDERLY AS THEY ARE CHAOTIC

Aside from human beings and stellar bodies, one of the clearest and simplest examples of open systems may be seen in the *Belousov-Zhabotinsky reaction* (Metaphorm 6a) which first interested Prigogine. The reaction is prompted by mixing four chemicals in a shallow dish at a select temperature. The chemicals rapidly form all sorts of pulsating waveforms—from concentric circles to spirals and more—as though they're alive and self-transforming. They live far from equilibrium, as self-organizing systems. Prigogine's contribution to the study of "order through *fluctuation*," as he called *chaos*, was founded on his analysis of

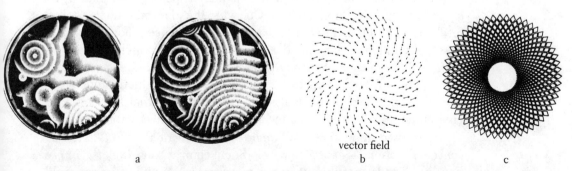

vector field

a   b   c

METAPHORM 6. Manifestations of Order. *The concept of* order *is an expression of our glances in looking at organized patterns of energy and assuming that they're forever orderly. Similarly, the concept of* chaos *is an expression of our ineptness at seeing non-chaotic patterns of energy, information, and ideas.*

the dynamics of this reaction. His primary inference was that the reaction actually decreased energy.

The Belousov-Zhabotinsky reaction, which is also known as *Bernard instabilities*, is presented in neurocosmology as a metaphorm. It represents every "creative" action of open systems, such as biological cells and galaxies. This metaphorm is particularly relevant to the covert actions of cell-assemblies. In a much broader sense, it is applicable to physical, biological, psychological, social, and symbolic systems—and their interactions. To the degree that all open systems experience conditions or situations involving chaos, they can grow in complexity as creative entities. As the process philosopher Henri-Louis Bergson speculated, evolution itself is a nonmechanistic, creative process stimulated by—or powered by—what he called "élan vital" (or vital impulse).

Other examples of "order-out-of-chaos" metaphorms include *gauge fields* (Metaphorm 6b; note the definition of gauge fields in Chapter 8). Gauge fields are generally used in theoretical physics to study the forces of interaction between elementary particles. They're presented here to suggest the "forces of interaction" in the brain. The skeletal image of the sunflower pattern (Metaphorm 6c) reveals another organization of "creative, integrative processes."

FEEDBACK IN CYBERNETIC OPEN SYSTEMS

In trying to remedy our world energy, pollution, population, and health problems, we demonstrate the principle of feedback. The more prosaic definition of the term "feedback" refers to the act of applying the output of a system to control or influence its performance. Feedback systems are evident in various forms of living matter. Although they are considerably less evident in forms of inanimate matter, perhaps feedback principles are operable nonetheless. Consider the possibilities of "stellar feedback"—in relation to biofeedback. If the inference holds that physical systems, such as stars, are open systems, then the principles of *cybernetics* (which means "steersman" in Greek) apply to them as well. The "cybernetics of stars" refers to the output of a star's production of energy which is controlled or influenced by either the external forces of the environment and gravity affecting a star or its internal environment.

Expressed another way: *Human control functions reflect "celestial control functions"—or the reverse.* (Such comparisons *literally imply* how our "heads are in the stars and vice versa"—though not necessarily how the stars control our heads.)

THE FLOWING NATURE OF FEEDBACK

The mathematician Norbert Wiener, who pioneered what are called "goal-seeking mechanisms," defined cybernetics as "the study of control and communication in man and machine." Wiener's classic book *Cybernetics: Or, Control and Communication in the Animal and the Machine* (1948) explored ways in which servo-mechanisms* and governors work in electronics. The most common example is the feeding of some output from, say, an amplifier back to the input in order to control things like noise, distortion, or instability. The popular term *negative feedback* refers to the reduction of, say again, an amplifier's output for the purpose of improving its performance by stabilizing the input. A loud example of *positive feedback* is the annoying crescendo and crackle from a loudspeaker or microphone mechanism that is destabilized.

The general principle of feedback is provided in the right-hand part of Metaphorm 2. Each of these examples is meant to prompt you to think of the relationship between the feedback systems in our brains and those in the stars. Gravity might be the source of "input-output information" influencing a star's actions by affecting its "servo-mechanisms and governors' performance." The action or performance is measured in terms of energy conversion or production. By "information" I mean what the anthropologist Gregory Bateson meant when he defined this phenomenon as "any difference which makes a difference in some later event." My only modification is the word "later," which I would omit, since it introduces a conceptual element that may be more illusory than tangible—namely, time.

THE UNIVERSE FEEDING BACK TO THE MIND AND THE REVERSE

In *Mind and Nature: A Necessary Unity*, Gregory Bateson theorized about sets of relationships that affect every aspect of our lives. He examined, for example, not only the nature of mind but also its connections with other minds and with the environment together with nature at large. In describing the properties of the human mind, Bateson offers

---

* There are two types of servo-mechanisms. In one type, the target or goal is known. The aim is to reach the target. In the other type, the target or goal is unknown. The aim is to discover it. The human nervous system functions by "knowing," discovering, and fulfilling its goals. Perhaps there is a third type which has no specific target or goal. Or rather, its "goal" is so integrated in its being that it's senseless to speak of targets as servo-mechanisms securing its existence or survival. This third "virtual" type may be the means by which stars live.

the notion that "*a mind is an aggregate of interacting parts or components. The interaction between parts of mind is triggered by difference,* and difference is a nonsubstantial phenomenon not located in space or time; difference is related to negentropy and entropy* rather than to energy."[13] Here, the presence of the universe is intimated by the terms *negentropy* and *entropy*.

I recall the first time I really thought about this passage and these terms as they relate to our conglomerate minds. In untangling this ponderance, I sensed that the ecology of the universe influences the ecology of our minds. This was one of the agendas of his book, I thought: The "interacting components" of our minds have something to do with the principles by which the universe interacts with its "aggregate of components." Although Bateson doesn't explicitly state this connection, it is written in the wind, water, earth, and fire of what he conjectures in this book. His principal conjecture is that "mental function is immanent in the interaction of differentiated 'parts.' 'Wholes' are constituted by such combined interaction."[14]

Bateson fascinated over these relationships between parts. He even developed a strapping list of criteria for distinguishing the characteristics of minds and thoughts from non-minds or simple material events. After exploring this list I realized that the "details of mind" still grazed in some mysterious land—one that is as wide open as all of our interpretations of mind, universe, culture, and other ephemeral quantities of matter and information.

Bateson's thinking is a source of fuel for anyone interested in traveling far enough to see how the languages of mind trip over themselves in interpreting the languages of the universe. He discusses the problems of language in *Mind and Nature,* in which he analyzes the flaws of ordinary language (meaning English rather than mathematics) in describing nature. Bateson writes: "It is the difference between talking in a language which a physicist might use to describe how one variable acts upon another and talking in another language about the circuit as a whole which reduces or increases difference."[15] As these problems of language remain largely unresolved, I can't imagine we will soon be able to smoothly discuss comparatively the cybernetics of brains and stars. And yet, we need to. Such discussions challenge our assumptions about *mind as nature.*

---

* Bateson defines *entropy* as the "degree to which relations between the components of any aggregate are mixed up, unsorted, undifferentiated, unpredictable, and random. The opposite is *negentropy*, the degree of ordering or sorting or predictability in an aggregate. In physics, certain types of ordering are related to quantity of available energy" (p. 250).

THE UNIVERSE ENFOLDING
WITHIN THE FOLDS OF HOLONOMY

The concepts of *holonomy* and *holomovement* proposed by the theoretical physicist David Bohm provide other examples of metaphorms for neurocosmologists to further explore.

In the early 1960s, it occurred to Bohm that quantum theory had not fully considered the process (and movement) by which physical states are continually transformed—one into the other and one by means of the other. Because of this oversight, he felt that quantum theory offered a partial description of reality. As long as the universe is contemplated in terms of distinct sets of phenomena (or physical states), the wholeness of the phenomena is missed. Part of the wholeness includes the observing instruments, along with the physicists and engineers who construct and operate these instruments.

Bohm was intent on forming a broader approach to representing the whole process of quantum reality. This subatomic reality includes the *quantum potential*. The term refers to the potential collective behavior of particles. Bohm reasoned: "Even in an 'empty' space in which there is no classical potential, the particle can be acted on by a quantum potential that does not fall off with the distance; one is now able to explain the well-known wave particle duality properties of matter."[16] The target of this point is that the quantum potential can involve non-local connections. These connections depend on the whole state of a system, which is, as Bohm expressed, "not reducible to a preassigned relationship among the parts."[17]

MOMENTARILY DROPPING ORDER

One day Bohm saw a BBC broadcast in which a drop of ink was dispersed in a cylinder filled with glycerine and then, as if by magic, the ink drop reappeared again in its whole form, "reconstituted" as it were. With his prepared mind, Bohm recognized the relevance of this demonstration. In his words: "When the ink drop was spread out, it still had a 'hidden' (i.e., non-manifest) order that was revealed when it was reconstituted. On the other hand, in our usual language, we would say that the ink was in a state of 'disorder' when it was diffused through glycerine. This led me to see that new notions of order must be involved here."[18]

Soon after, he made the connection between the ink-drop phenomenon and a hologram, in which, as Bohm realized, "the entire order of an object is contained in an interference pattern that does not appear to have such an order at all." He recognized that what they had in common was that "an order was enfolded; that is, in any small region of space there may be 'information' which is the result of enfolding an extended order and which could be then unfolded into the original order."[19] Bohm likened this dynamic to the enfolding-unfolding motions of the universe, which began from this ponderance: "Perhaps the movement of enfoldment and unfoldment is universal, while the extended and separate forms that we commonly see in experience are relatively stable and independent patterns, maintained by a constant underlying movement of enfoldment and unfoldment."[20] This was the concept of *holomovement* —a concept that conveyed something of the nature of the implicate order and of the quantum potential.

REBUILDING ORDER

Later on, Bohm peered into the limitations of these analogies. He sensed that something else was needed to complete his holographic model. The whole was conceived of as constantly enfolding into and unfolding from different regions of an electromagnetic field. This whole field could no longer be conceived as a self-contained totality. The dynamics of the whole depend crucially on the *super–quantum potential* for particles. To complicate things, this potential depends on the "wave function of the universe," in the same way the quantum potential for particles depends on the wave function of a system of particles. Recall that the wave function is the key mathematical construct used by physicists for exploring the possible interactions between you, the observer, and the phenomena observed. The motion picture we are led to envision about wave functions and the implicate order shows the flow of incoming and outgoing waves organized into independent sub-wholes.[21] We are left to determine how these sub-wholes are integrated into the whole. Bohm writes: "The explicate order itself may be obtainable from the implicate order as a special and determinate sub-order that is contained within it."[22]

There is no final decision on this conjecture. Paradoxically, Bohm's vision of our implicate-explicate universe resonates with the present concepts of chaos in which the turbulent world of matter is shown to have an orderly side, and vice versa. When you put together these insights into the oddities of order, a new image of the cosmos emerges, one that points to the connectedness of everything in quantum nature—the nature of processes—beyond proportions.

REBUILDING MEANING

The more open-minded we remain in searching for meaning in these speculations, the less limited we are in metaphorming our exploratory thoughts. We are also freer in interpreting our experiences of the world and in evaluating the residue of knowledge that is titrated from these experiences.

Of the hundreds of interpretations I've read that give insights into the nature of reality and the reality of nature, none has touched me as being complete. There always seems to be something missing to a philosophy of reality, a theory of nature, an observation of life. It's either too materially oriented or too spiritually oriented or too nebulous to relate to in a personal way. Our systems for knowing nature are knotted in the languages that comprise these very systems. Too often, we devote more energy to attacking or defending these systems of thought and beliefs than we give to looking at what they're pointing out. We just never get around to discussing the relationship between ideas. Instead, we argue over how we arrived at an idea or theory—specifically, what our hypotheses are based on, what our methodology consisted of, and related details. These arguments are important providing they don't stop there.

I keep thinking that there must be an easier—more natural—way to effectively communicate. Look at porpoises and whales. How can these and other advanced mammals pack so much into their conversations with their parsimonious language? How can we make our languages more fuel efficient and coherent, like the languages of our cousins? What are we missing (or what do we possess) that blocks us from achieving this efficacy?

Why do our languages—and attitudes—always get in the way? Could this be the way of communication in nature, this unresolved conflict? Or is it simply a temporary facet of human nature, before it evolves in the direction of parsimony? In posing these questions, I'm reminded of two expressions in traditional Indian philosophy which are used to describe reality: *neti-neti* ("not this, not that")—but "this *and* that"); and *sive-sive* ("this as well as that"). The words "as well as" can also mean *is*, as in "this *is* that." *This* (model, artform, construct, or theory) *is that* (a part of nature to which it directly or indirectly refers). Reality is all that we create—and recognize as both real and virtual—in nature.

I think that only daring speculation can lead us further and not accumulation of facts—Letter of ALBERT EINSTEIN *to Michele Besso, October 8, 1952*

Branches #3

EXPANDING VISIONS
OF OUR NATURAL INTELLIGENCE

In the early years of artificial intelligence (or "AI") some 35 years ago, critics of this new field of computer science prepared a repertoire of questions regarding AI's purpose. Two of the more tenacious questions have survived to this day: Why do we need it? What's it really good for?

I recall that the first time someone asked me a similar question about neurocosmology, my response was instinctively swift: "Hypothesis formation," I retorted with a sort of knee-jerk response. One of its "purposes" is to provide new ways of forming hypotheses through which we test our assumptions for their logical or empirical consequences. Our existence depends on the stream of assumptions and hypotheses we form in order to live and to make sense of our lives. In thinking about it further, I realized that even more important than hypotheses is the enjoyment that comes from making new conceptual connections. Unlike AI, the true goals and delights of neurocosmology are its aesthetics—as in the pleasure one experiences in discovering the ways of art in science and how the sciences essentially formed from the arts—through metaphorming and experimentation. This provision of sensual-cerebral pleasure is as much a nourishment as food, water, and other life-sustaining substances.

At first, AI seems unnecessary or, as the word *artificial* suggests, "false," "not real," or some other disparaging synonym applied to reflections of intelligence. But with some reading into its fascinating history, no one I know really teeters undecidedly about the usefulness of the AI endeavor nor do they question its many accomplishments. According to the three principles of neurocosmology, not only is AI useful in hypothesis formation but the computer languages it has developed based on its theories have provided significant insights into human mental processes. More radically, a neurocosmologist is inclined to recognize the term *artificial* as a misnomer, insisting that computer languages are as much a part of human psychical languages as all the other symbolic languages the mind has created. They are all *natural* languages, even though our machines "express" themselves—in however limited a capacity—using these constrained creations.

AI broadly addresses our mental capacities to acquire and to apply knowledge. It has succeeded at re-creating and codifying some aspect of the human thought process in its directive to "simulate thinking." Since its inception, it has emphasized designing computer systems in the image of human intelligence—where *intelligence* includes learning, discovering, reasoning, solving problems, and a host of other higher-order mental activities we associate with thought processes. AI is as much concerned about modeling the properties of thought as it is about applying its models to all aspects of education and industry. Proponents of AI contend that computing machines can symbolically model information in much the same way the human mind appears to. Just as we engineer knowledge, our computers may be taught "knowledge engineering"[1]— applying all sorts of expertise towards problem-solving.

Because neurocosmologists work with diverse forms of information —symbols, names, sentences, theorems, etc.—and not just numerical data, there's an affinity with AI, which works with similar information. AI programs were originally designed to speak in terms of symbols other than numbers. AI researchers have focused their energies on building better human/machine dialoguing devices, where computer programs can speak to humans and to fellow computers in ordinary language. The emphasis in neurocosmology is not on building these or any other devices. Rather it emphasizes understanding how all of our "devices" resemble the minds that have devised them. Neurocosmologists question whether this knowledge of "resemblances" will help improve our communications systems and provide a more thorough knowledge of our inventions. If computers, for example, can create a "linguistic mirror" of human languages—to invoke psycholinguist Noam Chomsky's important expression[2]—then presumably other devices with artificial languages also reflect our mental processes. To perfect our systems of communication means first understanding how *we express ourselves* through language and how we impart our knowledge of language. In this direction, AI seems to be both long on imagination and long on practical application—at least, as it may be applied to neurocosmology.

ENVISIONING THE SUM OF OUR LANGUAGES

AI's formative stages and studies might serve, then, as one of a number of models for neurocosmology. Many of its methods for cross-referencing and combining different sources of information are useful. Neurocosmology investigates our symbol-using capacity in exploring the world. Unlike AI, neurocosmology applies the conceptual tools and unlimited resources of the arts in its explorations. Also, many neurocosmologists are less concerned with "explaining" the phenomena of life and more interested in exploring information-related phenomena—often without regard to final goals or applications. For instance, I'm content with exploring the phenomena of life often without regard to final goals. The sum experience of this exploration, you could say, is an art without a specific artform.

Something "explained" is only temporarily secure insofar as one can always see that explanation in a broader context, or from a new angle of analysis, thus altering its meaning. In the early development of AI, two pursuits preoccupied its explorers: Discovering a mathematical basis for reasoning; and theorizing about the dynamics of reasoning, incorporating phenomena such as pattern recognition and analogy. Tackling the nonprogrammable, free-wheeling process of intuition was too much to handle. This particular process has a reputation for sabotaging schemes of codification and measurement, such as those employed by AI in its search to understand the way humans reason in representing knowledge.

AI has developed many specializations or categories of research, some of which are especially important to neurocosmology. Aside from the work in knowledge representation, research in learning and induction (which encompasses inference, reasoning, and heuristics [rules of practice]), is pertinent. Understanding the process of metaphorming and model-building is most relevant to neurocosmology. Model-building includes the activity of searching and selecting ideas or research directives; it stimulates us to think about our thinking nature. It also helps us understand the universes we compose with our thoughts, and how these *thought universes* map onto, or relate to, the brain.

As neurocosmology is saturated with different perspectives, languages, data, and other forms of information, some unifying principle is needed in bridging people, ideas, and tools. Next to pencil and paper, or chunk of clay, stones, and steel, computers are the best tools that can assist us in our studies. To understand how computer science, and AI specifically, might contribute to the development of neurocosmology, it may be wise to quickly sketch (in a paragraph or two) the kinds of problems AI confronted and continues to tackle.

Before creation of the computer software "LISP" (list programming), which was to become the first programming language of AI, not only were computing devices cumbersome to operate but there were experts in computer systems in nearly every field who were all talking in different languages. The combination of the cumbersomeness of tools and the diversity of languages made it oppressively difficult to communicate with specialists outside these languages or without the benefit of these tools. There seemed to be as many worlds of thought as there were languages to represent these worlds. Through the trials of AI, we have learned something about reconciling our "worlds" of representations.

SYMBOLS EXTENDING THE LIMITS OF NUMBERS

This cumbersomeness gradually changed with the invention of software (operating systems, programs, languages, applications) called *IPL*, or "Information Processing Language."* IPL used symbols instead of numbers in its data structures. This marked a major shift in consciousness from the strictly numeric orientation of data to the symbolic use of ordinary words and symbols. Manipulating networks of symbols was more desirable than managing networks of numbers. The idea was to portray information as lists of things. Hence the expression *list program-*

---

* IPL was created by three computer scientists, Herbert Simon and Alan Newell of Carnegie-Mellon University and Cliff Show of the Rand Corporation. The concept of IPL was presented at the first workshop on artificial intelligence conducted at Dartmouth University in 1956. The "LISP" software was created in 1958 by John McCarthy of M.I.T., who had attended this workshop.

ming language, which is commonly called *LISP.* The soul of the idea was to devise a means of searching, sorting out, and interpreting symbolic representations of information. This means of searching and representing the world, AI proponents claim, is remarkably similar to the integrative and discriminative processes by which our minds know and represent something. This neurocosmologist would argue that it is more likely similar to only one or two ways (out of many) by which the human mind processes information and represents what it processes.

This is about as far into the woods of AI I care to take you. The inclusion of AI in this discussion on the foundations of neurocosmology is not meant to imply that these are the only researchers whose work may guide neurocosmologists. There are many other groups in the arts, humanities, philosophy, and physics that are also important. The use of *LISP,* for example, and other AI tools in neurocosmology seems appropriate. To my mind, we need all the insights, analytical techniques, and methods we can assemble for studying the brain's relationship with the universe. There are not enough chefs and too many "dataists" handling data that our current disciplines cannot possibly prepare, serve, and digest what they make by themselves.

MANAGING INTEGRATED WORLDS OF SYMBOLS, NUMBERS, AND IMAGES

Other novel solutions to the management of information which may be useful in the work of neurocosmology include "information refineries." "Information-refining programs," as they're called by computer software specialists, are transforming the way we think about, and *think with,* information. By the expression "think with" I imply forming new connections between sources of ideas.

Researchers in the computer sciences have developed new kinds of programs that act as "machines within machines." Instead of simply instructing some piece of computer hardware to process a list of instructions (like conventional programs), scientists have created "software machines" that consist of the same substance as they transform. Simply put: *They are what they create.* They're encoded numbers and characters, or information. Through the actions of information refineries facts are swiftly converted into knowledge.

In a *Scientific American* article titled "The Metamorphosis of Information Management" (August 1989), computer scientist David Gelernter discusses how he and his colleagues at Yale University are creating new information-refining programs. The group is exploring the principles of so-called smart, or intelligent, software that is used to drive computers that process information simultaneously (or in parallel) rather than sequentially. The thrust of the research on parallel-processing computers is to develop new ways of managing information.

The concept of "information refineries" is one of these ways. In Gelernter's words, the goal involves "refining numbers or signals into useful information." At the same time, the parallel processors are increasing the speed of information processing. As is always the case, no matter how quickly we send or receive information electrically—or even optically—it always seems too slow.

Moving away from details such as the time (in processing) and the size (in megabytes) of information to be handled, what's important is the information itself. And also, the way information can be brought together to create novel syntheses. We now have the opportunity to coordinate—and, in fact, superimpose—two colossal sources of information about the physical world and our changing perceptions of this world. The parallel-programming system called *"Linda,"* for example, which Gelernter's group has created and applied toward the medical arts, makes this exchange of information a reality through this new development in communication.

ON METAPHORMING INFORMATION

The relevance of information-refining programs to this book converges at the following point. Neurocosmologists need some means of *filtering and refining* the potentially infinite number of metaphorms relating our thought processes to their physical realizations. In addition, we need to filter the various metaphorms relating brain processes to cosmic processes. We also need "smart" data bases to help us reason about the similarities and differences between data on both processes. In effect, the neurocosmologist supplies the metaphorms and the parallel-programming systems (such as *Linda*) form speculations and conclusions about information presented in a comparative sense. The conclusions are based on a description of some artifact or data whose properties are known. In this way increasingly complex hypotheses can be formed and tested.

Currently, two variations of parallel-software machines are being explored. One is called the "information filter" and the other is called the " 'smart' data base." According to Gelernter, the former transposes streams of incoming data into "higher-level knowledge." The latter sifts out "interesting patterns of information from records of similar objects and events." These objects and events can be anything or about anything, implying that they can *relate* any aspect of our brain processes and behavior to the world at large.

TOWARDS ANONYMOUS AND
"UNCOUNICATION" COMMUNICATION

Before leaving this subject of "computer-aided metaphorming" on the theme of the brain-universe, I want to briefly describe the most fascinating aspect of these information-refinery schemes, which is, to my mind, symbolic of the way all human beings may communicate in the future—"*Linda* style." I'll explain. Or rather, I'll let Gelernter explain the process by which his group's parallel-programming system works. "Linda's approach to communication is to simplify as much as possible," writes Dr. Gelernter. "Rather than setting up intricate communication paths among the parts of a software machine and complex protocols by which two or more parts can synchronize their actions, Linda makes communication anonymous and uncoupled. *A component that produces data need not know who will use it or when. Components that require data need not know who produces it*" (my italics).[3]

The metaphorical interpretation of this passage is "positive," providing we are generally *all held responsible* for the production and use of data. The way I interpret this description positively is as follows: Whether you're an artist (composer, painter, sculptor, musician, poet, etc.) or a scientist (biologist, psychologist, physicist, etc.) or a layperson interested in mathematics, you—as a "component"—need not know who will use the data (the artwork, or work of science, or concept) you produce. Nor do you need to know when someone will use it. And if you "require data," you "need not know who produces it." The versatile information is there to be used. Again, all anyone can hope for is that the "data" you produce and use as well as the data others use productively are of the highest quality. Moreover, that they are used with integrity and conscience—*regardless of the varying contexts or situations in which these data are applied.*

The example Gelernter presents as a way of introducing the "Linda-style communication" is a group of astronauts who are—at their own pace—constructing a space station. His plan is clear and simple: "An astronaut who has finished using a wrench lets go of it—sets it adrift. Another astronaut who needs the wrench can reach out and grab it. The same goes for any other tool, for a list of tasks that need doing or for some piece of information that might be shared by several astronauts. *Whoever produces the information simply releases it where anyone who needs it can look at it. Individual astronauts don't know who has something they need or who needs what they have; they just pick up tools or set them afloat. Nor do they synchronize their actions*" (my italics).[4]

The way I see it, we are all "astronauts," equipped with different gifts and skills, producing things—such as information and ideas—in one form or another. We do not know who will ultimately need what we produce, or when someone else will need it. Nor do we need to know what context or circumstance this information we produce will be applied in. We simply live the tireless assumption that someone will need what we produce at some time. And so, we produce—creating as we communicate and conveying through our creations.

RESIDUAL AMBIGUITY

Einstein reportedly remarked: "The hardest thing to understand is why we can understand anything at all." You may wonder if he was being too generous with this assessment. I'm often amazed at how *anything* is ever communicated by means of the complex systems of thinking and symbol-making in both the arts and sciences. Regardless of how tightly constrained these systems are, there's always some residue of ambiguity. For many in the arts and literature, this residue is the essence of the poetics of art. Discovering the richness of ambiguity, and how we gravitate towards certain associations and meanings, is the play of metaphorms and pleasure of contextures. Ambiguity is a process that teaches us to unlearn what we've been taught in order to see what we've learned.

When I speak of the "getting of wisdom" in neurocosmology and provide the mortar-and-pestle metaphorm, I mean that some neurocosmologists methodically collect and pulverize data in their disciplinary bowls. The data are then used—after "mortaring"—as a source for speculation, inference, and extrapolation. In dealing with metaphorms, you can almost count the number of misunderstandings beforehand, depending on their degree of ambiguity. The more explicit the metaphorm, the more its power to confound our senses decreases.

"EQUALITY" WITHOUT EQUIVALENCE

The ambiguous, rhetorical equations *Art≈Freedom*; *Freedom≈Power*; *Art≈Power*, where ≈ means *related to* rather than *equal to*, suggests one thing. Both in its practices and in its creations, art is unbound by any laws of expression. The equation is meant to be interpreted with the same sense of security in deductive reasoning as the standard syllogism "All A is B; all B is C; therefore all A is C." The metaphorm expresses this syllogism.

In effect, the power of art is inversely proportional to its capacity for freedom—and its expressions of this capacity. One of the most progressive thinking artists in the past 25 years, Joseph Beuys, defined art by using a very different equation. He declared that "Art = Creativity." I've always been bothered by this proposition in that it implies that art alone, as opposed to science, for example, possesses the unconditional privileges of creative freedom. Or that science must be artistic to be creative. This is as much a poor reflection on scientific thought as it is a limited perception of creativity. Most scientists and mathematicians would mutually claim that Science = Creativity. Both equations cast the process of creativity in too narrow terms. I suspect that the more broadly *creativity* is defined, the closer it is to the truth. This statement can be as meaningful or empty of meaning as you choose to understand it, like my statement: *Creativity is any unconditioned response or interpretation.*[5] Any unconditioned experience is aesthetic and creative in nature.

This statement implies that novelty evolves out of conditioning of material and mind. (See Chapter 8, page 168.) The statement on creativity intimates that this activity applies to the whole natural world, where "unconditioned responses" include creative, evolutionary processes.

EXPERIENCING EQUALITY

I said that both art and science are about experience. The American philosopher and educator John Dewey observed that "experience is equivalent to art." In his seminal book *Experience and Nature*, Dewey elaborates on how we came to separate the experience of art from that of science. He attributes this disparagement, where the ways of one are weighed against the means of the other, to experience and reason:

> The ground for depreciation was not that usually assigned in modern philosophy; it was not that experience is "subjective." On the contrary, experience was considered to be a genuine expression of cosmic forces, not an exclusive attribute or possession of animal or of human nature. It was taken to be a realization of inferior portions of nature, those inflicted with chance and change, the less *Being* part of the cosmos. Thus while experience meant art, art reflected the contingencies and partialities of nature, while science—theory —exhibited its necessities and universalities. Art was born of need, lack, deprivation, incompleteness, while science—theory—manifested fullness and totality of Being. Thus the deprecatory view of experience was identical with a conception that placed practical activity below theoretical activity, finding the former dependent,

impelled from outside, marked by deficiency of real being, while the latter was independent and free, complete and self-sufficing: that is perfect.[6]

In fact, what seems almost counter-intuitive is the activity of science: Its self-imposed constraints; its tightly policed languages and representations; its reserved protocols. The practice of science is frequently regarded as the authority of logical method over the intuitive leap, rather than the reverse. Its insistence on guiding the mind—on always "getting a grip on itself"—is perceived by many scientists as necessary for maintaining a sense of control, for filtering out the sense from nonsense. (By contrast, art is usually seen as a tool that loosens any grip, no matter how tightly affixed.) To exercise perfect control is to be perfectly godly. To control nature is to fulfill our godhood. If, as Dewey expounded, "art reflected the contingencies and partialities of nature," then to include art in science was to include all these contingencies and partialities. Such "imperfections" are the stuff that can disrupt methodologies and potentially nullify scientific interpretations. To fully embrace the arts in the sciences would be an admission of our inability to control through science with its assistant, technology.

WITHOUT QUALIFYING EXPERIENCE

Instead of embracing the most complex languages ever conceived, the scientific community has until very recently chosen not to deal with the arts—certainly not as an equal partner, directly incorporating the artistic process as a method of inquiry into the "necessities and universalities" of nature. It has elected, instead, to deny itself the experience of art, segregating such experiences from the body of science. And so many scientists remain bemused by the concerns or forms of contemporary art, rather than being impressed by artists' abilities to resuscitate or redirect our consciousnesses with philosophically and socially poignant concepts.

No doubt there are many stout-hearted explanations (justifications) for maintaining this distinction of sensibilities. At times we plainly act as isolated tribes of thinkers, creators, and policy-makers in virtually every discipline—tribes who have succeeded in compartmentalizing, and thus confusing, the *integrated matrix* of human activities essential to the evaluation of *any* idea, in *any* field, *all* the time. In partitioning our sensibilities, in favoring one approach to researching and interpreting the world to the exclusion of other approaches, we miss what many understand to be two obvious facets of nature: its contiguity and unity. Meanwhile our interactions remain "primitive" like those Charles Darwin described in *The Descent of Man* regarding a party of wild Fuegians who seemed tolerant of members of their group and "merciless to everyone not of their own small tribe."

FROM THE INWARD TOWARDS THE OUTWARD

Concerning the arts and sciences: Both are inward-moving outward processes. Both artistic and scientific imaginings are biologically rooted in brain processes. They're cerebral in the most poetically liberated and literal sense.

Art teaches science to go inward for questions; to explore ambiguity; to trust uncertainty; to exercise one's freedom to communicate, in searching and creating without fear. Science teaches art to go outward for answers; to challenge its conclusions; to exploit precision and consistency in exploring its insights. Art questions. Science answers. The two are mirror images of one another. The properties of one are opposite and sometimes reverse the properties of the other, although both move in relation to one another according to the mirror metaphorm.

Like an object standing firm in front of a plane triptych mirror, science stares at its triple reflections without questioning the reality of these virtual images that seem to stare back at it questioningly. Those images are art questioning the reality of the object, or the deceptive reality of the form and its surfaces. To separate art from science is to misunderstand the process of questioning and of answering in this reflexive relationship. To integrate science *And* art is to understand the unity of this process. It is to *experience* this unity in its purest state *(in your brain)* through intuition. Intuitions seem to answer their own questions before they pose them.

Contemporary art has shown with tidal-wave force that no information is too obscure, too removed from reality, too personal, or too abstruse to be explored in a meaningful way. Science has demonstrated that no information or phenomenon is too trivial to be investigated with important consequences. To believe in the principles of neurocosmology is to believe that everything can be connected and interrelated. And I mean *everything*—from the phenomena studied and represented to the disciplines doing the studying and representing.

IN EVERY WAY

There is one note of ill fate, regarding the works of artscience explored in neurocosmology. Because they're often *perceived* against the backdrop of our old (current) definitions of art which are understood apart from science, the first impulse any critic or viewer experiences in *looking* at them is defining their backdrop and stage props. Without this information, many people feel uncomfortable contemplating something. In order to appreciate the work, we usually indulge our impulse to name the context in order to clarify our interpretation of the content. This process determines whether or not the work is art-oriented or science-bound or technology-based. Unfortunately, few people may consider the consequences of continuing to contexture these artscience works such that they never sit permanently on one context. The metaphorm here is that of musical chairs—that "childhood game" we spend the rest of our lives playing in one way or another, in one career or another. Accordingly the music, as an expression of change, never stops playing. We just get tired of moving intellectually and so we sit—some of us never getting up again to look from a new context. Our imaginations, like muscles, atrophy. Also, our reasonings and sense-perceptions weaken. It rarely occurs to us that these works of art *And* science may be created to be interpreted from the vantage point of all perspectives and contexts—playing to all music.

Presently, we have no neutral category of creations which are neither one nor the other, but both—belonging to the *history of ideas* in general. As some neurocosmologists suggest by their actions, the visual arts' most promising future will present itself with the insight that our notions of "visual" and of "surface" need to be supplanted by a new definition and vision. A vision of nature with a new philosophy of mind that sees beyond all forms—including the forms we use to represent the formless processes of nature.

The multitude of the wise is the welfare of the world.
—*Wisdom of Solomon, VII, 24*

Afterword

Throughout this book, I've been talking about neuro-cosmology as an intellectual tool. But it can also serve to inform a social philosophy. We must recognize that our development and survival ride on our ability to experience life as an integrated community of explorers —providing we grant ourselves this privilege. We need to understand the creative potential of people, processes, and information living connected and exercising our gifts as wise, ethical creatures. The moment we forget that we are connected as caring, conscientious, or ethical human beings with emotions—the instant we dismiss our collective responsibilities to one another in sharing our knowledge—we descend without insight into the entropic, primordial world from which we emerged. We become slaves to our ignorance. We misunderstand that to be *human* is to be *nature at its best.* It is to live in nature's way. To be *inhuman* is to be *nature at its worst.* It is to live in the way of nature, like a disease obstructs the path of a healthy life. Human beings are forms of nature's being. Where nature connects with every detail of its creations, *we too are connected with every detail of our creations.* We are one of nature's details which can be grasped only by connecting our interpretations. Without some respected union of interpretations, we lose ourselves in the mayhem of meaningless actions and ecological anarchy. Our world closes.

I was taught to believe that stellar life and human life are distinctly different. The thought of these lives intermingling was intimated only by some ancient source of mystical wisdom that soared beyond my comprehension. I later learned to question my early teachings and to respect the flight of these complex wisdoms that trust intuition the way a child trusts loving parents. It's not that the parents know any better, it's just that they have more experience in seeing how the options of belief and vision are forever open to debate: No school of thought or apparatus of analysis is secure from change and criticism.

Celestial matter and minds both create space and inhabit the spaces they create. (Although one may take up more room than the other—the human mind, that is.) When we say that a person is "as inventive as nature," for example, are we not touching on some *archetypal thought* that's true to nature? What are these exaggerated expressions founded on? Are they really hyperboles or are they hyperrealistic statements?

The idea that one's imagination and ego can be as large as the heavens may once have been a startling thought that has since been pulverized into a cliché. To revive its original surprise, ask yourself how the meaning of this metaphor would change if this expression was understood to be more literal than figurative or "meta-literal"—if it was understood as a fact of our physiological makeup rather than of our fantasy's makeover. Or consider how the meaning of this expression would change if it was spoken into the deafening silence of deep space. Imagine how peculiar this boastful remark would sound, as you drifted with trepidation in an embryonic spacesuit—unattached to your mother, a lonely spacecraft in the distance. Feeling minuscule and mortal, would this conceit ever be thought or uttered? Ask yourself how this and other related metaphorms will change as we eventually inhabit this unfamiliar setting, as we journey alone into that frigid, black silence where even our thoughts are overwhelmed by the energies of the cosmos. To use a neurocosmological expression: Our thoughts will seem to undergo a process of *hydrostatic equilibrium*, like stars. The inward gravitational attraction from the weight of matter (our recognition of the truth) will balance the outward pressure forces at every point (our will to defy the truth). Perhaps, at this later date, we'll understand these counterbalancing tendencies of the human will.

Will we revert to primal-like expressions to empower us? Or will we readjust the scale of these comparisons as our confidence wanes—as our dreams retreat to the Platonic caves from whence they came? Perhaps, even as we confront these foreign circumstances, we will still stake out our place in the universe or, more ambitiously, *put ourselves on par with* the universe. Meaning that we'll continue to speak through anthropic principles and untempered visions. We'll continue to connect the celestial sky and mind through metaphorms.

The story of the brain-universe has been authored by all who have fascinated over this unity and have sought to express their fascinations. We are all storytellers when we describe how this relation will end as our species ends, or how it began when we began. To be privy to this knowledge is to understand the essence of nature which nature itself has not privileged us by demonstration. And may never.

Out of sheer nervousness, we might still broadcast our wishful thoughts of space conquests, the way we whistle in the dark when we're scared beyond composure or shriek when terrified! Our aspirations seem indomitable. It's unlikely that we'll be squelched even when confronted with the worst of horrors—whatever these are that each of us imagines. For me, it is the engulfing realization that we may be more alone and unique than we ever imagined. Without peers in space, we have only ourselves to share the experience of our total obscurity amidst the void. We invent aliens the way some children create sprawling war fields full of nameless soldiers dressed to kill in their battle fatigues. In a similar vein, we drive species after species of life forms into extinction, either by accident or by intractable will charged by some deadly conviction. At our present rate of development, the course of our creativity will look like a blunt and stubby branch atop the tree of evolution. Unless, of course, we think first—with foresight—about the roots and soil of our nature, about our minds in the garden of the universe.

IN THE PROCESS OF PROBING REALITY—A BALL OF MERCURY—
OUR PHILOSOPHIES OF REALITY BECOME CONTAMINATED   .
BY INCREASING COMPLEXITY.

Nature's Processes...

Facts...

Truths...

Realities.

Glossary

Some key terms coined by the author which the reader will encounter in Breaking the Mind Barrier:

BIOMIRROR or BIOLOGICAL MIRROR—*n*. **1**: theoretically, a bioelectric-magnetic-chemical mirror in the human brain constructed from certain patterns of electromagnetic activity and activated by a number of neurochemical systems involving the coordinated actions of both cerebral hemispheres and their subcortical systems. **2**: a specific process in the central nervous system that initiates unified electrochemical activity which is coordinated across the medial plane of the cerebral hemispheres down to the brain stem. **3**: the convergent actions of the whole brain. See *cerebral fusion*.

BRAIN/MIND UNION, ⲯ–*n*.: a transformation of the Greek symbols *psi*, ψ (mind), and *phi*, φ (body), signaling a transformation in philosophy whereby the brain and mind, matter and nonmatter, are integrated or considered one and the same thing. See *reflectionism*.

CEREBRAL FISSION—*n*. **1**: moments of analytical reasoning. **2**: a proposed splitting of brain processes. **3**: theoretically, an electrochemical disunity between cerebral hemispheres; one part of

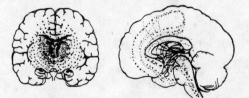

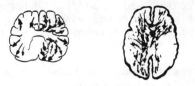

the brain, e.g., one hemisphere, is temporarily more forceful or assertive than the other.

CEREBRAL FUSION—*n*. **1:** an instant of intuition. **2:** a proposed merging of brain processes. **3:** theoretically, an electrochemical unity between and below the cerebral hemispheres; for tens of milliseconds or less, the two hemispheres function as one sphere. **4:** *cerebral parity*—a proposed equiv-

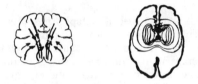

alence between functions of cerebral hemispheres and their integrated, subcortical systems.

CEREBRALISM—*n*. **1:** a visionary activity of neurocosmology, involving all disciplines and fields of knowledge, in creating a common language and system of communication. **2a:** a form of combined art and science which potentially comprises all other artforms and forms of science. **2b:** *cerebral art*: works of art and science which explore the nature of thinking, feeling, experiencing, and creating. **2c:** artscience works which do not emphasize a particular manner of expression and are eclectic in their representations. **3:** *cerebralists* are artists and scientists, writers and scholars, composers and musicians, performers and researchers who, in William James's words (1890), "combine the sensual and the intellectual" in their creations.

CEREBRARIUM—*n*. **1:** a model and apparatus for interpreting human brain/mind processes. **2:** a proposed multi-media, artificial-intelligence environment which will consist of hundreds of movable, computer-controlled video and film projectors and moving rearview projection screens that will receive a wide range of audio-visual material; the projected images are meant to stimulate processes of analogical thinking through metaphorms. The *Cerebrarium* is de-

signed to poetically transform current research in brain science, psychology, cognitive science, and physics, among other disciplines, in exploring the connectedness of the human mind and physical reality. **3:** a metaphoric brain theater of mental imagery.

CEREBREACTIONS—*n*. **1:** a proposed sudden reversal of feeling, thought, action, or tendency; a thought reversed by a contrasting feeling or impulse. **2a:** an outburst of emotional thoughts accompanying a sense of surprise or inspiration. **2b:** a creative impulse free from the anguish of indecision. **3:** moments of intuition balanced by reason; the presence of rationality in some instinctive movement.

CEREBREACTORS—*n*. **1a:** imaginary particle accelerators, nuclear fusion and fission reactors designed after the human brain and nervous system. **1b:** depictions of the human central nervous system as an organic linear accelerator, with the spinal cord as conduit and the brain as a reactor chamber, where atomic and subatomic particles (like nerve-cell interactions) generate energy fields that shape the pattern and substance of thought. **2a:** artistic constructions which suggest connections between two sciences that are ordinarily thought of as quite separate: neuro-

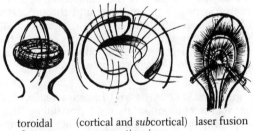

toroidal   (cortical and *subcortical*)   laser fusion
fusion   magnetic mirror
          fusion

physiology and nuclear physics. **2b:** metaphorical models that relate the mechanisms of nuclear and atomic reactors and the mechanisms of the human brain that generate and transform energy.

FACTUALIZE—*v*.: to make into fact or regard as fact; the complement of fictionalize.

METAPHORM—*v*. and *n*. **1:** an object, image, idea, or process that we compare to something else. **2:** all forms of metaphor which include allusion, allegory, analogy, symbolism and trope or figure of speech and which can involve all of the physical and psychological senses. **3:** the concept that *every* object, image, idea, or process is intrinsically metaphorical, whether or not one uses it metaphorically and regardless of the context in which the thing or process exists. **4:** to simultaneously compare and transform information.

METAPHORMING—*v.* and *n.* **1a:** the act or process of forming metaphors. **1b:** the transformative process of metaphorms. **2:** the interrelating, interweaving, comparing, assembling, or joining of parts into a connected whole, as in the process of *contexturing* or *contextualizing* information and matter; seeing something in a different context—from different points or angles of view—through metaphorms.

MYTHODOLOGY—*n.*: methods of myth-making in art and in science; applied "mythods" (serious double entendres).

NEUROCOSMOLOGY—*n.* **1:** the conceptual combining of neuroscience, cosmology, and art (among other disciplines) to study the nature and connectedness of the brain and the universe it explores and creates. **2a:** a vision of mind *as* nature where mind and body are understood in relation to the cosmos. **2b:** beyond the "micro-macrocosm" relation of structures, proportions, etc.: a study of the relationship between the processes of the universe and the processes of the things the universe creates. **2c:** a study of the relationship between the processes of the human brain and the things the brain creates. See *processmorphology* and *processmorph.*

NEURO-EXPRESSIONISM—*n.* **1a:** all visualizations or representations of thought, including those that emphasize subjective experience and those that are objectively or analytically composed; variations: *neuro-impressionism, neuro-surrealism,* among other art-related activities understood in relation to brain processes. **1b:** forms of *cerebral art* (see *cerebralism*).

NEUROSPHERE—*n.* **1:** a hypothetical sphere of processes related to the human nervous system; the neurosphere is to the nervous system what the biosphere is to the Earth. **2a:** variations: an imaginary, infinite *neurospace* that is related to the physical neuro-environment. **2b:** also called the *unisphere,* as when the two cerebral hemispheres are combined through the process of intuition in overcoming their complementarity to form one functional "sphere."

PROCESSMORPH—*adj.* and *n., biol.* and *phys.*: different in structure or appearance but sharing the same process. Also, *processmorphosis*—as in *process + metamorphosis.*

PROCESSMORPHISM—*n., pl.* **1:** the state or property of being processmorphic, alike in process but unlike in form. **2:** the de-emphasis of scale and proportion, hierarchies and levels, in the study of processmorphic relations.

PROCESSMORPHOLOGY—*n.* **1:** a comparative study of processes in like and unlike systems (biological, physical, symbolic, psychological, social, etc.). **2:** the combined branches of art, biology, physics, mathematics, and philosophy, among other disciplinary fields, dealing with the comparison of process-phenomena in natural and artificial systems, such as machines.

PSYCHOMIRROR—*n.* **1:** a metaphorm representing the mental mirror or conceptual mirror we create each time we separate or imagine dividing things that represent two different aspects of one and the same thing, such as mass and energy, space and time. **2:** theoretically, a state of mind in which one reflects on something or some thing reflects something else.

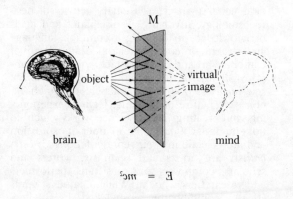

REFLECTIONISM—*n.* **1a:** the monistic/dualistic philosophy of neurocosmology that incorporates both Western and Eastern dialectics. **1b:** a belief system which holds that nature is one integrated whole, even though it can be described by many different perspectives; no perspective or representation of nature possesses absolute truth.

THOUGHT-ASSEMBLY—*n., pl.* -BLIES. **1:** an association of ideas or mental images. **2:** an artistic and/or scientific reconstruction of various perceptions and expressions. **3:** an account of the interactions of thought processes over the course of a specific time-space. **4:** an interpretation of a moment of creative or directed thought, visualizing the elements of imagination. **5:** the complement of cell-assembly (-blies).

TRANSCIENCE—*n.* **1:** a study of the transience of science or transient nature of science-making, i.e., paradigm shifts, variances in methodology, strategies, languages, etc. **2:** an examination of the transient tools (including mathematics) for studying the transient phenomena of nature.

End Notes

CHAPTER ONE

1. H. Butterfield, *The Origins of Modern Science* (New York: Macmillan, 1957), p. 75.

Kepler was very much inspired by the Pythagorean Society of Thought which believed in the supremacy of numbers, mixing mysticism and mathematics. The mathematician Tobias Dantzig relates that "the source of this mystic philosophy of the Pythagoreans, which left such a deep impression on the speculations of all Greek thinkers including Plato and Aristotle, is still a controversial question. To the modern mind steeped in rationalism the pompous number-worship may appear as *superstition erected into a system*.

When we view it in historical perspective, we are inclined to take a more charitable attitude. Stripped of its religious mysticism, the Pythagorean philosophy contained the fundamental idea that only through number and form can man grasp the nature of the universe. Such thoughts are expressed by Philolaus, Pythagoras' ablest disciple. . . . 'All things which can be known have number; for it is not possible that without number anything can be either conceived or known' (Philolaus)." (From T. Dantzig, 1954, p. 43.)

I also recommend Jamie Kassler's article "Music as Model in Early Science," *History of Science*, Vol. 20, pp. 103–139.

2. For an elaboration on this thought, read Nelson Goodman's *Ways of Worldmaking* (Indianapolis; Cambridge: Hackett Publishing Company, 1978).

Goodman's thoughts prompt me to recall the writings of the Nobel Prize dramatist Luigi Pirandello who explored a related subject—namely, the ways of reality and illusion—regarding our pictures of the world and ourselves. His play *It Is So! (If You Think So)* (1922), which was originally titled *Right You Are!*, resonates with Goodman's reflection on "worldmaking" specifically and with the reflections in *Breaking the Mind Barrier* more generally.

3. For more on this subject, I suggest reading Max Black's *Models and Metaphors* (Ithaca, N.Y.: Cornell University Press, 1962), Chapters 3 and 13. Also, for a synoptic overview of the different types of metaphors and analogies, there is Mary B. Hesse's *Models and Analogies in Science* (Notre Dame, Ind.: University of Notre Dame Press, 1966).

4. Arthur Miller, "Visualization Lost and Regained: The Genesis of the Quantum Theory in the Period 1913–27," in Judith Wechsler (ed.), *On Aesthetics in Science* (Cambridge, Mass.: M.I.T. Press, 1978), p. 73. An important book on this subject is Rudolf Arnheim's *Visual Thinking* (Berkeley: University of California Press, 1969); see Chapter 15, "Models for Theory."

5. From the preface to Arakawa and Madeline H. Gins, *The Mechanism of Meaning* (New York: Harry N. Abrams, 1979).

6. Jacob Bronowski, *Science and Human Values* (New York: Harper, 1956), p. 27.

CHAPTER THREE

1. Quoted in P. M. Knudtson, "Santiago Roman y Cajal: Painter of Neurons," *Science 85*, September 1985, Vol. 6, pp. 66–7.

2. *The American Heritage Dictionary*, 2nd college ed. (New York: Dell, 1985), p. 550.

3. As a parenthetical note: In the medical arts/sciences this distinction between process and structure is not so pronounced. For example, internist and pulmonary specialists alike will discuss a disease of the endocrine system in terms of the malfunctioning processes of the liver and not its structure (unless the incidence of finding shows that this organ is physically damaged).

4. Lewis Thomas, *The Lives of a Cell: Notes of a Biology Watcher* (New York: Bantam Books, 1974), p. 4.

5. Michael Zeilik and Elske van Panhuys Smith, *Introductory Astronomy and Astrophysics*, 2nd ed. (New York: Saunders College Publishing, 1987), p. 273.

6. Stan Woosley and Tom Weaver, "The Great Supernova of 1987," *Scientific American*, August 1989, p. 33.

7. Ibid., p. 33.

8. Zeilik and Smith, *Introductory Astronomy and Astrophysics*, p. 274.

9. Ibid., p. 180.

10. This quote is by astrophysicist Virginia Trimble of the University of California at Irvine. It appeared in an article by Marcia Bartusiak entitled "Coming Home," *Discover*, September 1988, p. 31.

11. "Aphorism XXIII" in Francis Bacon's *The Novum Organum* (1620), from Fulton H. Anderson (ed.), *Bacon: The New Organum* (New York: Bobbs-Merrill, 1960).

12. Gregory Bateson, *Mind and Nature: A Necessary Unity* (New York: Bantam Books, 1980), p. 18. Bateson writes: "I remember the boredom of analyzing sentences and the boredom later, at Cambridge, of learning comparative anatomy. Both subjects, as taught, were tortuously unreal. We could have been told something about the pattern which connects: that all communication necessitates context, that without context, there is no meaning, and that contexts confer meaning because there is classification of contexts."

CHAPTER FOUR

1. My notes are based on Dr. Samuel Schacher's text "Determination and Differentia-

tion in the Development of the Nervous System," in Eric R. Kandel and James H. Schwartz, *Principles of Neural Science* (New York: Elsevier/North-Holland, 1981), p. 504.

2. Ibid., p. 504.

3. Ibid., p. 510.

4. Quoted in Aniela Jaffe's article "Symbols in an Individual Analysis," in C. G. Jung, *Man and His Symbols* (New York: Dell Publishing, 1964), p. 306.

5. Michael Zeilik and Elske van Panhuys Smith, *Introductory Astronomy and Astrophysics*, 2nd ed. (New York: Saunders College Publishing, 1987), p. 278.

6. Ibid., p. 352. The description here of the birthing process of massive stars is best read in the original. I've tried to relate the gist of this process, adhering to Zeilik and Smith's account.

One fine introductory text—which is both scientific and poetic—is Dr. Martin Cohen's *In Darkness Born: The Story of Star Formation* (New York: Cambridge University Press, 1988). Cohen is a professor of astronomy at the University of California, Berkeley, and a researcher at NASA–Ames Research Center.

7. Zeilik and Smith, *Introductory Astronomy and Astrophysics*, p. 352.

8. In supporting his speculations, Carl Sagan cites the research of Dr. Leigh van Valen, an evolutionary biologist from the University of Chicago. You may want to read van Valen's "Brain Size and Intelligence in Man" in *American Journal of Physical Anthropology* (Vol. 40, 1974, pp. 417–424) to further appreciate Sagan's views on this issue raised in *The Dragons of Eden: Speculations on the Evolution of Human Intelligence* (New York: Ballantine Books, 1977), pp. 34–41.

Sagan specifically stated that "differences of many hundreds of grams may be functionally unimportant." I tend to agree. However, I think it's important to consider not only the areas of the human brain these grams of matter are either missing or removed from but also the age and relative health of the individual. If we're talking about the removal of cortical tissue and limiting ourselves to the region of the cerebral hemispheres, then I'm in complete agreement. If, on the other hand, we're speaking of subcortical tissue—such as removing a few hundred grams of gray matter (for example, from the subthalamic nuclei, a small mass of gray matter in the diencephalon [the "between brain"], or from the substantia nigra, another small though important mass in the midbrain)—I disagree. This means that even the most subtle subtractions of matter from one area of the brain can make a critical difference in the cognitive, affective, and motor behavior of a human being. And yet, the modest subtraction of neocortical matter can have minimal effects on cognition and the continuity of movement, or action—two key aspects of intelligence.

There is one potential trap regarding this quantification scheme. It may be found in the argument of the American psychologist Karl Lashley, put forth earlier in this century, that collective brain mass alone—and not neuronal connectivity or architecture—is critical for learning and intelligence. Lashley's pioneering theory of *mass action* (or *aggregate field* as it was also called) deemphasized the importance of neural connections and individual neurons and nuclei. Need I say this theory has since undergone some major revisions, due to the forceful evidence of a number of important neurophysiologists (among them Wilder Penfield and Norman Geschwind) who studied the localization of functions in the brain.

9. Zeilik and Smith, *Introductory Astronomy and Astrophysics*, p. 213.

10. Ibid., p. 213.

11. Ibid., pp. 222, 223.

12. I recommend James Gleick's *Chaos: Making a New Science* (New York: Penguin Books, 1987), p. 28. The history of this important science is documented in a lively and well-informed manner.

13. Hofstadter's quote is from James Gleick's *Chaos*. It is, perhaps, one of the most ubiquitous quotes, summing up the strangeness of chaotic phenomena.

CHAPTER FIVE

1. For further information on the functional components of neurons, read Dr. Eric R. Kandel's "Nerve Cells and Behavior," in Eric R. Kandel and James H. Schwartz, *Principles of Neural Science* (New York: Elsevier/North-Holland, 1981), pp. 14–23.

2. You may find this issue of *virtual structures and processes* as perplexing as I do. Why would nature show us only half of itself in the form of the finite physical world—a world of tangible limits—taunting us with the idea of a world without limits? Could this really be the world of mind, of brain? Are the ways of nature that paradoxical and mischievous? Are contradictions the means by which nature conceals its inner secrets? There appears to be no way of showing these virtual processes of mind (or matter) by physical means. Thus far, they've resisted full scrutiny by our most sophisticated and ingenious technologies, although we can observe their "dimensions" in a limited and speculative way by studying the mind's secondary and tertiary effects—namely, the artifacts of our creations and *their* effects. Despite our attempts we still cannot grip the primary effects of virtual processes.

3. Carl W. Cotman and James L. McGaugh, *Behavioral Neuroscience* (New York: Academic Press, 1980), p. 112.

4. For more detailed information on the nature of action potentials in neuronal signaling, you may want to read Bernard Katz's classic book *Nerve, Muscle, and Synapse* (New York: McGraw-Hill, 1966).

5. Cotman and McGaugh, *Behavioral Neuroscience*, pp. 26, 27.

6. For a sensitive account of brain damage involving sensory and motor aphasias, ataxias, etc., I recommend the neurologist Oliver Sacks's fascinating book *The Man Who Mistook His Wife for a Hat: And Other Clinical Tales* (New York: Harper & Row, 1987).

Another excellent account of the effects of brain damage as reported by a pioneer in the field is Howard Gardner's *Art, Mind and Brain: A Cognitive Approach to Creativity* (New York: Basic Books, 1982), Part IV: "The Breakdown of the Mind."

There is an avalanche of unanswered questions concerning the degradation and breakdown of the nervous system. This avalanche is slowly losing speed and will hopefully settle in the valley of the neurosciences in the next century as we learn much more about brainways—and as we render this knowledge into some coherent, unified form with the assistance of both mathematics *and* the arts, among other disciplines.

7. Michael Zeilik and Elske van Panhuys Smith, *Introductory Astronomy and Astrophysics*, 2nd ed. (New York: Saunders College Publishing, 1987), pp. 71, 72.

8. My point about gravity and gravitons as the "language of galaxies" should be understood in the context of Newton's notion that gravity is a force. Einstein's view, by contrast, represented gravity as a manifestation of the curvature of the space-time continuum. If I were to adopt this view in speaking metaphorically about how galaxies "communicate" with one another, I would associate the phenomenon of space-time curvature with the "grammar" of stellar communication—that is, its formal features and "use of language and rules" of communicating. You might say a galaxy exercises its powers of communication, or "gravitational influences," as it interacts with other galactic systems or communities.

Einstein's special and general theories of relativity (1905 and 1915, respectively) established the science of what is called *relativistic cosmology*, on which most cosmological models are based. Where the special theory combined the concepts of time and space to create a four-dimensional weave of time-space, the general theory applied the principles from special relativity to gravitational phenomena.

9. Donald Hebb, *The Organization of Behavior* (New York: John Wiley & Sons, 1949), p. 88.

10. D. R. Hofstadter and D. C. Dennett, *The Mind's I* (New York: Basic Books, 1981), p. 200.

11. Zeilik and Smith, *Introductory Astronomy and Astrophysics*, pp. 71, 72.

12. This quotation is from an article by David L. Chandler in the "Sci-Tech" section of *The Boston Globe*, Monday, March 20, 1989, pp. 27, 28.

13. For an informed and extended account of electrical currents in relation to the human body, I suggest you read Dr. Robert O. Becker and Gary Selden's book *The Body Electric: Electromagnetism and the Foundation of Life* (New York: William Morrow, 1985).

CHAPTER SIX

1. G. S. Kirk and J. E. Raven, *The Presocratic Philosophers* (Cambridge: Cambridge University Press, 1957). Also M. C. Stokes, *One and Many in Presocratic Philosophy* (Cambridge, Mass.: Harvard University Press, 1971).

2. Quoted from C. N. Yang, "Nobel Prize Lecture," in Heinz R. Pagels, *The Cosmic Code: Quantum Physics as the Language of Nature* (New York: Bantam Books, 1982), p. 256.

3. From a chapter titled "The Uniformity of Biochemistry" in Francis Crick's *Life Itself: Its Origin and Nature* (New York: Simon & Schuster, 1981), pp. 42, 43.

4. For a discussion on "philosophy without mirrors" and a full treatment of this subject, I strongly recommend Richard Rorty's *Philosophy and the Mirror of Nature* (Princeton, N.J.: Princeton University Press, 1979), p. 357. The author offers a cogent polemic against the seventeenth-century systematic philosophers, pointing out the strengths of those Western "edifying" philosophers (such as [the early] Wittgenstein, Heidegger, Dewey, Sellars) who avoided what Rorty calls "the self-deception which comes from believing that we know ourselves by knowing a set of objective facts."

5. Ibid., p. 357.

6. Ibid., p. 357.

7. For an in-depth analysis of this subject, I recommend Mario Bunge's scholarly *Mind-Body Problem* (New York: Pergamon Press, 1980), p. 6. In Bunge's history of this "problem," we learn of several contemporary philosophers, such as Herbert Feigl, who straddled between notions of the identity theory, expressing the belief that "no matter how much the concepts of psychology may differ from those of neurophysiology, they have the same referents." This statement from Feigl appears in Paul K. Feyerabend, ed., *Mind, Matter, and Method* (Minneapolis: University of Minnesota Press, 1966).

8. Stephen Hawking, *A Brief History of Time* (New York: Bantam Books, 1988), p. 11.

9. Betty Heimann, *Facets of Indian Thought* (New York: Schocken Books, 1964), pp. 118, 119.

10. For a synoptic description and finer illustration of J. Richard Gott's "Tachyon Universe," see Timothy Ferris, *Galaxies* (New York: Stewart, Tabori & Chang, 1982), pp. 174, 175.

CHAPTER SEVEN

1. William Aspray and Arthur Burks, eds., *Papers of John von Neumann on Computing and Computer Theory* (Cambridge, Mass.: M.I.T. Press; Los Angeles: Tomash Publishers, 1987). Also, read John Von Neumann, *The Computer and the Brain* (New Haven: Yale University Press, 1958).

2. For more on Turing, I suggest you read Andrew Hodges's biography *Alan Turing: The Enigma* (New York: Simon & Schuster, 1983).

In the context of this discussion, consider Turing's (1952) theoretical study of morphogenesis, the development of pattern and form in living organisms. His main goal was to show how a uniform, symmetric structure could grow and develop, as a result of diffusion, into a strongly unsymmetric structure with a definite pattern.

3. From Todd Siler, *Cerebreactors* (New York: Ronald Feldman Fine Arts, Inc., 1980), p. 12.

4. Ibid., p. 12.

5. Note that the "mirror machine reactors" employed in plasma fusion are not to be confused with the "mirror machine" used by British physicist-psychologist C. Maxwell Cade, among other

researchers, who continue to apply this machine in measuring and exploring EEG patterns in biofeedback experiments. Most of the experiments concentrated on inducing what Cade calls "lucid awareness," or "awakened mind," as in a state of heightened mental activity.

Although no reference is made in the literature connecting these devices, a neurocosmologist might make such a connection. She might remark, for example, that the inventions of both devices are born from "the plasma fusion state of mind"—the fourth state of mind and matter—like plasma. In this heightened state of awareness, and intuition, *the mind momentarily sees itself for what it is.*

6. J. D. Lawson, *The Physics of Charged-Particle Beams* (New York: Clarendon Press, 1977).

7. I recommend Gerard 't Hooft's article "Gauge Theories of the Forces between Elementary Particles" in Richard A. Carrigan, Jr., and W. Peter Trower, eds., *Particle Physics in the Cosmos: Readings from Scientific American Magazine* (New York: W. H. Freeman, 1989), pp. 78–105.

CHAPTER EIGHT

1. For further reading on the diencephalon, in particular the limbic system and its role in emotion, read J. W. Papez, "A Proposed Mechanism of Emotion," *American Medical Association Archives of Neurology and Psychiatry*, 1937, Vol. 83, pp. 725–743. On this subject, you may also find it interesting to read *The Mitchell Beazley Atlas of Body and Mind* (London: Mitchell Beazley, 1976), which discusses the workings of the limbic system.

2. For an in-depth (but somewhat complicated) look at the theory of "personality fields," read K. Lewin, *A Dynamic Theory of Personality: Selected Papers by Kurt Lewin* (New York: McGraw Hill, 1935).

3. Wolfgang Köhler, *Gestalt Psychology* (New York: Liveright Publishing, 1947). For a discussion on the nature of insight, note the work of Max Wertheimer, *Productive Thinking* (New York: Harper, 1959).

4. Aside from the studies in neuromagnetics, there is major research in the field of medical imaging that utilizes positron emission tomography. This device allows scientists to actually see the organic changes in states of brain as they correspond to various cognitive processes. I refer you to W. D. Hess and M. E. Phelps, eds., *Positron Emission Tomography of the Brain* (New York: Springer-Verlag, 1983).

5. For further reading: R. W. Sperry, "Lateral Specialization in Surgically Separated Hemispheres," in F. O. Schmitt and F. G. Worden, eds., *The Neurosciences: Third Study Program* (Cambridge, Mass.: M.I.T. Press, 1974), pp. 5–19. One good introduction to this research: Michael S. Gazzaniga's article "One Brain—Two Minds?," *American Scientist*, May–June 1972, Vol. 60, No. 3, pp. 311–317. For a unique overview on split-brain research is Robert E. Ornstein's book *The Psychology of Consciousness* (San Francisco: W. H. Freeman, 1975).

6. A classic: C. Wernicke, "The Symptom-Complex of Aphasia," in A. Church, ed., *Diseases of the Nervous System* (New York: Appleton, 1908), pp. 265–324.

7. Other important research: C. Trevarthen, "Analysis of Cerebral Activities that Generate and Regulate Consciousness in Commissurotomy Patients," in S. J. Dimond and J. G. Beaumont, eds., *Hemisphere Function in the Human Brain* (New York: Wiley, 1974), pp. 235–263.

8. Suggested readings on the issue of laterality and cerebral dominance: G. Berlucchi, "Cerebral Dominance and Interhemispheric Communication in Normal Man," in F. O. Schmitt and F. G. Worden, eds., *The Neurosciences: Third Study Program* (Cambridge, Mass.: M.I.T. Press, 1974), pp. 65–69; N. Geschwind, "Disconnexion Syndromes in Animals and Man: Parts 1 and 2," *Brain*, Vol. 88 (1965), pp. 237–294 and 585–644. Also: M. S. Gazzaniga, "On Dividing the Self: Speculations from Brain Research," in W. den Hartog Jager, G. Bruyn, and A. Heijstee, eds., *Neurology* (Amsterdam: Proceedings of the 11th World Congress of Neurology, International Congress Series No. 434), pp. 233–244.

9. An important book on this subject is C. P. Snow's *The Two Cultures* (Cambridge: Cambridge University Press, 1961).

10. M. E. Phelps, J. Mazziotta, and S. C. Huang, "Study of Cerebral Function with Positron Computed Tomography," *Journal of Cerebral Blood Flow Metabolism*, Vol. 2, pp. 113–162.

11. H.-L. Teuber, *Perception* (New York: Springer-Verlag, 1978). Also: B. Milner, "Hemispheric Specialization: Scope and Limits," in F. O. Schmitt and F. G. Worden, eds., *The Neurosciences: Third Study Program* (Cambridge, Mass.: M.I.T. Press, 1974), pp. 65–69.

12. Revisiting the subject of laterality: R. Puccetti, "The Case for Mental Duality: Evidence from Split-Brain Data and Other Considerations," *The Behavioral and Brain Sciences*, Vol. 4 (1981), pp. 93–123; also H. A. Whitaker and G. A. Ojemann, "Lateralization of Higher Cortical Functions: A Critique," *Annals of the New York Academy of Sciences*, Vol. 299 (1977), pp. 459–473.

CHAPTER TEN

1. There are a number of companies that now manufacture biomagnetic imaging devices that are being used in medicine to visualize the functional activity of the brain and body for diagnostic purposes. The neurological applications will have increasing significance as more detailed studies on the health effects of exposure to X-rays (as in computer tomography, or "CT") and high-intensity magnetic fields (as in magnetic resonance imaging, or "MRI") are done.

Also, note: H. S. Burr, "The Meaning of Bioelectric Potentials," *Yale Journal of Biological Medicine*, Vol. 16 (1944), p. 353. Also by Burr on this subject: *The Fields of Life* (New York: Ballantine, 1972); and see A. S. Presman, *Electromagnetic Fields of Life* (New York: Plenum Press, 1970).

2. Donald Hebb, *The Organization of Behavior* (New York: John Wiley & Sons, 1949), p. 80.

CHAPTER ELEVEN

1. The Isaac Asimov quote is from Marvin Minsky's *The Society of Mind* (New York: Simon & Schuster, 1985), p. 24.

2. Ilya Prigogine and Isabelle Stengers, *Order Out of Chaos: Man's New Dialogue with Nature* (New York: Bantam Books, 1984).

CHAPTER TWELVE

1. Heinz Pagels, *The Cosmic Code: Quantum Physics as the Language of Nature* (New York: Bantam, 1982).

2. Paul C. W. Davies, *The Accidental Universe* (New York: Cambridge University Press, 1982).

3. For a broad and clear discussion of this subject there is John D. Barrow and Frank J. Tipler's book *The Anthropic Cosmological Principle* (Oxford; New York: Oxford University Press, 1986). More recently: John Gribbon and Martin Rees, *Cosmic Coincidences: Dark Matter, Mankind, and the Anthropic Cosmology* (New York: Bantam Books, 1989).

4. See Davies, *The Accidental Universe*, p. 121.

5. The physicist Brandon Carter is one of the first scientists to have stated the weak and strong versions of the anthropic principle, exploring their implications with regard to astrophysical processes. Carter's article is included in M. S. Longair, ed., *Confrontation of Cosmological Theories with Observation* (Dordrecht: Reidel, 1974).

6. Some art historians may charge that reappropriating this image and its content is inappropriate, or even irrelevant, to the art itself; meaning that this artwork wasn't intended for such use. However, as a cosmological image, it is so rich in its visual associations that it naturally stimulates these sorts of connections with the science of cosmology. Not to recognize these associations is to miss this art's message or underestimate its ability to inform this science. I could reverse the situation and take the driest, most analytically composed "image" of either an isotropic or anisotropic universe and convert it into a sensuous, visually flowing image comparable in scope to the picture of our world. To amplify a statement I made earlier: I'm not taking things out of context for the sake of the taking. The way I see it is, *I'm putting things back in*

their original context—as they were perhaps in the "first act of unity" occurring in the mind and cosmos alike.

7. See Davies, *The Accidental Universe*, p. 81.

8. Fred Alan Wolf, *Star*Wave: Mind, Consciousness, and Quantum Physics* (New York: Macmillan, 1984).

9. From Todd Siler, "Reality" (M.I.T. Master's thesis, 1981), pp. 17, 18.

10. Ibid., p. 18.

11. Paul Davies elaborates on a similar point in describing how elementary particles must be understood as a "network of relations." For me, the concept of *network* emphasizes the combination rather than integration, of things and processes. Things can be combined through acts of networking, while remaining unintegrated. By using the term *union*, I mean to accent this issue of a world of fully integrated processes. For further reading: Paul Davies, *Superforce: The Search for a Grand Unified Theory of Nature* (New York: Simon & Schuster, 1984).

12. Note that Minsky's explorations of the nature of mind consider only the mind's physical aspects. As he has stated, *"Minds are simply what brains do."* In the scheme of reflectionism, Minsky would be a reflectionist monist. Consequently, many of the issues examined in neurocosmology—such as those relating physics or celestial mechanics to the mind, and similar speculations—are not entertained in his models.

One belated note: I believe we need to work toward creating socie*ties* of minds—not a society of mind which is composed of "mindless agents" (to use Minsky's phrase)—whose collective activities add up to intelligent actions.

13. Marvin Minsky, *The Society of Mind* (New York: Simon & Schuster, 1988), p. 288.

14. Ibid., p. 319.

15. Ibid., p. 319.

16. Peter Medawar, *The Limits of Science* (New York: Oxford University Press, 1984.)

17. Karl Lashley devoted his life's work to discovering the nature of memory. After more than 40 years of neurobiological research, he concluded that memory is not situated in any specific place in the brain but exists as a collective property of brain processes. This insight was critical for the neuroscientist Karl Pribram, who studied with Lashley and who later developed his holographic memory metaphorm based on this work and on physicist David Bohm's holographic universe metaphorm. See Karl S. Lashley, "Persistent Problems in the Evolution of Mind," *Quarterly Review of Biology*, Vol. 24 (1949), pp. 28–42; also, Karl S. Lashley, "In Search of the Engram," *Symposia of the Society of Experimental Biology*, Vol. 4 (1950), pp. 454–482.

18. From Douglas R. Hofstadter's "Prelude . . . And Fugue" in *Gödel, Escher, Bach: An Eternal Golden Braid* (New York: Basic Books, 1979), p. 283.

CHAPTER THIRTEEN

1. Ludwig Boltzmann quote is from Arthur I. Miller, *Imagery in Scientific Thought Creating 20th-Century Physics* (Cambridge, Mass.: M.I.T. Press, 1984).

CHAPTER FOURTEEN

1. My notes supporting this point are drawn from physicist Frank Wilczek's article "The Cosmic Asymmetry between Matter and Antimatter," in Richard A. Carrigan, Jr., and W. Peter Trower, eds., *Particle Physics in the Cosmos: Readings from Scientific American Magazine* (New York: W. H. Freeman, 1989), pp. 165, 166.

2. Fred Alan Wolf, *Star*Wave: Mind, Consciousness, and Quantum Physics* (New York: Macmillan, 1984), p. 137.

3. My explanatory notes here are informed by Alan H. Guth and Paul J. Steinhardt's article "The Inflationary Universe," in Richard A. Carrigan, Jr., and W. Peter Trower, eds., *Particle Physics in the Cosmos: Readings from Scientific American Magazine* (New York: W. H. Freeman, 1989), p. 181.

4. Ibid., p. 179.

5. Todd Siler, *The Biomirror* (New York: Ronald Feldman Fine Arts, Inc., 1983).

CHAPTER FIFTEEN

1. From Todd Siler, "Architectonics of Thought: A Symbolic Model of Neuropsychological Processes" (M.I.T., Ph.D. dissertation, Interdisciplinary Studies in Psychology and Art, 1986).

2. From Todd Siler, "Reality" (M.I.T., Master's thesis, 1981).

3. This quote is from Edward Hoffman's *The Right to Be Human: A Biography of Abraham Maslow* (Los Angeles: Jeremy P. Tarcher, 1988), p. 208.

BRANCHES #1

1. Note Gyorgy Kepes, "The Visual Arts and the Sciences: A Proposal for Collaboration," *Daedalus*, Vol. 94 (No. 1), Winter 1965. Also, for more superb reading on issues concerning arts' confluence with science and technology, there are Kepes's many books, among them: *Language of Vision* (Chicago: Paul Theobald, 1944); also *The Vision + Values Series* (New York: George Braziller), which includes *The Education of Vision* (1965), *Structure in Art and Science* (1965), *The Nature of Art and Motion* (1965), *Module, Proportion, Symmetry, Rhythm* (1966), *The Man-Made Object* (1966), *Sign, Image, Symbol* (1966), and *Arts of the Environment* (1972—"Art and Ecological Consciousness," pp. 1–12).

2. Quoted from the dedication catalogue of the Center for Advanced Visual Studies at M.I.T., Gyorgy Kepes, 1967.

3. Otto Piene and Heinz Mack, eds., *Zero, 1, 2, 3* (Cambridge, Mass.: M.I.T. Press, 1973). The artists in Group Zero included, among other renowned innovators, Yves Klein, Jean Tinguely, and Pol Bury. In the mid-twentieth century, the creations of Group Zero (in Düsseldorf), Gruppo T (in Milan), and Groupe de Recherche d'Art Visuel (in Paris) mark one of many outstanding periods in art's evolution.

4. *Centerbeam* was included in the international art exhibition *documenta 6* at Kassel, Germany, June 24 through October 2, 1977, and was exhibited on the Mall in Washington, D.C., in the summer of 1978. For a more current overview of Sky Art explorations, see Elizabeth Goldring, "Desert Sun/Desert Moon and the SKY ART Manifesto," *Leonardo*, Vol. 20 (No. 4), 1987.
The history from which this machine-technology-oriented work emerged includes the Russian Constructivist Tatlin's "Memorial to the 3rd International," Moholy-Nagy's "Manifesto of the Dynamic-Constructive Power System," Marcel Duchamp's and the Dadaists' "machines," and—in more recent times—the technological works of Experiments in Art and Technology (or E.A.T.), founded by the engineer Billy Kluver and artist Robert Rauschenberg in the mid-1960s; see Billy Kluver, *Pavilion* (Toronto, Vancouver: Clarke, Irwin, 1972). Other important developments in this direction are documented in Maurice Tuchman's *Art & Technology: A Report on the Art & Technology Program of the Los Angeles County Museum of Art, 1967–71* (New York: Viking Press, 1972).

5. From Otto Piene and Elizabeth Goldring, eds., *Centerbeam* (Cambridge, Mass.: M.I.T. Press, 1980), p. 24.

6. Nancy Holt, ed., *The Writings of Robert Smithson: Essays with Illustrations* (New York: New York University Press, 1979).

7. Diana Shaffer, "Nancy Holt: Spaces for Reflections or Projections," in Alan Sonfist, ed., *Art in the Land: A Critical Anthology of Environmental Art* (New York: E. P. Dutton, 1983), p. 174; Nancy Rosen, "A Sense of Place: Five American Artists," *Studio International*, Vol. 193, March-April 1977, p. 119; also Ted Castle, "Nancy Holt, Siteseer," *Art in America*, Vol. 70, March 1982.

8. Nancy Holt, "Sun Tunnels," *Artforum*, Vol. 15, April 1977, p. 35. For a sensitive overview on Holt's work in relation to other major works with a similar focus, read Nicholas J. Capasso, "Environmental Art: Strategies for Reorientation in Nature," *Arts Magazine*, Vol. 59, January 1985, pp. 73–77.

9. Edward Fry, "Robert Morris; The Dialectic," *Arts Magazine*, Vol. 49, September 1974, pp. 22–23; Robert Morris, "Observatory," *Avalanche*, Fall 1971 (No. 3), pp. 30–35.

10. From an interview with Janet Saad-Cook in "Touching the Sky: Artworks Using Natural Phenomena, Earth, Sky and Connections to Astronomy," *Leonardo*, Vol. 21 (No. 2), 1988, p. 124; La Jolla Museum of Contemporary Art, *The Substance of Light: Sunlight Dispersion, the Solar Burns, Point/Star Space: Selected Works of Charles Ross*, exhibition catalogue, La Jolla, California, 1976.

11. Janet Saad-Cook with Charles Ross, Nancy Holt, and James Turrell, "Touching the Sky: Artworks Using Natural Phenomena, Earth, Sky and Connections to Astronomy," *Archeoastronomy*, Vol. 8 (Nos. 1–4), January-December 1985, pp. 118–141.

12. Walter De Maria, "The Lightning Field," *Artforum*, Vol. 18, April 1980, pp. 52–59; Melinda Wortz, "Walter De Maria's 'The Lightning Field,' " *Arts Magazine*, Vol. 54, May 1980, pp. 172–173.

13. From the project description *The Van Sant GeoSphere: The Earth Situation Room*, issued by Eyes on Earth, a nonprofit corporation founded and directed by Tom Van Sant in Santa Monica, California.

14. Jonathan Benthall, "Haacke, Sonfist, and Nature," *Studio International*, Vol. 181, March 1971; *Hans Haacke, unfinished business* (New York: New Museum of Contemporary Art; Cambridge, Mass.: M.I.T. Press, 1986).

15. Robert J. Horvitz, "Nature as Artifact: Alan Sonfist," *Artforum*, Vol. 12, November 1973, p. 32.

16. Carol Hall, "Environmental Artists: Sources and Directions," in Alan Sonfist, ed., *Art in the Land: A Critical Anthology of Environmental Art* (New York: E. P. Dutton, 1983), pp. 26–29. One of Sonfist's works which most notably explores ecological issues is his public collaborative project *Time Landscape* (1965–1978). See Jonathan Carpenter, "Alan Sonfist's Public Sculptures," in *Art in the Land*, pp. 142–154; Harold Rosenberg, "Time and Space Concepts in Environmental Art," in *Art in the Land*, pp. 191–216; and Corinne Robbins, "Alan Sonfist: Time as Aesthetic Dimension," *Arts Magazine*, Vol. 51, October 1976, pp. 87–88.

17. Quoted in Robert J. Horvitz, "Nature as Artifact," p. 32.

18. Quoted in Nancy Foote, ed., "Situation Esthetics: Impermanent Art and the Seventies Audience," *Artforum*, Vol. 18, January 1980, p. 29.

19. Kim Levin, "Helen and Newton Harrison: New Grounds for Art," *Arts Magazine*, Vol. 52, February 1978, pp. 126–129; Peter Selz, "Helen and Newton Harrison: Art as Survival Instruction," *Arts Magazine*, Vol. 52, February 1978, pp. 130–131; and Kristine Stiles, "Helen and Newton Harrison: Questions," *Arts Magazine*, Vol. 52, February 1978, pp. 131–133.

20. *Panamarenko Multiples 1966–1988* (Antwerp: Galerie Ronny Van De Velde & Co., 1988); also Panamarenko, *U-Control III* (Antwerp: Galerie Ronny Van De Velde & Co., 1972).

21. Jean-Christophe Bailly, *Piotr Kowalski* (Paris: Fernand Hazen, 1988); also Jean-Christophe Bailly, ed., *Alea*, No. 3 (Paris: Christian Bourgois, 1982), pp. 45–50.

22. *Piotr Kowalski, Time Machine + Projects*, Centre Georges Pompidou. December 16, 1981, to February 6, 1982, art exhibition catalogue; also, Félix Guattarí, "Interview Piotr Kowalski," in *L'Art et la Ville: Town-Planning and Contemporary Art* (Paris: Secrétariat Général des Villes Nouvelles, 1990), pp. 121–33.

23. Graig Adcock, "Perceptual Edges: The Psychology of James Turrell's Light and Space," *Arts Magazine*, February 1985, pp. 124–128; also read Lawrence Weschler, *Seeing Is Forgetting the Name of the Thing One Sees: A Life of Contemporary Artist Robert Irwin* (Berkeley: University of California Press, 1982), p. 131.

24. James Turrell's statement, "space-looked-out-onto," is excerpted from an interview with the environmental artist Janet Saad-Cook in "Touching the Sky: Conversations with Four Contemporary Artists," *Archeoastronomy*, Vol. 8 (Nos. 1–4), January-December 1985, pp. 118–141.

25. Patricia Failing, "James Turrell's New Light on the Universe," *ARTnews*, April 1985, p. 71; also see James Turrell, *Occluded Front* (Los Angeles: Fellows of Contemporary Art; Larkspur Landing, Calif.: Lapis Press, 1985).

26. Lowry Burgess, *The Quiet Axis* (Montreal: Editions du Trecarre, 1987), p. 25; also see Miriam Seidel, "Artist for the Cosmos," *New Art Examiner*, March 1987, pp. 25–28; Ken Sofer, "Dante and the Orbiting Asteroids," *ARTnews*, February 1982, p. 148.

27. Ibid., *The Quiet Axis*, pp. 39, 41.

28. Notes on the *Model of the Solar System* from *Chris Burden: A Twenty-Year Survey* (Newport Harbor, Calif.: Newport Harbor Art Museum, 1988). The book accompanied the exhibition Chris Burden: A Twenty-Year Survey, organized by Paul Schimmel and Anne Ayres of the Newport Harbor Art Museum.

29. *Geneviève Cadieux: Canada XLIVth Biennale di Venezia*, with text "Language Is a Skin" by Chantal Pontbriand (Montreal, Quebec: The Montreal Museum of Fine Arts and Parachute, 1990).

30. *Kristin Jones and Andrew Ginzel: "Antithesis,"* Kunsthalle Basel, August 20 to October 29, 1989. (Exhibition catalogue translated in German and English; photographs by T. Charles Erickson; see p. 15.)

31. *Kenneth Snelson: The Nature of Structure* (New York: The New York Academy of Sciences, 1988), p. 12. The book accompanied the exhibition of Snelson's work at the New York Academy of Sciences, organized by Joelle Burrows.

32. Robert S. Root-Bernstein, "Beauty, Truth, and Imagination: A Perspective on the Science and Art of Modeling Atoms," in *Kenneth Snelson: The Nature of Structure*, pp. 15–20.

33. Ibid., pp. 12, 14.

34. Edwin Schlossberg, *Word: Nerve Tidal Gestures* (published by Edwin Schlossberg, 1990); Edwin Schlossberg, *Einstein and Beckett: A Record of an Imaginary Conversation with Albert Einstein and Samuel Beckett* (New York: Links, 1973); also, Edwin Schlossberg, *The Pirated Edition of How I Learn How I Learn* (Zurich: Princelet Editions, 1984).

35. Arakawa and Madeline H. Gins, *The Mechanism of Meaning* (New York: Harry N. Abrams, 1979); *Arakawa* (Padiglione D'Arte Contemporanea, Milano, Via Palestro, 1984), p. 14 (art exhibition catalogue translated in French, Italian, and English); Arakawa/Madeline Gins, *Pour Ne Pas Mourir: To Not Die* (translated in French by Francois Rosso) (Paris: Editions de la Différence).

36. Mark Rosenthal, *Anselm Kiefer*, exhibition catalogue organized by Mark Rosenthal, Philadelphia Museum of Art, and A. James Speyer, Art Institute of Chicago, distributed by Prestel-Verlag, 1987.

37. *Mia Westerlund Roosen*, Joseloff Gallery, Harry Jack Gray Center, August 16 to September 24, 1989 (Hartford, Conn.: University of Hartford, 1989).

38. From a conversation with Deborah Menaker, Associate Curator of Exhibitions. Quoted in *Artworks: Adrian Piper*, Williams College Museum of Art, January 20–March 25, 1990.

39. *Adrian Piper Reflections 1967–1987* (New York: Alternative Museum, 1987; Reprint, John Weber Gallery, New York, 1989.)

40. See exhibition catalogue *Lorna Simpson*, September 8–October 14, 1989. (New York: Josh Baer Gallery, 1989); also, Eleanor Heartney's review "Lorna Simpson at Josh Baer," in *Art In America*, November 1989, pp. 185, 186; also read Regina Joseph's "Lorna Simpson interview" in *Balcon Magazine*, Spring 1990, pp. 35–39. As a point of contrast and perspective, you may want to examine the work of conceptual artists Barbara Kruger and Antonio Muntadas, who use text and photographs, video installations, and other media in exploring with great force and clarity a wide range of social and politically oriented issues.

41. Quoted from Adrian Piper's text "Ways of Averting One's Gaze," in Laurence Alloway and Lucy Lippard, *Adrian Piper* (Washington: University of Washington, 1989). This superb text demonstrates how Piper has managed to skillfully apply her earlier doctoral studies in analytic phi-

losophy towards art—not by presenting the content of these studies, but by using her knowledge of philosophy to help inform and articulate her artistic inquiry into racism. This writing is also important in that it documents the process by which Piper creates, selects, and evaluates her work. The process involves overseeing all stages and aspects of the work's production, including its interpretation by the public.

42. From Steinman's text for *Cenotaph*, in Claude Gosselin, ed., *Lumières: Perception—Projection*, Les Cent Jours d'Art Contemporain Montréal 86, Centre International D'Art Contemporain, August 1–November 2, 1986. Also Bruce Ferguson, "The Art of Memory," *Vanguard*, Vol. 18 (No. 3), Summer 1989, pp. 10–15, and James D. Campbell, "History Hurts: Barbara Steinman and Installation," *C*, June 1989, pp. 14–19.

43. *Nancy Spero*, catalogue copublished by the Institute of Contemporary Art, London/Fruitmarket Gallery, Edinburgh/Orchard Gallery/Foyle Art Project; also, *The Village Voice*, "Art Special," October 6, 1987, Vol. III, No. 2 (cover and p. 17); Robert Taylor, "Nancy Spero's Work Filled with Liberating Strength," *The Boston Sunday Globe*, October 23, 1988, p. 80.

44. *Ida Applebroog, Nostrums Paintings*, catalogue published by Ronald Feldman Fine Arts, Inc., New York, 1989; also, *Applebroog*, catalogue with text by Ronald Feldman, Carrie Rickey, Lucy R. Lippard, Linda F. McGreevy, and Carter Ratcliff, published by Ronald Feldman Fine Arts, Inc., New York, 1987.

Other books which may be of interest in the context of the artworks and individuals discussed here include:

Duane and Sarah Preble, *Artforms*, 4th ed. (New York: Harper & Row, 1989).

Lucy Lippard, *Overlay: Contemporary Art and the Art of Prehistory* (New York: Pantheon Books, 1983).

Alan Sonfist, ed., *Art in the Land: A Critical Anthology of Environmental Art* (New York: E. P. Dutton, 1983).

John Beardsley, *Probing the Earth: Contemporary Land Projects*, exhibition catalogue, Hirshhorn Museum and Sculpture Garden, Washington, D.C., 1977.

P. C. Vitz and A. B. Glimcher, *Modern Art and Modern Science: The Parallel Analysis of Vision* (New York: Praeger, 1984).

Ezra Orion, "Sculpture in the Solar System: From Geologically Based Earthworks to Astro-Sculpture," *Leonardo*, Vol. 18, 1987.

Frank J. Malina, "On the Visual Arts in the Space Age," *Leonardo*, Vol. 3, 1987, p. 323.

Frank Popper, *Techno-science Art, the Art of the 21st-Century: The New Immaterialists* (Paris, forthcoming).

BRANCHES #2

1. Carl G. Jung, *Man and His Symbols* (New York: Dell, 1964), p. 310.

2. Robert Snyder, *R. Buckminster Fuller: An Autobiographical Monologue/Scenario* (New York: St. Martin's Press, 1980).

3. To my knowledge, *synergetics* marks one of the first concerted efforts to exchange perspectives and ideas in the interborough of scientific disciplines. Physicists talk with biologists who talk with sociologists who, in turn, speak with chemists speaking, full circle, with physicists. Somehow, in this lattice of communication, artists and scholars were omitted from all formal conversations or rather their views were unsolicited.

4. Hermann Haken, ed., *Synergetics: A Workshop* (Proceedings of the International Workshop on Synergetics at Schloss Elmau, Bavaria, May 2–7, 1977) (Berlin, Heidelberg, New York: Springer-Verlag, 1977), pp. 2–9 *passim*. Hermann Haken, ed., *Synergetics of the Brain* (Berlin, Heidelberg, New York: Springer-Verlag, 1978), pp. 3, 4.

5. Charles Hampden-Turner, *Maps of the Mind: Charts and Concepts of the Mind and Its Labyrinths* (New York: Macmillan, 1981), p. 158.

6. Ervin Laszlo, *The Systems View of the World: The Natural Philosophy of the New Developments in the Sciences* (New York: George Braziller, 1972), p. 22.

7. Ibid., pp. 20, 21.

8. Quoted in Alain Danielou, *The Gods of India: Hindu Polytheism* (New York: Inner Traditions International, 1985), p. 23. (Quotation from Karapatri, "Lingopasana-rahasya," p. 155.)

9. Ibid., pp. 23, 24. (Quotation from Maitrayani Upanishad, 5.2)

10. One complementary text to Prigogine and Stengers' *Order Out of Chaos* (New York: Bantam Books, 1984) I would call traditional Hindu philosophy! Perhaps physicists, chemists, and biologists will one day have as *required readings*—not as optional, supplementary studies—these great works of philosophy. If natural scientists and mathematicians allowed themselves to creatively interpret and metaphorm these special sources, I tend to believe they would produce broader artistic/scientific paradigms or more complete theories. It seems we're using the same palette and paints—but different brushes, brush strokes, media, and aesthetics—to paint our "unique portraits of one reality, one universe, one nature, with its many "whole" parts or entities.

11. From Pamela Weintraub, ed., *The Omni Interviews* (New York: Ticknor & Fields, 1984).

12. Danielou, *The Gods of India*, p. 229. (Quotation from Gopinatha Kaviraja, "Lingarahasya, Kalyana, Shiva anka," p. 476.)

13. Gregory Bateson, *Mind and Nature: A Necessary Unity* (New York: Ballantine, 1979), p. 102.

14. Ibid., p. 104.

15. Ibid., p. 120.

16. David Bohm, "Hidden Variables and the Implicate Order," in B. J. Hiley and F. David Peat, eds., *Quantum Implications: Essays in Honor of David Bohm* (London, New York: Routledge & Kegan Paul, 1987), pp. 36, 37. For additional reading: Ken Wilbur, ed., *The Holographic Paradigm and Other Paradoxes: Exploring the Leading Edge of Science* (Boulder, London: Shambhala Press, 1982), Chapter 5, "The Enfolding-Unfolding Universe: A Conversation with David Bohm."

17. Hiley and Peat, eds., *Quantum Implications*, p. 38.

18. Ibid., p. 40.

19. Ibid., p. 40.

20. Ibid., p. 40.

21. Ibid., p. 43.

22. Ibid., p. 44.

BRANCHES #3

1. William Arnold and John Bowie describe "knowledge engineering" in the clearest of terms in their introductory book *Artificial Intelligence: A Personal, Commonsense Journey* (Englewood Cliffs, N.J.: Prentice-Hall, 1986), p. 17. If you're interested in the subject of AI, I recommend Patrick Winston's excellent textbook, *Artificial Intelligence* (New York: Addison-Wesley, 1984). Winston was a student of Minsky and Papert and he directed the AI Lab at M.I.T. One book that bridges AI and psychology in modeling human thought processes is Roger Schank and Kenneth Colby, eds., *Computer Models of Thought and Language* (New York: W. H. Freeman, 1973).

2. In using the expression "linguistic mirror" I invoke the renowned psycholinguist Noam Chomsky's notion that language mirrors the mind. Chomsky maintains that as we come to understand how language is used or how it *works*, we will learn how our unconscious and conscious minds work in using language and how our minds know the world and organize this knowledge for purposes of communication. Read Noam Chomsky's important book *Reflections on Language* (New York: Pantheon Books, 1975).

3. David Gelernter, "The Metamorphosis of Information Management," *Scientific American*, August 1989, p. 68.

4. Ibid., p. 68.

5. When I coined this phrase many years

ago, I was searching for a way to broadly define the nature and origin of creative impulses. I thought that the broader the definition, the more truthful and useful. Broadness is more useful in this case, as opposed to specificity, because it encompasses not just our species' thought processes but those that are so often overlooked by our homocentric views regarding the "creativity" of other species.

Also, this definition was influenced by my interest in patenting inventions. When writing a patent, your invention must be described in the broadest terms, after which your claims must de- tail and support each statement of the invention. Note that all claims must be reduced to practice before they are considered patentable. That's the "filtering" process between theory and practice. A patent is as much a pedagogical tool as it is a legal device for protecting an invention. The clearer and broader its teachings, the easier it is to use "for one skilled in the arts."

6. John Dewey, *Experience and Nature* (Chicago, London: Open Court Publishing, 1926), p. 355.

Bibliography

The following books bear upon one or more aspects of neurocosmology—its scope, philosophy, practice, practitioners, and artifacts. I suggest that they be read in the context of all that has been discussed in this book.

Abbott, Edwin Abbott. 1884. *Flatland: A Romance of Many Dimensions*. Reprint. New York: Barnes & Noble, 1983.

Arguelles, Jose. *Transformative Vision*. Boulder: Shambhala Press, 1975.

Armstrong, David M. *What Is a Law of Nature*. Cambridge; New York: Cambridge University Press, 1983.

Arnheim, Rudolf. *Visual Thinking*. Berkeley: University of California Press, 1969.

Bachelard, Gaston. *Poetics of Space*. New York: Orion Press, 1964.

Bateson, Gregory. *Steps to an Ecology of Mind*. New York: Ballantine Books, 1975.

———. *Mind and Nature: A Necessary Unity*. New York: Bantam Books, 1979.

Becker, Robert O., M.D., and Gary Selden. *The Body Electric: Electromagnetism and the Foundation of Life*. New York: William Morrow, 1985.

Bertalanffy, Ludwig von. *General Systems Theory*. New York: George Braziller, 1968.

Black, Max. *Models and Metaphors: Studies in Language and Philosophy*. Ithaca, N.Y.: Cornell University Press, 1962.

Bohm, David. *Wholeness and the Implicate Order*. London and Boston: Routledge & Kegan Paul, 1981.

Bronowski, Jacob. *The Visionary Eye: Essays in the Arts, Literature, and Science*. (Selected and edited by Piero E. Ariotti in collaboration with Rita Bronowski.) Cambridge, Mass.: M.I.T. Press, 1978.

Calder, Nigel. *Violent Universe: An Eyewitness Account of the New Astronomy*. New York: Viking, 1969.

Campbell, Joseph. *The Masks of the Gods: Creative Mythology*. New York: Viking, 1968.

Capra, Fritjof. *The Turning Point: Science, Society and the Rising Culture*. New York: Simon & Schuster, 1982.

Carrigan, Richard A., and W. Peter Trower, eds. *Particle Physics in the Cosmos: Readings from Scientific American*. New York: W. H. Freeman, 1989.

Carroll, Lewis. 1872. *Through the Looking-Glass*. Reprint. New York: Random House, 1946.

Casti, John L. *Paradigms Lost: Images of Man in the Mirror of Science*. New York: William Morrow, 1989.

Chaisson, Eric. *Cosmic Dawn*. Boston: A Berkeley Book/Little, Brown and Company, 1981.

Chomsky, Noam. *Language and Mind*. New York: Harcourt Brace Jovanovich, 1972.

Chorover, Stephan. *From Genesis to Genocide*. Cambridge, Mass.: M.I.T. Press, 1983.

Churchland, Patricia. *Neurophilosophy*. Cambridge, Mass.: M.I.T. Press, 1988.

Coomaraswamy, Ananda K. *The Transformation of Nature in Art*. Cambridge, Mass.: Harvard University Press, 1934.

Danielou, Alain. *The Gods of India: Hindu Polytheism*. New York: Inner Traditions International, 1985.

Davies, Paul. *The Accidental Universe*. Cambridge: Cambridge University Press, 1982.

———. *Superforce: The Search for a Grand Unified Theory of Nature*. New York: Simon & Schuster, 1984.

Eccles, John Carew. *The Neurophysiological Basis of Mind: The Principles of Neurophysiology*. Oxford: Clarendon Press, 1953.

———. *The Human Mystery*. Berlin and New York: Springer-Verlag, 1979.

Eckstein, Gustav. *The Body Has a Head*. New York: Harper & Row, 1970.

Eisler, Riane. *The Chalice and the Blade*. New York: Harper & Row, 1987.

Eliade, Mircea. *The Sacred and the Profane*. (Translated by William Trask.) New York: Harcourt, Brace, and World, 1959.

———. "Time and Eternity in Indian Thought," in J. Campbell, ed., *Man and Time: Papers from the Eranos Yearbooks*, Bollingen Series XXX, Vol. 3, pp. 173–200. Princeton, N.J.: Princeton University Press, 1983.

Faser, J. T., F. C. Haber, and G. H. Muller, eds. *The Study of Time*. Berlin: Springer-Verlag, 1972.

Ferguson, Marilyn. *The Aquarian Conspiracy: Personal and Social Transformation in the 1980s*. Los Angeles: J. P. Tarcher, 1980 (distributed by St. Martin's Press, New York).

Ferris, Timothy. *Galaxies*. New York: Stewart, Tabori & Chang, 1982.

Feynman, Richard. *The Character of Physical Law*. Cambridge, Mass.: M.I.T. Press, 1967.

Fuller, R. Buckminster. *Synergetics: Explorations in the Geometry of Thinking.* New York: Macmillan, 1975.

Gardner, Howard. *Art, Mind and Brain: A Cognitive Approach to Creativity.* New York: Basic Books, 1982.

———. *Frames of Mind: The Theory of Multiple Intelligences.* New York: Basic Books, 1983.

Gardner, Martin. *The Ambidextrous Universe.* New York: Basic Books, 1964.

Ghiselin, B., ed. *The Creative Process.* Berkeley: University of California Press, 1952.

Gleick, James. *Chaos: Making a New Science.* New York: Penguin Books, 1987.

Gombrich, E. H. *Art and Illusion: A Study in the Psychology of Pictorial Representation.* New York: Pantheon Books, 1960.

———. "The Sky Is the Limit," in R. B. MacLeod and H. L. Pick, eds., *Perception: Essays in Honor of James J. Gibson.* Ithaca, N.Y.: Cornell University Press, 1975.

Goodman, Nelson. *Ways of Worldmaking.* Indianapolis: Hackett, 1978.

Gould, Stephen J. *The Mismeasure of Man.* New York: W. W. Norton, 1981.

Gribbin, John, and Martin Rees. *Cosmic Coincidences: Dark Matter, Mankind, and Anthropic Cosmology.* New York: Bantam, 1989.

Hadamard, J. *The Psychology of Invention in the Mathematical Field.* Princeton, N.J.: Princeton University Press, 1945.

Haken, Hermann. *The Science of Structure: Synergetics.* (English translation by Fred Bradley.) New York: Van Nostrand Reinhold, 1981.

Hampden-Turner, Charles. *Maps of the Mind: Charts and Concepts of the Mind and Its Labyrinths.* New York: Macmillan, 1982.

Hanson, Norwood R. *Patterns of Discovery: An Inquiry into the Conceptual Foundations of Science.* Cambridge: Cambridge University Press, 1958.

Harrison, Edward. *Masks of the Universe.* New York: Macmillan, 1985.

Hawking, Stephen. *A Brief History of Time.* New York: Bantam Books, 1988.

——— and G. Ellis. *The Large Scale Structure of Space-Time.* Cambridge: Cambridge University Press, 1973.

Hebb, Donald O. *The Organization of Behavior.* New York: John Wiley & Sons, 1949.

Hilbert, D., and S. Cohn-Vossen. 1938. *Geometry and the Imagination.* Reprint. New York: Chelsea, 1952.

Hofstadter, Douglas R. *Gödel, Escher, Bach: An Eternal Golden Braid.* New York: Basic Books, 1979.

——— and Daniel Dennett. *The Mind's I: Fantasies and Reflections on Self and Soul.* New York: Basic Books, 1981.

Humboldt, Alexander, Freiherr von. *Kosmos,* Vols. 1–4. Stuttgart: Cotta, 1845.

James, William. *Principles of Psychology.* Vols. 1 and 2. Reprinted. New York: Dover, 1890.

Jantsch, Erich. *The Self-Organizing Universe: Scientific and Human Implications of the Emerging Paradigm of Evolution.* Oxford: Pergamon Press, 1980.

Jung, Carl G. *Man and His Symbols.* New York: Dell Books, 1964.

Kandel, Eric R., and James H. Schwartz. *Principles of Neural Science.* New York: Elsevier/North-Holland, 1981.

Kaufmann, William J. *Relativity and Cosmology.* New York: Harper & Row, 1973.

Kepes, Gyorgy. *The New Landscape in the Arts and Sciences.* Chicago: Paul Theobald, 1956.

———. *Education of Vision.* Chicago: Paul Theobald, 1965.

Klee, Paul. *The Thinking Eye.* (Edited by J. Spiller.) New York: G. Wittenborn, 1969.

————. *The Nature of Nature*. (Edited by J. Spiller and translated by H. Norden.) New York: G. Wittenborn, 1973.

Koestler, Arthur. *The Act of Creation*. New York: Macmillan, 1964.

Kuhn, Thomas. *The Structure of Scientific Revolution*, 2nd ed. Chicago: University of Chicago Press, 1970.

————. *The Essential Tension*. Chicago: University of Chicago Press, 1976.

Langer, Susanne Katherina. *Philosophy in a New Key: A Study in the Symbolism of Reason, Rite, and Art*, 3rd ed. Cambridge, Mass.: Harvard University Press, 1957.

Lévi-Strauss, Claude. *The Savage Mind*. Chicago: University of Chicago Press, 1966.

Lewin, Kurt. *Field Theory in Social Science: Selected Theoretical Papers*. (Edited by Dorwin Cartwright.) New York: Harper, 1951.

Malone, Adrian, and Steven Talley. *The Secret*. Boston: Houghton-Mifflin, 1984.

Maslow, Abraham, ed. *New Knowledge in Human Values*. New York: Harper, 1959.

Maturana, Humberto R. *The Tree of Knowledge: The Biological Roots of Human Understanding*. Boston: New Science Library (distributed in the United States by Random House), 1987.

McCulloch, Warren S. *Embodiments of Mind*. Cambridge, Mass.: M.I.T. Press, 1988.

McLanathan, R. B. *Images of the Universe. Leonardo Da Vinci: The Artist as Scientist*. Garden City, N.Y.: Doubleday, 1966.

Miller, Arthur I. *Imagery in Scientific Thought: Creating 20th-Century Physics*. Cambridge, Mass.: M.I.T. Press, 1987.

Minsky, Marvin. *The Society of Mind*. New York: Simon & Schuster, 1988.

Monod, Jacques. *Chance and Necessity: An Essay on the Natural Philosophy of Modern Biology*. New York: Knopf, 1971.

Morrison, Philip. *Ring of Truth*. New York: Random House, 1987.

Munitz, Milton K. *Theories of the Universe: From Babylonian Myth to Modern Science*. New York: The Free Press/Collier Macmillan, 1957.

Neisser, Ulric. *Cognition and Reality: Principles and Implications of Cognitive Psychology*. San Francisco: W. H. Freeman, 1976.

Ornstein, Robert E. *The Psychology of Consciousness*. San Francisco: W. H. Freeman, 1975.

Pagels, Heinz. *The Cosmic Code: Quantum Physics as the Language of Nature*. New York: Simon & Schuster, 1982.

Penfield, Wilder. *The Mystery of the Mind: A Critical Study of Consciousness and the Human Brain*. Princeton, N.J.: Princeton University Press, 1975.

Piaget, Jean. *Biology and Knowledge: An Essay on the Relations Between Organic Regulations and Cognitive Processes*. Chicago: University of Chicago Press, 1971.

Pirsig, Robert M. *Zen and the Art of Motorcycle Maintenance: An Inquiry into Values*. New York: Bantam Books, 1974.

Plato. "The Republic," in *The Dialogues of Plato*. (Translated by B. Jowett.) New York: Random House, 1937.

Popper, Karl R. *Conjectures and Refutations*. New York: Basic Books, 1962.

————. *Objective Knowledge: An Evolutionary Approach*. Oxford: Clarendon Press, 1972.

Pribram, Karl. *Languages of the Brain*. Englewood Cliffs, N.J.: Prentice-Hall, 1971.

Prigogine, Ilya, and Isabelle Stengers. *Order Out of Chaos: Man's New Dialogue with Nature*. New York: Bantam Books, 1984.

Redfield, Robert. *The Primitive World and Its Transformations*. Ithaca, N.Y.: Cornell University Press, 1953.

Richter, Irma A., ed. *Selections from the Notebooks of Leonardo Da Vinci* (with commentaries). Oxford and New York: Oxford University Press, 1977.

Root-Bernstein, Robert. *Discovering.* Cambridge, Mass.: Harvard University Press, 1988.

Rorty, Richard. *Philosophy and the Mirror of Nature.* Princeton, N.J.: Princeton University Press, 1979.

Rothenberg, A. *The Emerging Goddess: The Creative Process in Art, Science, and Other Fields.* Chicago: University of Chicago Press, 1979.

Rucker, Rudy. *Infinity and the Mind.* Boston: Birkhauser, 1982.

Sagan, Carl. *The Dragons of Eden: Speculations on the Evolution of Human Intelligence.* New York: Ballantine Books, 1977.

————. *Broca's Brain: Reflections on the Romance of Science.* New York: Ballantine Books, 1979.

Sambursky, S. *The Physical World of the Greeks.* (Translated from the Hebrew by Merton Dagut.) London: Routledge & Kegan Paul, 1963.

Sandage, Allan. *The Hubble Atlas of Galaxies.* Washington, D.C.: Carnegie Institute of Washington, 1961.

Schrödinger, Erwin. *What Is Life? The Physical Aspects of the Living Cell.* Cambridge, England: Cambridge University Press, 1943.

————. *Mind and Matter.* Cambridge, England: Cambridge University Press, 1958.

Siler, Todd. *Metaphorms: Forms of Metaphor.* New York: The New York Academy of Sciences, 1987.

————. *The Art of Thought.* Montreal, Quebec: Saidye Bronfman Centre, 1987.

————. "Architectonics of Thought: A Symbolic Model of Neuropsychological Processes." M.I.T. Ph.D. dissertation, Interdisciplinary Studies in Psychology and Art, 1986.

————. "Neurocosmology: Ideas and Images Towards an Art-Science-Technology Synthesis," *Leonardo,* Vol. 18 (No. 1), 1985, pp. 1–10.

————. *The Biomirror.* New York: Ronald Feldman Fine Arts, Inc., 1983.

————. *Cerebreactors.* New York: Ronald Feldman Fine Arts, Inc., 1980.

Sivaramamurti, Calambur. *The Art of India.* New York: Harry N. Abrams, 1977.

Smullyan, Raymond. *5000 B.C. and Other Philosophical Fantasies.* New York: St. Martin's Press, 1983.

Sperry, Roger W. "The Great Cerebral Commissure," *Scientific American,* Vol. 210 (1964), pp. 42–52.

————. "Hemisphere Deconnection and Unity of Conscious Awareness," *American Psychologist,* Vol. 23 (1968), pp. 723–733.

Thomas, Lewis. *The Lives of a Cell: Notes of a Biology Watcher.* New York: A Bantam Book/Viking, 1974.

Thompson, D'Arcy. *On Growth and Form,* abridged ed. Cambridge: Cambridge University Press, 1961.

Thorne, Kip. "The Search for Black Holes," in D. Gingerich, ed., *Cosmology + 1.* San Francisco: W. H. Freeman, 1977.

Toffler, Alvin. *Future Shock.* New York: Random House, 1970.

Von Neumann, John. *The Computer and the Brain.* New Haven: Yale University Press, 1958.

Weyl, Hermann. *Space Time Matter.* (Translated from the German by Henry L. Brose.) New York: Dover, 1952.

Whitehead, Alfred North. *The Concept of Nature.* Cambridge: Cambridge University Press, 1920.

————. *Process and Reality.* New York: Macmillan, 1929.

Wiener, Norbert. *Cybernetics: Or, Control and Communication in the Animal and the Machine.* Amsterdam and New York: Elsevier, 1948.

————. *The Human Use of Human Beings: Cybernetics and Society*. Garden City, N.Y.: Doubleday/Anchor Books, 1954.

Wilber, Ken. *The Holographic Paradigm and Other Paradoxes: Exploring the Leading Edge of Science*. Boulder and London: Shambhala Press, 1982.

Wittgenstein, Ludwig. *Tractatus Logico-Philosophicus*. (Translated by D. F. Pears and B. F. McGuiness with an introduction by Bertrand Russell.) London: Routledge & Kegan Paul, 1921.

Wolf, Fred Alan. *Star*Wave: Mind, Consciousness, and Quantum Physics*. New York: Macmillan, 1984.

Zeilik, Michael, and Elske van Panhuys Smith. *Introductory Astronomy and Astrophysics*, 2nd ed. New York: Saunders College Publishing, 1987.

Permissions

Graph of the mass luminosity relationship for binary systems from Zeilik and Smith, 1987. Reprinted by permission of the publisher.

Diagram of brain mass versus body mass. Copyright © 1980 by Carl Sagan. All Rights Reserved. Reprinted by permission of the author.

Three drawings illustrating a contact binary star system from Zeilik and Smith, 1987. Reprinted by permission of the publisher.

Drawing of star clusters adapted from R. B. Tully in *Astrophysical Journal*, 257:389, 1983. Reprinted by permission of the author.

Two views of the spiral galaxy, NGC 1097, from *The Radiant Universe*, by Michael Marten and John Chesterman, Macmillan Publishing Co., Inc., 1980. Courtesy CTIO. Reprinted by permission of the publisher.

Microphotograph of the Hippocampus from *Fundamental Neuroanatomy*. By Walle J. H. Nauta and Michael Feirtag. Copyright © 1986 by W. H. Freeman and Company. Reprinted by permission of Cell Press.

Photograph of a vela supernova remnant copyright © 1949 Royal Observatory, Edinburgh. Reprinted by permission.

Two microphotographs of neural tissue from *Fundamental Neuroanatomy*. By Walle J. H. Nauta and Michael Feirtag. Copyright © 1986 by W. H. Freeman and Company. Reprinted by permission of Cell Press.

Photograph of colliding galaxies from *The Radiant Universe* by Michael Martin and John Chesterman, Macmillan Publishing Co., Inc., 1980. Reprinted by permission of the National Optical Astronomy Observatories, Tucson, AZ.

Photograph of lightning at Kitt Peak Observatory, copyright © Gary Ladd, 1972. Reprinted by permission of Gary Ladd.

Excerpt from *Galaxies* by Timothy Ferris; Stewart, Tabori and Chang, Publishers. Copyright © 1982 by Timothy Ferris. Used by permission.

Excerpt from *Life Itself: Its Origin and Nature* by Francis Crick, Simon and Schuster, 1981. Reprinted by permission of the publisher.

Drawing of Toroidal plasma fusion and adapted drawing of *Cerebreactor* theatron from *Nuclear Fusion* by R. Hulme and M. B. Collier, Taylor & Francis publishers, Philadelphia, 1969 (formerly, Wykeham Publications Ltd.). Reprinted by permission of the publishers.

Drawings of magnetic fields from *Driven Magnetic Fusion Reactors* by B. Brunelli, ed., Pergamon Press, 1978. Reprinted by permission of the publisher.

Drawing of the brain from *Basic Neurology* by J. P. Schade and D. H. Ford, American Elsevier Co. Publishers, 1967. Reprinted by permission of the publisher.

Diagram of Argonne Reactor from *Principles Fuels of Energy Conversion*, American Society, A.N.S. News. Courtesy of Argonne National Laboratory.

Photograph of assembled torus from "Operation of the ORMAK Fusion Device in a Cryogenic Vacuum Environment" by W. Halchin, S. M. Decamp, S. O. Lewis, D. C. Lousteau and J. D. Bylander in *Fifth Symposium on Engineering Problems of Fusion Research*, Institute of Electrical and Electronics Engineers, Inc., 1973. Reprinted by permission of IEEE, Inc.

Drawing from "The Technology of Mirror Machines—LLL Facilities for Magnetic Mirror Fusion Experiments" by T. H. Batzer in *Seventh Symposium on Engineering Problems of Fusion Research*, Institute of Electrical and Electronics Engineers, Inc., 1977. Reprinted by permission of IEEE, Inc.

Nova Laser photos, target chambers. Courtesy of the Lawrence Livermore National Laboratory, Livermore, CA.

Photograph of linear electron reactor from "Accelerating Structure in Technology" by A. L. Eldrege, A. V. Lisin, V. G. Price in *Linear Accelerators*, P. M. Lapostolle and A. L. Septier, eds.;

Index

Page numbers in italics *refer to illustrations.*